Mining Heritage and Tourism

Many former mining areas have lost their industrial function and are now turning to tourism for regional revitalization and community economic development. The transformation process of these industrial, and in some cases derelict, mining sites and landscapes into an area of interest for tourists is a major challenge both for planners and for tourism managers. It involves complex consideration to both the preservation of the physical site and community mining heritages as well as the health, safety and environmental factors inherent in opening these vast sites to the public.

Mining Heritage and Tourism includes contributions from internationally recognised authorities and is the first book to focus on these issues, challenges and potentials in redeveloping mines as cultural heritage attractions which are explored thematically throughout the book. It draws on multidisciplinary research to consider the dichotomy between heritage preservation and tourist development goals for mining heritage sites as well as to explore the practical challenges of developing these sites. These themes are illustrated by case studies from a vast range of geographical locations around the globe to offer operational insights into the planning and management of these sites for both heritage and tourism purposes, as well as innovative site management techniques.

There has never before been a more comprehensive book on mining heritage tourism representing the latest developments in strategy, policy and practices. This book serves as an invaluable guide for students, researchers, academics and practitioners in the areas of Tourism and Heritage Management.

Michael V. Conlin is a Professor in the Okanagan School of Business in Kelowna, BC, Canada. He also teaches for the University of Guelph and Royal Roads University. His academic interests include island tourism and his current work includes an investigation of the use of industrial infrastructure for touristic purposes in British Columbia and elsewhere.

Lee Jolliffe is an Associate Professor of Hospitality and Tourism at the University of New Brunswick, Saint John. Her academic work explores the use of culture for tourism in a variety of contexts that include rural and island tourism situations.

Routledge Advances in Tourism
Edited by Stephen Page, London Metropolitan University, London.

17 Drive Tourism
Trends and emerging markets
Bruce Prideaux and Dean Carson

18 Tourist Customer Service Satisfaction
Francis P. Noe, Muzzafer Uysal and Vincent P. Magnini

19 Mining Heritage Tourism
A global synthesis
Edited by Michael V. Conlin and Lee Jolliffe

20 Tourist Experience
Richard Sharpley and Phillip Stone

Forthcoming:

Tourism Supply Chain Management
Haiyan Song

Sustainable Tourism in Rural Europe
Donald Macleod and Steven Gillespie

Critical Turn in Tourism
An academy of hope
Nigel Morgan, Irena Atelkevic and Annette Pritchard

New Perspectives in Tourism Geographies
Julie Wilson

Mining Heritage and Tourism

A global synthesis

Edited by Michael V. Conlin and Lee Jolliffe

Routledge
Taylor & Francis Group
LONDON AND NEW YORK

First published 2011
by Routledge
2 Park Square, Milton Park, Abingdon, Oxfordshire OX14 4RN

Simultaneously published in the USA and Canada
by Routledge
711 Third Avenue, New York, NY 10017

First issued in paperback 2014

Routledge is an imprint of the Taylor & Francis Group, an informa business

Typeset in Times New Roman by
Pindar NZ, Auckland, New Zealand

British Library Cataloguing in Publication Data
A catalogue record for this book is available from the British Library

Library of Congress Cataloguing in Publication Data
Mining heritage and tourism / Edited by Michael V. Conlin and Lee Jolliffe.
p. cm.
Includes bibliographical references and index.
1. Heritage tourism. 2. Mines and mineral resources. I. Conlin, Michael V. II. Jolliffe, Lee.
G156.5.H47M56 2010
622.074—dc22 2010008765

ISBN13: 978-1-138-88069-6 (pbk)
ISBN13: 978-0-415-56090-0 (hbk)

Contents

List of illustrations

Figures

Tables

Map

Acknowledgments

As investigations into the tourism phenomena progress over time, there are still a few topics that have not been investigated in detail, and mining heritage in relation to tourism is one of them. The genesis of this book comes from a discussion the editors had at the 2005 TTRA Canada Conference at which one co-editor presented a related paper 'Mines as Heritage Attractions in Atlantic Canada'. The paper was warmly received by participating tourism academics, many expressing a peripheral interest in the subject area. And while the subject area has not been completely unexplored in the literature, this book builds on and acknowledges the pioneering work in this area by a number of other authors.

We appreciate the interest shown by a large number of academics around the world in this project and in particular, the active participation of the cadre of 37 authors who eventually contributed to the book. We thank everyone involved for their patience with us as we reviewed their work and asked for the necessary input to finalize the manuscript. In particular we are grateful to a number of authors who joined the enterprise during its latter stages, increasing the global coverage from North America, Europe and Australia, to include Asia (Vietnam) and South America (Brazil).

The book has been four years in the making, a time during which we had the opportunity to visit a number of mining attraction sites including those located in and around Butte and Anaconda, Montana, Greenwood, British Columbia, and perhaps most importantly, the British Columbia Museum of Mining located at the Canadian Heritage listed Brittania Beach mine site near the head of Howe Sound on Canada's west coast. Particular thanks are extended to the Museum's Director, Kirsten Klausen, who has been generous with both her time and experience in helping to develop our understanding of the real challenges and opportunities which practitioners face as they work to preserve mining heritage in an increasingly competitive tourism environment.

We wish also to extend recognition to Okanagan College which provided a Grant-In-Aid to support the field research undertaken by the editors in preparing the proposal for this book. In particular, we wish to thank Dr Jayne Brooks, the Dean of the Okanagan School of Business who has been consistently supportive of this project since its inception, both in terms of funding and other forms of institutional support. The support and interest of colleagues and institutions in Atlantic Canada is also appreciated including Dr Keith Dewar at the University of New Brunswick, Dr Randy Miller at the New Brunswick Museum and Dr Kim Lemky at the Cabot Trail Working Association.

On a personal level we appreciate the support of our respective spouses, Roxi Alix and Darrell Tschirhart who recognized our need to be constantly involved in the project, particularly during its latter stages. Roxi was also very helpful in assembling the final manuscript for us, a task that isn't very exciting but is nonetheless very important.

Thanks are also due to our publisher, the Routledge Advances in Tourism series. In particular, the series editor Professor Stephen Page who was very helpful in reviewing our original proposal and who provided the suggestion of the title. Emma Travis, the Commissioning Editor for Tourism at Taylor and Francis effectively managed the proposal review process and Michael Jones and Faye Leerink, the Editorial Assistants for Tourism and Geography, have been patient and supportive of the proposal especially through the challenging period of assembling the final manuscript. We thank them all for their support.

Michael Conlin
Kelowna, British Columbia

Lee Jolliffe
Saint John, New Brunswick

List of contributors

Deborah Che is an Affiliate Assistant Professor and Senior Fellow, School of Public Policy, George Mason University. She received her PhD from Clark University in 2000. Her research interests include rural and community development, natural resource-based tourism (i.e. agritourism, ecotourism, hunting) development and marketing, cultural/heritage tourism, and arts-based economic development strategies. As a result of her numerous publications and funded research projects, she has been invited to present in international geography and tourism conferences in the USA, Europe, Asia, and Canada. Dr Che is on the editorial board of *Tourism Geographies* and has served as Chair of the Recreation, Tourism, and Sport specialty group of the Association of American Geographers. deborah.che@gmail.com

Nancy Chesworth is an Associate Professor in the Department of Business and Tourism and Hospitality Management at Mount St. Vincent University in Halifax, Nova Scotia. Her research interests and publications focus on ecotourism, demographics, niche markets and climate change. Nancy.Chesworth@msvu.ca

Michael Conlin is a Professor at the Okanagan School of Business, Okanagan College in Kelowna, British Columbia, Canada. He also holds teaching appointments at the University of Guelph, Ontario and Royal Roads University in Victoria, British Columbia. He has been an elected Board Member with International CHRIE and the Canadian Chapter of the Travel and Tourism Research Association. His research interests have included island tourism and he was the co-editor (with Dr Tom Baum) of the first book on the subject, *Island Tourism: Management Principles and Practices* by Wiley. In addition to industrial heritage tourism and transportation infrastructure attractions, his current research focuses on sustainable building technologies and education in the tourism field in conjunction with Okanagan College's innovative Centre of Excellence in Sustainable Building Technologies and Renewable Energy Conservation in Penticton, British Columbia. When completed in 2011, the Centre will be one of the world's first LEED Platinum research and academic instruction facilities and include an incubator for companies engaged in research and innovation relating to sustainable technologies. mconlin@okanagan.bc.ca

Keith Dewar began his academic career in 1992 after 20 years with Parks Canada. With Parks Canada, he served as park naturalist and visitor services officer at the Point Pelee, Jasper, Rideau Canal and Saint Lawrence Island parks. After leaving Parks Canada, he taught tourism and environmental management at Massey University in New Zealand for 10 years before returning to Canada to teach at University of New Brunswick-Saint John.

He has also taught in China. His major research interests are natural and cultural heritage management. kdewar@unbsj.ca

Angela Cabral Flecha is Professor of the School of Tourism, Federal University of Ouro Preto, Minas Gerais, Brazil and a PhD candidate in Production Engineering. She teaches marketing, tourism, hospitality and events disciplines and coordinates two research groups. The first one focuses on tourism studies in general – Research and Advanced Studies Tourism Group, NUPETUR – where she has a special interest in the studies on tourism networks. The second one focuses on impacts of tourism events – she coordinates the Events Studies Group. She was responsible for the organization of the most important event about tourism in Brazil in 2008 for specialists from three continents on the theme tourism management. aflecha@turismo.ufop.br

Kathryn Flynn is a Senior Lecturer in the Swansea Business School at Swansea Metropolitan University. Her main teaching duties are with the School's undergraduate portfolio in leisure and tourism management. Her teaching specializations relate to leisure operations management, culture and arts management and student work based learning activities. She completed her MPhil in cultural regeneration and tourism and continues to research within this area. kathryn.flynn@smu.ac.uk

Elspeth Frew is a Senior Lecturer in Tourism Management at La Trobe University in Melbourne, Australia. Dr Frew moved to Australia from the UK in 1991 to take up a position at Monash University, Melbourne. From there she moved to Victoria University, Melbourne before joining La Trobe University in 1998. Her research interest is in cultural tourism, with a particular focus on festival and attraction management, industrial tourism and personality and tourism. Her PhD focused on the managerial aspects of operating industrial tourism attractions, including the interpretative aspects of operating these cultural heritage sites, and the extent to which personality type can distinguish tourists and reflect their tourism choices. In recent years, she has received several competitive research grants, particularly from the Australian Cooperative Research Centre for Sustainable Tourism. These grants considered the following: innovation and change management among small to medium sized tourism enterprises, success factors among nature based tour operators, interpretation of cultural heritage in national parks and, professionalism and excellence among small business tourism operators. E.Frew@latrobe.edu.au

Warwick Frost teaches tourism at La Trobe University, Melbourne, Australia. He has written extensively on tourism and heritage, ranging from ethnic groups to popular culture icons. Much of his research has focused on the Victorian Goldfields. He was the Convenor of the Organising Committee for the 150th Anniversary of the Victorian Gold Rushes Festival in 2001. His latest publication is *Tourism and National Parks: International Perspectives on Development, Histories and Change* (co-edited with C.M. Hall, Routledge 2009). W.Frost@latrobe.edu.au

Mary Beth Gouthro is Lecturer and Programme Leader at the School of Services Management at Bournemouth University (Dorset, UK). She currently leads the programme delivery for the BA (Honours) in Events Management programme at Bournemouth University, and lectures at the first and final year of the degree programme. Her research interests span the experience of consumption in areas that include tourism, events and leisure settings. She has published in the areas of event satisfaction in festivals, tourism consumption in heritage settings and perspectives on qualitative research approaches. Other research interests include festival & event volunteering, event motivation and qualitative research methodologies adopted in the scope of services management. Dr Gouthro's industry and consultancy experience spans the charity sector, corporate organisations, small business and government. mgouthro@bournemouth.ac.uk

Kevin Hannam is Professor of Tourism Development at the University of Sunderland, UK. He is a founding co-editor of the journal *Mobilities* (with John Urry and Mimi Sheller). He is Chair of the ATLAS Backpacker Research Group and has published widely on aspects of cultural, heritage and ecotourism development in India and Scandinavia. He holds a PhD in geography from the University of Portsmouth. kevin.hannam@sunderland.ac.uk

Cuong Nguyen Duc Hoa was born in Hanoi, Vietnam and is a Lecturer in the Faculty of Management and Tourism at Hanoi University specializing in teaching tourism planning and sustainable tourism. He completed a Master of Science in Tourism Management at CERAM Graduate Business School in Sophia Antipolis, France in 2002. For several years, he worked as a tourism researcher for Vietnam National Administration of Tourism where he participated in numerous tourism planning projects throughout the country. He has also been providing consulting services to local government authorities and various international development organizations and projects in Vietnam, in particular the Asia Development Bank (ADB) or Netherlands Development Organization SNV. He has written several dozen articles and planning documents that were published in tourism magazines and publications of international development organizations in Vietnam. cuongndh@hanu.edu.vn

Mary Hollick is Deputy Head of School, Undergraduate Programs and Senior Lecturer in Tourism, in the School of Business, University of Ballarat, Victoria. She has considerable experience as an educator, commentator and consultant in tourism policy and planning. She has a strong background in museum services combined with extensive experience in the fields of staff development, training and visitor services, specializing in social history and Australian heritage management. Dr Hollick has provided consultant services in destination management, attractions management and areas of social and built history conservation and presentation. Her areas of interest and expertise include: tourism attraction management; pathology of tourism business failure; sustainable heritage management; micro tourism enterprise skills development; community capacity building and service quality measurement. Her specific research interests include regional tourism planning, heritage management and visitor learning in museums. m.hollick@ballarat.edu.au

Stuart Howard is a Senior Lecturer in Social History at the University of Sunderland, UK. He helped found the North East England Mining Archive and Research Centre (NEEMARC). NEEMARC is home to important mining information and houses trade union records, health and safety information, technical reports and legal records relating to mining legislation. stuart.howard@sunderland.ac.uk

Jana Jablonská is an Associate Professor at the Technical University of Košice, BERG Faculty. She is a mining geologist and earned her PhD in 1992 researching the geological implications of zircon morphometry. Later, she devoted her attention to environmental geology and related areas. She has also been doing research work at the University of Nice and the University Ecole de Mine de Paris. Jana.Jablonska@tuke.sk

Lee Jolliffe is an Associate Professor in the Faculty of Business at the University of New Brunswick, Saint John, Canada and a Visiting Professor at Hanoi University, Vietnam and the University of the West Indies, Barbados. With an academic and professional background in museum studies, in 1997 she turned her attention to cultural tourism, first working on tourism research for the North Atlantic Islands Program at the University of Strathclyde in Glasgow, Scotland. Since then her research has focused on cultural and heritage tourism, seasonality in tourism, and various forms of culinary tourism. She is a member of the editorial board for *Annals of Tourism Research* (Resource Editor for museums) and the author of two edited volumes: Tea and Tourism: Tourists, Traditions and Transformations (2007) and Coffee Culture, Destinations and Tourism (in press for 2010) both published by Channel View Publications. ljolliff@unbsj.ca

Andrew Jones is a Senior Lecturer at Swansea Business School, Swansea Metropolitan University. He has professional experience in planning and tourism at international, regional and local levels and has been an enthusiastic contributor to the academic tourism community within the UK and Europe. In this respect he has been an active practitioner, researcher and teacher in planning, conservation, regeneration and tourism planning since 1981. In his present post as Director for both postgraduate and undergraduate tourism programmes Dr Jones specializes in Town and Country Planning, Cultural Regeneration, Tourism Development and Environmental Policy. He has carried out numerous consultancy projects relating to tourism and leisure and has published a number of industry and academic articles, reports and conference papers. In addition, he has contributed to international conferences in Europe, North America, Asia and Australasia and is currently editing two books. andrew.jones@smu.ac.uk

Marcos E.C.G. Knupp is a member-researcher of NUPETUR – (Research and Advanced Studies Tourism Group) where he developed several research projects and publications in the area of tourism, with special attention to issues of complex networks, events and tourism. He was one of the organizers of the 1st International Meeting on Tourism Management held in 2008 in Ouro Preto. His research interests include policies on tourism, tourism planning, complex networks, social networks, organizational networks, environmental education and events. marcos.knupp@gmail.com

Jennifer Laing is a Lecturer in Tourism in the Tourism Research Unit, Department of Management, Monash University, Australia. She has degrees in law and management, and worked in legal, publishing and marketing roles before returning to academia. Her PhD was on tourist behaviour, focusing on motivations behind niche travel experiences. Her research interests include heritage tourism, the tourist experience and links between tourism and well-being. jennifer.laing@buseco.monash.edu.au

Jane Legget joined Auckland University of Technology in 2008 as Associate Director (Cultural Heritage) with the New Zealand Tourism Research Institute and Senior Lecturer in Tourism Studies within the School of Hospitality and Tourism, specializing in cultural heritage tourism. She has worked in a number of museums and heritage organizations in Britain and New Zealand, most recently with the Museum of New Zealand Te Papa Tongarewa. Her PhD research on museums and accountability was supported by a Bright Futures Top Achiever Doctoral Research Fellowship from the Foundation for Research, Science and Technology. She is an elected Fellow of the Museums Association for Great Britain and edits *Te Ara – Journal of Museums Aotearoa*. Her research interests include museums and tourism, museum and heritage development, partnerships with community stakeholders, quality standards in museums and art galleries, public accountability in the heritage sector, cultural heritage policy, and women and heritage. jane.legget@aut.ac.nz

Kim Lemky has worked in both government and academic agencies for the last 25 years. She studied tourism in Ecuador as part of her Master's research. More recently, she has focused on visitor management and hiking trail management in protected areas (e.g. Cape Breton Highlands National Park of Canada). Since graduating with a PhD in Geography from the University of Waterloo in 2002, she has taught geography and recreation across Canada, from Memorial University of Newfoundland to the University of Northern British Columbia. Currently, she is the Tourism Coordinator for the Cabot Trail Working Association, a non-profit organization. She is involved in re-creating the Cabot Trail as a scenic destination area, working on projects from developing a way-finding system, entry signs, and interpretive panels as well as managing the www.cabottrail.travel website. She lives near the Cabot Trail, in Cheticamp, Nova Scotia. cabottrail.tourism@ns.sympatico.ca

Antonio Liccardo is a Professor at the Federal University of Paraná in Brazil. As a geologist he works with general geology, gemstone mineralogy, mining history and especially geotourism. His research into this new area of geology includes science divulgation and interface actions with different areas, including tourism. He is the author of a dozen books including *Geotourism in Curitiba, Mining History of Paraná and Gemstones of Minas Gerais.* His current research focuses on the implantation projects in Geoparks in Brazil, specifically Quadrilátero Ferrífero Geopark (Minas Gerais) and Campos Gerais Geopark (Paraná), together with other researchers. liccardo@geoturismobrasil.com

Gui Lohmann is a Lecturer in Transportation and Tourism at the School of Tourism and Hospitality Management, Southern Cross University, Australia. He is also a visiting professor in the Masters Program in Culture and Tourism at Santa Cruz State University, Brazil. He is also the tourism series co-editor for Aleph, the largest tourism publishing house in Brazil, and the director of national and international relations of the Brazilian Tourism Research Association (ANPTUR). His research interests include tourism transportation geography, tourism marketing, theoretical aspects of tourism and leisure and tourism for families. lohmann.gui@gmail.com

Randall Miller is the Curator of Geology and Palaeontology at the New Brunswick Museum in Canada. He is also an Adjunct Professor in the Department of Geology at the University of New Brunswick and a Professional Geoscientist with the Association of Professional Engineers and Geoscientists of New Brunswick. Prior to his appointment at the New Brunswick Museum, he undertook contract research for the Geological Survey of Canada and the Canadian Museum of Nature. Dr Miller has published more than 60 scientific papers on a variety of topics with a focus on fossils or fossil sites. In January 2004, his work on fossil sharks was described in Discover magazine and featured on the CBC Radio show 'Quirks and Quarks'. He has also written a children's book, *Fossil Hunter – Will and the Giant Trilobite*, to accompany an exhibit of the same name, and contributed to a best-selling book about the geology of the Maritimes called *The Last Billion Years*, and co-authored *Rebuilt in Stone: Geology and the Stone Buildings of Saint John, New Brunswick*. Randall.Miller@nbm-mnb.ca

Catherine Palmer is Assistant Head of School – Head of Research for the School of Service Management, University of Brighton, UK. She is part of the School's Centre for Tourism Policy Studies. She is a member of the Association of Social Anthropologists and a Fellow of the Royal Anthropological Institute. Her research and publications focus on culture, materiality, identity, and heritage including: 'Royalty, National Identity, Heritage and Tourism', in P. Long and N. Foster (eds) *Royal Tourism: Excursions around Monarchy* (2008); 'Stalking the cannibals: photographic behaviour on the Sepik River', *Tourist Studies* (with J. Lester, 2007); 'An ethnography of Englishness: Tourism and national identity', *Annals of Tourism Research* (2005); 'From theory to practice: Experiencing the nation in everyday life', *Journal of Material Culture* (1998). Dr Palmer's teaching and supervision of research students reflect her research interests. C.A.Palmer@bton.ac.uk

Carlo Perelli is Research Assistant at CRENoS, University of Cagliari and University of Sassari, Italy. After an interdisciplinary MA in Mediterranean Studies between Venice, Montpellier and Madrid, he received his PhD at *The School for Advanced Studies in Venice Foundation*, for a thesis on the implementation of sustainability policies in mass tourism coastal destinations. His interests are in tourism geography, planning for tourism, Integrated Coastal Zone Management (ICZM) and Tourism Carrying Capacity Assessment (TCCA) mainly in Sardinia (Italy), Morocco, Tunisia and Algeria. He has been working for a year in the team developing the Socio Economic Analysis and the Executive Plan for the Sardinian Geopark. carloperelli@hotmail.com

Paola Pinna is Research Assistant at CRENoS, University of Cagliari and University of Sassari, Italy. She received a MBA in Economy and Marketing of Tourism at the Faculty of Political Science, University of Sassari, Italy and a MA degree in Cultural Management at Barcelona University, Spain. She worked on the guidelines for the Strategic Management Plan of the Archeological Park of Xoclán (Mexico). Her main research interests are in sustainable development, cooperation and development, cultural tourism, cultural heritage at local, national and international level. She has been working for a year in the team developing the Socio Economic Analysis and the Executive Plan for the Sardinian Geopark. papinna@gmail.com

María de los Ángeles Plaza-Mejía is a Senior Lecturer in managerial skills and strategic management at the Spanish University of Huelva where she is in charge of its Business Internationalization. She is an active member of the research group on "Innovation and Development Strategies in Tourist Firms" (GEIDETUR). Her research is mainly about industrial tourism. plaza@uhu.es

Nuria Porras-Bueno is a Senior Lecturer in business administration at the University of Huelva in Spain, where she is an active member of the research group on "Innovation and Development Strategies in Business Tourism" (GEIDETUR). She is the author of many papers published in scientific journals. porras@uhu.es

Bruce Prideaux is Professor of Marketing and Tourism Management and Director of the Centre for Tourism and Hospitality Research at the Cairns campus of James Cook University. He is actively involved in research and manages a number of long-term large scale research projects that examine aspects of marine and rainforest tourism in North Queensland, and climate change. His research unit currently employs three Postdoctoral researchers and three part time research assistants. In addition, he is principal supervisor for six PhD students. He has a diverse range of research interests including climate change, tourism transport, heritage, destination development, seniors, backpackers, crisis management, marine and rainforest research and more recently future studies. He has published over 200 book chapters, journal articles, conference papers and reports. His most recent publications in 2009 are two books, an edited volume on river tourism and a book that explores a range of issues about resorts and destinations. He is also a member of nine editorial boards of scholarly journals. bruce.prideaux@jcu.edu.au

Keir Reeves is a Monash Research Fellow co-housed in the Tourism Research Unit and the National Centre for Australian Studies at Monash University, Australia where he is working on a five-year project titled 'Heritage tourism and the historical landscapes of Australia, Asia and the Pacific'. This research investigates how cultural landscapes can be used as a research framework to evaluate heritage tourism sites. He is the current historian member of the Heritage Council of Victoria and his research has been closely connected to Australian and Asian history, world heritage studies, heritage tourism, mining heritage, sustainable heritage management and development tourism in Asia. With William Logan he is a contributing co-editor of *Dealing with Difficult Heritage: Remembering Places of Pain and Shame* (Routledge, 2009) a key outcome of an Australian Research Council Discovery grant. He is also the contributing co-editor to *Deeper Leads: New Approaches in Victorian Goldfields History* (BHS, 2007). keir.reeves@buseco.monash.edu.au

Pavol Rybár is a Professor at the BERG Faculty, Technical University Košice and Director of the Institute of Geotourism. He was the Dean of the same Faculty. In the late 1980s, he worked as a geologist in Havana. He is a foreign member of the Russian Mining Academy of Science, President of International Association for Geotourism, past president of the Society of Mining Professors, past deputy of the Slovak Republic in INTER OCEAN METAL – body for extraction of minerals from sea floor, past member of the Board of Directors, Federation of European Mining Programs. He had teaching assignments in the former Yugoslavia, France, Finland and the USA. His interest in geotourism has led also to publications on mining tourism related to Banska Stiavnica. Pavol.Rybar@tuke.sk

Giovanni Sistu is Professor of Geography at the Faculty of Political Science, University of Cagliari, Italy. He graduated with honours in Geological Sciences at the University of Cagliari in 1985. He received his PhD in Mineral Prospecting at the Faculty of Engineering at the University of Cagliari. He worked for the *Centre for Geo-mining and Mineralogical Studies* of the CNR (Italian National Research Council). He was Junior Lecturer in Political and Economic Geography and Environmental Policies at the Faculty of Political Science, University of Cagliari, between 1995 and 2005. His main research interests are in environmental policies and local development; cultural tourism and geography in the Maghreb; water and waste management in the Mediterranean Region. sistug@unica.it

Geza M. Timcak is an Associate Professor at the Institute of Geotourism, BERG Faculty, Technical University, Košice, Slovakia. As the first chairman of the Faculty Senate after the Velvet Revolution and a long-standing member of the University Senate, he supported the diversification of the teaching programme of the Faculty of Mining that eventually led to forming the Department of Geotourism. He was the founding director of the Institute of Lifelong Learning at the Technical University, Košice. He also taught at the College of Tourism and Hotel Management in Nicosia, Cyprus, at the Indian School of Mines in Dhanbad and at Imperial College, London. He is a member of the Hungarian Engineering Academy. He serves on the editorial board of *Geotourism, Tourism Today* and the *European Journal of Tourism Research*. In 2002, he established the GEOTOUR – International Conference on Geotourism. He has written a number of textbooks for students in his respective fields of interest. These include information systems in the tourism industry, strategic planning in tourism and human potential development in the tourism profession. His books on selected aspects of human potential development have been published in Slovak, Hungarian and German. gejza.timcak@tuke.sk

Dallen Timothy is Professor of Community Resources and Development at Arizona State University, and Director of the Tourism Development and Management Program at the same institution. He is also Visiting Professor of Heritage Tourism at the University of Sunderland, England, and Editor of the *Journal of Heritage Tourism*. Professor Timothy is commissioning co-editor of Channel View's Aspects of Tourism book series and serves on the editorial boards of twelve international social science journals. He has authored many books and articles on various aspects of tourism, including heritage, politics, international borders, participatory planning, and religious tourism. Dr. Timothy's current field work and data-collection includes projects related to political borders, heritage, tourist trails, religious tourism, nature conservation, and souvenir consumption in Israel, Belgium, the Netherlands, Canada, Costa Rica, Panama, Mexico, the Republic of Georgia, Bhutan, Nepal, India, and various parts of the United States. dtimothy@asu.edu

Alfonso Vargas-Sánchez is a professor of strategic management at the Spanish University of Huelva. He heads its research group on "Innovation and Development Strategies in Tourist Firms" (GEIDETUR). He is also the author of many papers published in journals such as *Journal of International Food and Agribusiness Marketing, European Journal of Information Systems, Human IT, Journal of High Technology Management Research, Technology Analysis & Strategic Management, Journal of Tourism and Development, International Journal of Business Environment, and the Journal of Travel Research*. vargas@uhu.es

Fiona Wheeler is a PhD student in the Tourism Research Unit at Monash University, Melbourne. Her PhD topic explores the process of marketing and branding regional (i.e. rural) tourism areas through a case study of North East Victoria, Australia. Fiona has degrees in commerce and tourism management and a strong marketing background prior to returning to academia, with experience in product and market development, customer experience management, and branding. Her research interests include: the regeneration of regional areas through tourism; tourism marketing, communications and branding; and cultural and heritage tourism. fiona.wheeler@buseco.monash.edu.au

Leanne White is a Lecturer in Marketing and a Research Associate of the Centre for Tourism and Services Research at Victoria University in Melbourne, Australia. She has taught Marketing, Public Relations, Communications, Politics and Australian Studies at universities since 1988. Leanne has also worked in the areas of public relations, research and policy in government and higher education. Her research interests include: advertising, national identity, commercial nationalism, tourism, sports marketing and sponsorship, and the Olympic movement. Leanne's doctoral thesis examined the manifestations of official nationalism and commercial nationalism at the Sydney 2000 Olympic Games. Leannek.White@vu.edu.au

PART I

Introduction

1 What happens when mining leaves?

Michael Conlin and Lee Jolliffe

Introduction

Around the world, mining has been a basis for the development of industrial societies. Now there is something new in terms of mining – the conversion of mining valued for industrial purposes to mining valued for its heritage and tourism aspects. As the societal and communal values related to mining change, a new type of tourism related to industrial heritage has developed including the establishment of visitor attractions related to mines. These developments have produced a wide variety of visitor attractions ranging from the opportunity for tourists to visit working mines, to the creation of mining based visitor attractions of varying complexity many of which are located at or near defunct mine sites, through to the more traditional formation of community based museums that preserve and celebrate local mining heritages and extending to the bundling of mining attractions into heritage routes. The development of cultural attractions related to mines and minerals is universal, growing and important from both historical and economic perspectives, as reflected by the establishment of the UNESCO Global Geoparks Network and other regional consortiums of geotourism activity such as the European Geoparks Network.

When mining leaves a community, there are invariably negative impacts, most notably in terms of the decline in economic activity and the resultant loss of employment. Transforming mines into attractions will never replace the level of economic benefits created by mining or replace all the jobs lost, but it can contribute to the development of tourism as an alternate economic activity and can also preserve mining heritages for the benefit of communities. The development of defunct mines into visitor attractions can result in direct and indirect revenue generation by tourists, funding by governments and other authorities for economic diversification activities and through the creation of alternative employment, albeit usually on a much smaller scale than was available through mining directly. Retired miners, for example, act as tour guides at the Hollinger Gold Mine in Timmins, Ontario which provides employment and, at the same time, lends authenticity to the tourism experience of visiting the decommissioned mining site (Kauremszky, 2009). This process can be found around the world at mining heritage visitor attractions and a number of examples are described later in this book.

Why then a book about mining heritage in relation to tourism? There is definitely a need to understand the complexities of why and how mining heritage and mining sites should be preserved, interpreted and presented to the public, be it the local mining community or the international community of tourists. The need stems as much from the physical nature of mining sites as large and formerly industrial sites as it does from the variety of meanings and narratives associated with the act of mining, for example, including but not limited to indigenous, colonial, industrial and now post industrial subtexts. Visiting sites related to mining can be of interest to the industrial enthusiast as well as the purposeful cultural tourist

(McKercher and du Cros, 2002). The activity of visiting areas of mineral occurrence and production can also be identified as a form of special interest or niche tourism (Novelli, 2005) called geotourism (Hose, 2005). While the book's title purports to provide a global synthesis of the state of the connection between mining heritage and tourism, it is in fact only a beginning, surveying how mining heritage is interpreted (Part II), how mines are transformed into heritage attractions (Part III), the development of mining heritage destinations (Part IV), the growth of mining heritage destinations globally (Part V) and finally identifying opportunities for the development of mining related tourism (Part VI).

The preservation vs tourism issue

The following chapters examine mines and mining heritage as the basis for the development of tourism. As we have discussed, this process produces a wide variety of visitor attractions and activities. However, the common feature of most mining attractions is some form of a museum. These museums range from vast complexes of actual mines, smelters, equipments, buildings and supporting infrastructure through to single buildings or rooms which contain collections of documents and small artifacts. The British Columbia Museum of Mining at Britannia Beach at the headwater of Howe Sound on Canada's West Coast, is a prime example of the former encompassing a large property with accessible mine shafts, numerous buildings housing extensive collections of documents and artifacts, a significant equipment collection, access to a small historic town site, and most notably the 26-storey tall heritage listed gravity-fed concentrator mill.

A good example of the other end of the scale is the collection of documents, photographs and small artifacts commemorating the smelter activities founded at Anaconda, Montana at the turn of the last century. This collection tended by a small group of volunteers is housed in a basement room in the old city hall, an historic property which was itself under threat of demolition in the near past. The smelter facilities at Anaconda processed the output of the large mines located at Butte, Montana. It is impossible when visiting the town not to know that it was a site of enormous mining related activity. After all, the 585 foot Anaconda Smelter Stack is visible from miles away and billed as the tallest free standing masonry structure in the world. The Stack is now on the US National Register of Historic Places and is a state park, thus assuring its preservation. However, the best any tourists can do is to view the Stack from a significant distance as access to the site is forbidden. Virtually, all other remnants of the town's former industrial glory have disappeared. The town shows significant signs of economic decay and the collection described above is susceptible to abandonment as the municipality has not formalized its heritage value.

At first glance, the term mining museum may seem somewhat oxymoronic. After all, museums usually bring to mind images of quiet, clean, often scholarly repositories of documents, pictures, works of art and human artifacts. Mines on the other hand evoke images of noisy, dirty and dangerous holes in or under the ground. Museums are often considered to be elite and for the leisure class while mines bring to mind serious industrial activity where the working class use their trade skills and labour and risk their lives to earn a living for their families. Notwithstanding the obvious fallacy of these stereotypes, the world of the museum and the world of the mine would seem to be very far apart in cultural, sociological and economic terms. However, these perceptions all change when mines either exhaust their primary extractive purpose or when their operators look to expand the opportunities presented by both the mining infrastructure and artifacts and the activity of mining itself to generate revenue, interest and goodwill from visitors who are for the want of a better term, tourists.

In these instances, the defunct mine site takes on significance as a symbol of cultural, social and economic heritage and the alternative activity provides the basis for the continuation or preservation of the mining site after its original economic *raison d'etre* ends. It is at this stage of a mine's life cycle that this book focuses its enquiry. The fundamental purpose of this enquiry is to examine why, when and how mines are transformed into visitor attractions, be they museums, tourism complexes or informal collections.

The injection of the concepts of tourism, heritage preservation and interpretation, and financial sustainability into a discussion about mines and museums provides an ideal foundation for the fusion of the two activities. And it is this fusion within the context of what can be called industrial heritage tourism that is the focus of this book. The chapters comprising this book describe, analyse and critique the development of mines and mining sites as visitor attractions. As we know, the process of developing these sites into visitor attractions is accomplished mainly through the creation of museums. However, as some of the chapters reveal, this process can and does take development well beyond the more traditional notion of a museum and into the realm of more contemporary tourist attraction as is the case in the development of the Sovereign Hill site in Australia.

That there are museums based on mines and mining should not be surprising when considered within the perspective of heritage preservation. The International Council of Museums in its standard definition of a museum makes it clear that heritage preservation is the fundamental objective of museums:

> A museum is a non-profit, permanent institution in the service of society and its development, open to the public, which acquires, conserves, researches, communicates and exhibits the tangible and intangible heritage of humanity and its environment for the purposes of education, study and enjoyment.
>
> (ICOM, 2004)

Therefore, it is not surprising that given the pervasiveness of mines and mining throughout history, their value as symbols of heritage has been recognized in ways which among others, includes the establishment of mines as museums at appropriate points in their life cycles. However, much of this book discusses the inevitable conflict between the more traditional roles of a museum as a means of preserving heritage for the future with the inevitable challenges of funding this preservation and turning what, in most cases, are now defunct economic activities back into sustainable opportunities for their communities.

The chapters in Part II of the book provide ample evidence of this process, from a historical perspective and also in terms of the modern challenges facing heritage preservation. Issues of interpretation from anthropological, sociological and political perspectives abound in the history of mining-based visitor attraction development. What becomes clear, however, from the contributions to this book is that these challenges of interpretation are impacted by more modern and arguably mundane challenges such as funding and future preservation efforts. In reality, however, these modern challenges are not mundane since they threaten the concept of the mining attraction if not resolved and discussions with practitioners in the mining museum field invariably highlight the competing perspectives of historical preservation and interpretation and current and future survival. Much of the discussion in the chapters found in Part III highlights this conflict and examines approaches to resolve the dilemma which preservation and funding produces.

In terms of tourism development, the concept of mines and mining museums is not new and the chapters in Part IV illustrate a number of these historical developments. However, this development is increasing and around the world, communities are realizing that their mining heritage has value, both culturally and economically. The chapters in Part V of the book illustrate how dramatic this development is. What becomes clear from this extensive discussion is that the fusion described above produces a very complex range of developments full of opportunities and fraught with challenges. Chapter 20 summarizes these opportunities and challenges in a typology of mining visitor attractions that provides lessons for development and sustainable management.

Outline of the book

The chapters in Part II (Chapters 2–5) deal primarily with issues related to historical interpretation. As such, Chapter 2 is an excellent launching point for consideration of the historical perspectives which have given rise to the formation of museums dealing with industrial heritage and in the case of the northeast of England, specifically coal mining. Howard and Hannam, in their comprehensive discussion of the founding of two mining museums, The John Bowes Museum and the Beamish Open Air Museum, highlight the sometimes conflicting interpretations which can be made of the historical development of the mining industry. The authors contrast the aristocratic (as they label it) motivation of the Bowes family in the nineteenth century, to provide an educational opportunity for workers in their mines, with the late twentieth century goal of a group of local government bodies of preserving and celebrating the local folk culture of the region. These two differing perspectives result in quite two different attractions, both of which have value and which, in these specific cases, are highly successful tourism attractions.

In Chapter 3, Reeves *et al.*, discuss, among other points, the primary issue addressed by Howard and Hannam, namely differing interpretations of the same history, in the context of the changing interpretation of the role of Chinese migrants in the development of mining in Australia, Canada, New Zealand and the United States. The authors discuss the shift in interpretation of the role of Chinese migrants from a focus on negative elements of racism and exploitation to a more humanistic perspective based on the development of relationships between races and classes in the mine fields of the Pacific Rim. They point to, among other consequences, the growth of vibrant 'Chinatowns' in the mining regions which, for the most part, have survived the disappearance of mining and which provide the basis for tourism development. Complimenting this discussion, the authors provide a very comprehensive discussion of the development challenges and responses related to visitor attractions based on the Pacific Rim gold mining industry including issues of excavation, preservation of relics, site and location accessibility and the desirability of employing a cultural landscape approach to interpretation of these sites in order to capture the full significance of them, both historically and from a tourism perspective.

Gouthro and Palmer, in Chapter 4, approach the issue of historical interpretation from a perspective grounded in the anthropology of tourism. They make the argument that visits to mining heritage sites can be considered as a form of pilgrimage. Their argument compares historical and contemporary pilgrimage with tourism, both of which involve travel and visitation to historically significant sites. In their discussion of The Big Pit Museum in Wales and the Cape Breton Miners' Museum in Nova Scotia, Canada, they use the results of research conducted at both sites to reinforce the notion that visits to such historical sites can, for families and individuals with a connection to the region or the mining industry, reinforce their cultural and national identity.

The first three chapters in Part II focused on the historical perspectives that have given context to the interpretation of mining attractions. The last chapter in this section, Chapter 5, deals with a completely modern influence, namely the impact of media on interpretation and its significance for visitor attraction development and management. Leanne White discusses the process of exploitation that took place following the accident at the Beaconsfield Mine in northern Tasmania in 2006. She chronicles the extensive news broadcasting that accompanied the disaster, the range of cultural by-products (songs, poems, plays and films) which have flowed from the incident, the role of national, regional and local governments in managing and ameliorating such events, the developments in the small town to compensate for the loss of employment caused by the accident and the impact which the event has had on tourism in the community, both negatively and positively.

Part III (Chapters 6–10) shifts the discussion from historical interpretation to the process of transforming mining sites into visitor attractions. Like many of the chapters throughout the book, Jane Leggett's discussion in Chapter 6 of challenges faced by two mining museums on New Zealand's North Island highlights the range of issues faced by visitor attraction developers. However, unlike many of the other examples in this book, the two museums examined by Leggett are small, community based ventures heavily dependent upon volunteer support and with limited access to resources. The chapter discusses the growth of volunteerism in the context of a desire on the part of some in their communities to preserve their heritage, and identifies the challenges faced by volunteer groups in securing funding, maintaining a cohesive vision for their projects and motivating their communities as a whole with the value of their projects. The impact of governmental policy on land use, funding and the management of historical and heritage sites is also discussed. The conflict between heritage preservation and visitor attraction development is also a theme throughout the chapter, as it is in many of the examples in this book.

Chapters 7 and 8 both look at the development of mining sites as visitor attractions alongside non-touristic operations, in one case, working mines and in the other, a multi-purpose and somewhat synergistic operation. Elspeth Frew uses two examples, PowerWorks in Victoria and the Super Pit in Western Australia, both the location of working mines, to identify the practical considerations which operators must face when contemplating the opening of their sites for visitation by tourists. These range from the obvious but very important issue of investment necessary to provide for tourists through to the possibility of unforeseen consequences on, for example, the public's perception of the mining process. Because of the nature of mining sites, there are inherent dangers which give rise to issues of visitor liability. Frew concludes that the two sites she examines provide positive lessons in how to manage what are two distinctly different activities on the same site. She further concludes that separating the two activities by developing separate purpose-built attractions on the site, sometimes done to address the conflicts between the two different activities, results in the loss of the 'real life' element that makes this challenging model of mining museum development so valuable.

In Chapter 8, Deborah Che, in her discussion of the newly opened Kansas Underground Salt Museum at Hutchinson, Kansas in the United States, illustrates how mines can be used for synergistic purposes alongside their primary function. The unique climate in the salt mine provides an ideal place for the storage of materials which would otherwise have been negatively impacted by the surface climate. The mine is now home to historical collections of films, costumes, data tapes and corporate papers. These collections, in turn, form part of the visitor attraction alongside the non-operational areas of the mine. The museum also forms part of a range of attractions in the region and its role, both in terms of financial impact on the local economy and as a visitor attraction, are discussed.

Returning again to Australia, Warwick Frost examines, in Chapter 9, the development of the Castlemaine Diggings National Heritage Park in Victoria which has the distinction of being the first national park established in Australia which explicitly celebrates the historical and cultural landscape created by the Australian gold rushes of the nineteenth century. Specifically, Frost addresses the complexity of constructing an interpretation plan which adequately addresses both the environmental and the cultural landscapes found in the park. In doing so, he focuses on the challenges posed by a lack of historical research relating specifically to the environmental impact of gold mining on the site and the absence of existing models to provide guidance to park managers in their developmental activities. He concludes the chapter by offering a range of options for the development of interpretation schemes which would address the complexities of the site.

The final chapter in Part III also looks at an Australian site, the Sovereign Hill Museum at Ballarat in Victoria, and offers lessons of a more commercial nature. Unlike many of the sites discussed throughout this book, Sovereign Hill is a genuine tourism success story when measured in terms of levels of visitation and revenue generation. It includes a range of activities which, for many heritage tourists, might be seen as somewhat frivolous. However, the site has been recognized by various national and regional bodies for its development and management as an attraction based on mining and its success brings the promise of future sustainability. Specifically, Mary Hollick examines the development and management of the Museum's most successful 'product', the 'Blood on the Southern Cross' sound and light show which chronicles the historically significant Eureka Rebellion of 1854. Hollick traces the development of the show approximately 15 years ago and follows its growth and rejuvenation through the application of management tools to ensure its relevance to the intended market. As such, the chapter is a valuable guide for managers seeking to expand the touristic component of their attractions.

The four chapters comprising Part IV (11–14) of the book offer a rich inventory of developed mining attraction and museum examples. Flynn and Jones, in their discussion of the development of mining attractions in South Wales, trace the growth of heritage tourism in the region starting in the late 1960s and culminating in the broad range of attractions which now form part of the Welsh tourism scene. Their research, outlined in the chapter, provides evidence of the contribution of attraction developments to employment and environmental improvements although these contributions, they conclude, were limited by the size of the attraction, the larger ones being more positive. These outcomes were, in turn, attributed to a higher level of governmental support for the larger attractions.

In Chapter 12, Lemky and Jolliffe in their examination of heritage tourism in the coal mining regions of Cape Breton on Canada's East Coast also point to the early efforts dating back to the late 1960s to exploit the opportunities presented by a growing interest in the history and significance of mining to the region. These developments mirrored the phasing out in the same period of coal mining in the region and in particular, the pressing need to find alternative employment for thousands of miners. They describe a complex environment of competing governmental authorities, conflicting visions about how to develop tourism in the region, and the resulting mixed set of outcomes in terms of economic development. In particular, the authors examine the impact of the Cape Breton Miner's Museum established in 1967, several railway attractions originally built to service the mining industry and the more recent development of the Colliery Route in 2002 as examples of this development.

Chapter 13 moves halfway around the world to examine the development of geotourism in Slovakia in Eastern Europe. Timcak *et al.*, point out that geotourism activities (trekking, mountaineering, etc.) have occurred in the region dating back many centuries. Following a detailed description of the geology of the region, the authors discuss the country's efforts to expand geotourism to include mining heritage tourism. They describe a rich inventory of sites based in and around a number of major cities for which tourism plans focusing on mining attractions have been developed and implemented. However, like many of the other authors, they point to challenges related to funding and the conflict between tourism advocates and heritage advocates and environmentalists as being barriers to ongoing development. Chapter 14 remains in Europe and focuses on the Riotinto Mining Park in Spain. Vargas-Sánchez *et al.*, identify the aspects of mining heritage in the Province of Huelva and examine the opportunities for expanding on the experience of the Riotinto Mining Park. Specifically, the chapter reports on the results of a survey of tourists visiting local mining attractions to determine visitor demographics which will contribute to planning further mining attraction development in the region. The chapter concludes that notwithstanding weaknesses in policy, funding and tourism planning, the future of mining heritage tourism in the region is positive given the range of mining heritage and geotourism assets and the opportunities for collaborative development with, among other segments, non-heritage mass tourism visitors to the region.

Part V (Chapters 15–19) will describe the emergence of mining attraction development around the world, notably in Vietnam, Brazil, Sardinia, Canada, Australia and the United States. In Chapter 15, Cuong Nguyen Duc Hoa *et al.*, trace the development of the Ha Tu Mine located in the northeast of Vietnam from its early history in the French colonial period through its use during the Japanese occupation in the 1940s to its transfer to the Government of Vietnam in 1955. Since then, mining at the site has expanded and continued notwithstanding bombing attacks by American forces during the Vietnam War. The nearby region of Ha Long Bay was named a UNESCO World Heritage Site in 1994 and this has provided a strong foundation for the development of tourism. In spite of conflict between mining and tourism, authorities have experimented with touristic activities which include mine tours and other activities designed to provide tourists with a greater understanding of the role of the mine and its history. The authors conclude their analysis by exploring three possible growth strategies for tourism at the mine. Chapter 16 shifts the focus to South America where Flecha *et al.*, examine the tourism potential in mining heritage in the Brazilian city of Ouro Preto. Designated as a UNESCO World Heritage Site, Ouro Preto is proposing the development of a UNESCO Geopark covering the region. The chapter identifies the various elements present in the region to support tourism generally and mining heritage tourism specifically. Sistu *et al.* in Chapter 17 discuss the experiences of Sardinia leading up to and following the establishment of a Geopark on the island. Not surprisingly, the authors point to solid development opportunities which are moderated by issues of funding, competing jurisdictions and conflict between environmental concerns and tourism.

The final two chapters in the section deal with the transition from stagnant or abandoned locales which contain geological and mining heritage potential and, therefore, provide a window into what the future of development in terms of mining attraction development holds. In Chapter 18, Dewar and Miller discuss the opportunities that exist in the Bay of Fundy region on Canada's East Coast. In particular, they point to the establishment of a UNESCO Biosphere Reserve in the region and the growth of the UNESCO Geopark Network as models for development in the region. As such, they link the potential success of mining tourism in the region to global models for preservation, growth and management and in this way, support the notion of a global synergy in the field. Finally, Prideaux and Timothy, in Chapter 19, present the Ghost Town Life Cycle model to help guide planners in reviving mining communities which have essentially died. Extrapolating on Butler's Tourism Area Life Cycle model, the authors provide a useful tool for identifying phases of growth and decline in mining communities and suggesting points within that cycle for intervention and rejuvenation. They illustrate the model through reference to two locales, Tombstone, Arizona in the United States and Ravenswood,

Queensland in Australia. In this way, they provide a theoretical framework for much, although not all, of the discussion throughout the book.

Conclusion

The chapters in this book present a comprehensive and compelling picture of the current state globally of mining heritage based visitor attraction development. As noted previously in this chapter, these developments can, and in many cases are, different in scope, location and vision. It is also the case, however, that they tend to share many of the same challenges. Given wealth of information and research, this book will present a typology of mining heritage based visitor attractions in Chapter 20.

Heritage tourism has influenced the nature and development of visitor attractions as noted by Williams (1998), introducing tourism into locations previously unknown as tourism destinations. Heritage sites related to mining, often located away from traditional touring routes fall into this category of visitor attractions. They include what Edwards and Llurdés (1996) described in their typology of industrial heritage attractions as productive, processing, transport and socio-cultural attractions. The typology presented in Chapter 20 will seek to expand on the more generic models identified above and in that way, provide a more detailed guide for practitioners and academics alike for the planning, development and management of mining heritage based visitor attractions.

References

Edwards, J. and Llurdés i Coit, J. (1996) Mines and quarries: Industrial heritage tourism. *Annals of Tourism Research*, 23(2): 341–363.

Hose, T. A. (2005) Geotourism: Appreciating the deep time of landscapes, in M. Novelli (ed) *Niche Tourism: Contemporary Issues, Trends and Cases*, Elsevier: Oxford, pp. 27–38.

ICOM News Thematic Files: The definition of the museum. *ICOM News*, vol. 57, 2004 no. 2.

Kauremszky, I. (2009) 'Subterranean: so this is where bling came from?' *Canadian Geographic Travel*, Summer, pp. 18–19.

Novelli, M. (2005) *Niche Tourism: Contemporary Issues, Trends and Cases*. Elsevier: Oxford.

McKercher, B. and du Cros, H. (2002) *Cultural Tourism: The Partnership Between Tourism and Cultural Heritage Management*. New York: Haworth.

Williams, S. (1998) *Tourism Geography*. London: Routledge.

PART II

Interpreting mining heritage

2 The making of two mining museums

Bowes and Beamish, North East England

Stuart Howard and Kevin Hannam

Introduction

The North East region of England boasts two of the most popular museums of their kind in Britain, the Josephine and John Bowes Museum in Barnard Castle and the Beamish Open Air Museum near Chester-le Street. Both Bowes and Beamish attract large numbers of visitors. After dropping down to around 60,000 visitors per annum in the 1990s, Bowes now attracts approximately 100,000 visitors per annum (Northern Echo, 2005), while Beamish attracts in excess of 320,000 visitors per annum (Beamish, 2003). Both of these institutions find their *raison d'etre* in that most potent symbol of North East England – deep coal mining. But in their origins, style and mission they are very different. The Bowes museum is a static, 'disciplined' museum, an institution that regulates 'the performative aspects of their visitors' conduct' (Bennett, 1995: 6). The other being a vast open air museum where the visitor is highly mobile and somewhat less regulated and disciplined (Figures 2.1, 2.2). However, these two museums are also very different in terms of the passions they evoke. In this chapter, we argue that Bowes and Beamish articulate economic and social differences which reflect the region's modern history and structure and ingrained patterns of regional identity and polity. In this sense at least, mining heritage in North East England is culturally significant and hotly debated.

Relatively little attention has been paid to the historical contexts which produced the institutions which mediate contemporary heritage and so often are the subject of contemporary analysis. Yet the history of Beamish and Bowes, for example, casts a different kind of light on the heritage debate, one that exposes the operation of historical forces upon the struggle for representation. Here, we are dealing with origins and we shall engage with the question, why were these institutions established? Here structural factors are of primary consideration, hence, this chapter will identify and analyse the historical forces which created, orientated and sustained Beamish and Bowes in their early years and in different and changing contexts.

John Urry (1990) has argued it is the service classes, i.e. the professional and managerial classes, which have led the contemporary drive towards heritage tourism. And it is this service-class of people who have been better able to exclude competing memories or visions of the past and impose their values upon other classes and social groups. One important way in which this dominance of representation has been accomplished is in museums. Museums often claim to be politically neutral displays of regional or national cultures that tell the stories of those cultures to both domestic and international visitors. However, some objects are selected while others are excluded sometimes in order to meet the regional political aims of particular museums. Museums, then, are far from neutral and are highly selective in terms of both what they communicate and what they exhibit (Bennett, 1995). They can and do exhibit only certain objects and exclude others, depending upon complex historical forces (Hannam and Knox, 2010).

Figure 2.1 Bowes Museum.

Figure 2.2 Beamish Open Air Museum.

The historical narratives transmitted through heritage, both public and/or private, are thus selective, partial and frequently biased and distorted (Johnson, 1999), because, to a significant degree, the narratives reflect political and social structural themes of contemporary people. The study of heritage tourism thus involves acknowledging the contextual power relations that underpin various representations and interpretations. Conceptualizing this as a series of contested meanings allows us to analyse the underlying conflicts and tensions and thus remain culturally sensitive (Graham *et al.*, 2000; Bruner, 2005). In this context Graham *et al.* (2000: 93) have put forward the idea of the dissonance of heritage representations, defined as:

> . . . the mis-match between heritage and people, in space and time. It is caused by movements or other changes in heritage and by migration or other changes in people, transformations which characteristically involve how heritage is perceived and what value systems are filtering these perceptions. The most pervasive source of heritage dissonance lies in the fundamental diversity of societies . . . The complexities of dissonance are further exacerbated by the contemporary expansion in the meanings and scope of heritage, and the concomitant multiplication of conflicts between its uses.

It is this dissonance that we explore in the historical context of North East England. Other chapters in this volume analyse how different mines have been developed into heritage attractions throughout the world by being re-opened to the public for the purpose of interpreting the industrial process. Such sites, of course, appeal to the growing numbers of 'heritage tourists' and 'industrial heritage enthusiasts' and their development for visitation involves complex considerations that impinge on preservation and conservation principles in terms of the physical sites and intangible mining heritages; as well as the health, safety and environmental factors in opening these sites to the public. However, in the cases presented in this chapter we analyse the complex histories behind two museums – Bowes that has its aristocratic origin in the money made from mining and Beamish that has its origins in a re-claiming working class heritage. Hence, this chapter seeks to look at the problems involved in preserving, presenting and interpreting community mining heritages and identities through these sites.

The North East historical context

For much of the Victorian era the North East of England housed a very small but vociferous middle-class concentrated, almost exclusively, in the suburbs of Newcastle and Sunderland. This group followed the self-improving cultural agenda of the nineteenth century middle-class as a whole, but some of its number also indulged in the construction of a mythical regional history based upon a romantic vision of Northumbria and Northumbrianess. However, this did not become the popular focus of regional identity since it was merely an empty ideological construct that failed to engage with the experience and lived culture of the vast majority of people which it, anyway, failed to represent. The majority of people in the region were, indeed, not Northumbrians, many were immigrants, and the largest industrial occupational group in the region – coal miners – were rarely represented in the cultural forms of the day. For much of the nineteenth century, Methodism and Liberalism were the dominant ideologies in North East England and these could be seen as conduits of shared meanings and values, but it must be remembered that both were class bound (Colls, 1977), a phenomenon revealing the primacy of class over other social and political forces during the nineteenth and much of the twentieth centuries. This is not to say that social mobility by way of position, education and social circle could not re-position visions and values. The region's most important labour leaders were often in tension with the grass roots as a result of their perspectives. Once elected to office the labour leaders often moved in lofty social circles and they frequently reflected, in their various contexts, what they saw as progressive thinking, which the traditionalist grass roots resisted.

The history of the Durham Miners Association reflects this ideological struggle between the leadership and the rank and file as does the post 1945 modernization movement in the region (Bulmer, 1978). It would be wrong to suggest that there was no variation in senses of belonging and identity in North East England, but overwhelmingly it was occupation and a sense of social class which lent to people of the region a sense of who they were. Deep coal mining was the key force in the making of the regional working class, its attendant class culture and class-based forms of identity which were derived by way of occupational identity (Williamson, 1982). 'Northeasterners', both old and new, working in, often related, industrial monoliths such as iron, later steel, and heavy engineering drew their cultural water from similar wells, premised as they were on company towns and villages, large workforces, heavy and dangerous manual labour, a strict sexual division of labour and extensive productive forms exemplified by the shift system. The roots of regional identity are thus neither mythical nor to be found in the sociability of the Newcastle crowd but were shaped by work in the context of stupendous economic and demographic growth in Durham and south east Northumberland during the nineteenth century (Colls and Lancaster, 2001). The people of North East England persistently represented their interests and celebrated their way of life, but they did so in the context of a fierce resistance engendered by the moral and social ideologues of industrial capitalism, and, arguably it is only within this dynamic that the making of both Beamish and Bowes can be understood.

The economic and social forces which Beamish and Bowes exemplify were generated by the expansion of the Durham Coalfield after the end of the Napoleonic Wars. Until this time, Durham had been little more than a semi-feudal backwater dominated by an omnipotent Prince Bishop and a small group of aristocratic landowners who exerted a hegemonic grip over the County's economic and social life. It was these elite which entered the coal trade as entrepreneurs, rentiers (someone whose income is from property rents) and industrialists as the sea coal trade grew and the region's mineral economy evolved.

As Beynon and Austrin (1994) have shown, the entrepreneurial aristocrats and their bourgeois imitators not only constructed the industrial infrastructure of the coalfield by building company towns and villages, but imposed upon them the social relations of the old order. The Third Marquess of Londonderry for example spoke of 'his pitmen' and his agent referred to Londonderry's men as his Lordship's 'little black family' (Beynon and Austrin, 1994: 27). Pease and Partners felt the need to stamp every brick in their colliery villages with their monograph (Moore, 1974). But the social relations of the Durham Coalfield were best characterised by the Bond, an instrument which bound the men to their masters for a year (after 1844 this was reduced to one month), but did not guarantee them employment. For the men, though not the masters, breach of the Bond was a criminal offence and offenders were incarcerated in Durham Jail.

The expansion of Durham's labouring population through the nineteenth century was staggering, moving from approximately 350,000 in 1801 to almost 2,000,000 in 1901. Occasionally, the county experienced twice the national average rate of growth and consistently achieved above average population growth in every decade between the 1820s and 1920s (House, 1954). This rapid rise in the labouring population had a profound and lasting effect upon the social structure and ultimately the culture of the region. County Durham quickly became a society polarised by social class. Durham's colliery villages and company towns were unambiguously class bound, characterised by tiny hegemonic industrial elites, a disproportionately small middle-class largely comprising a handful of managers and shop-keepers, and vast workforces uniformly engaged in some of the most dangerous and alienating occupations that industrialization has yet produced.

The class culture that evolved in County Durham was premised upon the confluence of these factors, that is to say a large migrant population dispersed in relatively small, isolated and industrially homogenized industrial communities, subject to an 'overloaded' paternal order (Beynon and Austrin, 1979). Work was an energising and shaping social force. Work in the 'occupational community' lent form and rhythm and value to everyday life and a powerful sense of role and identity to the individual. In Durham, as Jack Lawson (1932) has pointed out, a man was not a man until he attended the coal face. The social structure and constraints of the occupational community also conditioned the making of the formal institutions of the regional working class. In the industrial villages of the Durham Coalfield the masters demanded deference in exchange for economic and social patronage. Throughout the nineteenth and the first half of the twentieth century the Durham miners sought independence from paternalistic control and in doing so they created the institutions of labour, i.e. free chapels, the Durham Miners Association and the regional Labour Party. In Durham occupation, class, culture, identity and polity were almost indivisible.

In 1919, the representatives of the Durham miners formed the first Labour administration to run a County Council (Beynon and Austrin, 1994), and except for a brief break in the 1920s Labour has been in power ever since. The Durham miners who built Labour's political machine were intent on using it to promote their vision, their values and their way of life. The 'municipal socialism' they espoused and implemented was inward looking, regional in its ideology and in the orientation of its economic and social policy; it attempted to deal with specifically north-eastern problems not least the decline of the coal trade and attendant social deprivation and exclusion. The first Labour leader of Durham County Council, Peter Lee, epitomised Labour's project. Guided by the miners' long struggle for independence, Lee eschewed a Parliamentary career in favour of building a strong Labour Party power base dedicated to modernization and social regeneration on the coalfield. It was to be a model for future north-eastern Labour administrations, but in some ways it contradicted traditional sensibilities which were rooted in patterns of community based culture and identity. This tension led to an ideological contradiction in Labour's north-eastern cultural politics exemplified in the post-war period by the Durham Development Plan, the rise and ideology of T. Dan Smith and the establishment of the Beamish Open Air Museum, which was born out of the regional culture movement.

The 1951 Durham Development Plan was an explicit attempt to reverse spiralling industrial decline and attendant social deprivation and population loss by modernizing the County's infrastructure. The strategy was to concentrate urban and industrial development into key areas in order to break the historic pattern of scattered piecemeal development on the coalfield, and to improve infrastructure and the environment in order to attract new industries and other forms of inward investment (Durham County Council, 1951). Following post-war Government housing policy, the new towns of Newton Aycliffe and Peterlee were already being built and these were integrated into the Plan. Aycliffe and Peterlee featured futuristic housing and public art symbolic of the break with the past, but their populations were to be drawn from former pit villages. Some of these communities were deemed to have served their purpose and according to the Development Plan they were to be demolished and, as it were, erased from history. This radical attempt to modernize and re-brand the region demonstrated the dangers and difficulties of social and cultural engineering in a culturally distinctive area such as North East England since vigorous campaigns were fought by local people, throughout the 1960s and 1970s, to save the doomed communities (Bulmer, 1978).

The political career of T. Dan Smith parallels the period discussed above. Smith was the son of a Durham miner and like Peter Lee was a consciously regional politician, who rejected a Parliamentary career in favour of building a regional power base premised upon the Labour Party. Smith was an important figure because he articulated the ideology of post war regionalism in North East England, arguing for semi-autonomy and regional representation in national government by means of replacing the House of Lords with a House of Regions. During the 1960s Smith led Newcastle City Council, but more significantly he became Chairman of the Northern Economic Development Agency. Like many of the region's 'party bosses' Smith claimed, and was said to have, considerable influence within Labour Party and Labour Governments. In approach Smith was a modernizer recently characterised by Byrne (2001: 349–50) as 'just as much an elitist futurist as the (city and academic) planners . . . What drove him was a sense of 'vision' –vision not so much of social justice as of the "new" as "good" '. Smith had been a consultant on the project to develop the new town of Peter Lee, and he made grandiose predictions for the region claiming, for example, that he would transform Newcastle into the 'Milan of the North' through redevelopment and the promotion of cultural initiatives particularly in the arts (Potts *et al*., 1994: 7–26). Smith's modernizing project for the region and its 'capital city' proved every bit as unpopular among those it touched as the Durham Development Plan. Newcastle City Centre was redeveloped and became a tangle of brutal concrete fly-overs and often poorly built high rise blocks where former residents of the tightly knit working class communities endured the alienation and cultural exclusion of the modernist project.

In North East England modernization found a counterpoint in that other movement of the 1960s and 1970s – cultural politics (Colls and Lancaster, 1992). The period saw a heightening in class-based consciousness, confidence and pride expressed not only in the rise of trade unionism and socialism but in class-based literature, history, folk culture and media. Traditional values in the region had always found their most potent expression in the Durham Miners Gala. Here the massed banners and brass of the miners' lodges protested not only the political case but also the culture of the mining communities. The miners' leader Sam Watson articulated this sensibility when reviewing an exhibition by the pitman painter Tom McGuinness in 1964:

> He belongs to that type of miner that feels his own culture within himself . . . No one can go round this exhibition without getting a real insight into mining. The pit village has its place in the world of art and beauty along with the palaces and great buildings. Mr McGuinness has captured a great deal of beauty, culture, fellowship and companionship and displays great personal integrity.
>
> (McManners and Wales, 1997: 51)

It was around this core of cultural consciousness, the feeling of one's own culture within oneself that the regional culture movement developed, more often than not articulated by young working class men and women who were the first generation to have attended college or university. The folk song movement was an early and important example. The notes accompanying a record entitled 'The Colliers' Rant', released in 1962, gives an impression of its nature:

> This record has two aims. First, to jolt the memory of North-east miners; to bring back to their minds songs which have as much feeling, to be tender or humorous or down-right aggressive, as anything the 'pop' market can produce, and which are far more akin to

> the reality of their lives. The second aim is to let people outside the north east hear what miners think of themselves and their work.
>
> (Killen *et al.*, 1962)

Some modern regional histories reflected the mood, particularly the work of Robert Colls (Colls, 1977), as did the oral history movement which developed the 'from the bottom up' approach. In literature, working class regional writers had a renaissance and new publications such as 'Strong Words' which represented the thought and experience of 'ordinary people' proliferated. In media, the rise and celebration of the 'northeastener' began with programmes like the *Likely Lads*, moved to historical drama with *When the Boat Comes In* and back to comedy with *Auf Wiedersehen Pet*. The effect of the movement on popular culture in the region was striking. During the 1970s the celebration of all that was 'Geordie' reached new heights and is best exemplified by the work of local art critic Edward Scot Dobson, who produced the affectionately satirical, some say patronizing, *Larn Yersel Geordie* (Dobson, 1970), a text which spawned many imitations and variations (Howey, 1971).

The regional culture movement was popular because it connected with patterns of grassroots experience, consciousness and culture, something which the proponents of the modernization project palpably failed to do. Indeed it is fair to say that the modernization movement strove to obliterate the industrial legacy of the region and was quite blind to the historical and sociological dimensions of planning. The planners could see nothing in the villages but 'backwardness and decay' and nothing in the villagers' protests except an 'obdurate traditionalism' (Bulmer, 1978: 192–3). It was within this context of cultural struggle that the Josephine and John Bowes Museum fell into the care of Durham County Council.

The making of the museums

John Bowes (1811–85) typified the nineteenth century Durham aristocratic entrepreneur. He was the illegitimate son of the 10th Earl of Strathmore who, through his mother Mary Eleanor Bowes, had inherited vast estates in Durham and Yorkshire. The 10th Earl married Bowes' mother, the commoner Mary Miller, only hours before his death in order to secure his son's succession. However, the legacy was contested by the family and John Bowes was denied his father's title, but acquired the Durham estates after being schooled at Eton and Cambridge University. Although at times in his life the victim of aristocratic snobbery John Bowes had the best of the arrangement, since the Durham estates were rich in coal and he inherited and worked his measures at a time when demand for sea coal, i.e. coal bound for the London market, reached record levels. Bowes was a consummate entrepreneur, alive to the possibilities and shifts associated with industrialization. In 1852, for example, he pioneered the use of the steam collier which enabled regional producers to counter growing inland competition for the London trade.

Initially, John Bowes spent much of his money on racehorses. In the succeeding decades, however, Bowes dedicated more of his time and increasing wealth to his passion for all things French, but in particular French decor and *objets d'art*. In 1852, he had married the French actress Josephine-Benoite Coffin-Chevallier (1825–74), and it was she who conceived of the idea of creating a purpose built museum designed in the style of a French Chateau to house their growing collection of art on land near to the family's south Durham estate near Barnard Castle (Bowes Museum, 2009; Figure 2.1). It is claimed that their policy was largely educational, 'in order to introduce ordinary people to the wider world of art' (Bowes Museum, 2003), even if, as Kane (1995: 86) points out, 'no statement about the intended function of the museum, nor indeed any clues as to what prompted the Bowes to found a museum in the first place' appears in John Bowes correspondence between 1860 and 1880. Certainly, during the 1860s and the first half of the 1870s, they collected all manner of furniture, paintings, ceramics and tapestries especially for the museum. The foundation stone of Bowes Museum was laid by Josephine in 1869 but financial problems, which were to dog the project throughout its history, prevented its opening until 1892, by which time both John and Josephine were dead (Bowes Museum, 2003).

The Josephine and John Bowes museum project was perhaps the grandest gesture in the tradition of paternalistic symbolism and largesse practiced by the north-eastern aristocratic entrepreneurs during the nineteenth century. The Museum, if the contemporary Bowes establishment is to be believed, was for the benefit of ordinary people but in architecture and content it was a celebration of aristocratic status, cultural values and cultural power. That Bowes chose to spend his money this way rather than by improving wages and conditions in his collieries and colliery villages, or that the museum, as originally conceived, made no reference to the region or its people is, to judge by contemporary standards, not surprising. It is ironic however, that the foundation stone of Josephine and John's grand educational project was laid in the same year that Bowes' own men gathered with the other delegates of the Durham pitmen in the Market Tavern Durham City in order to improve their condition by forming the Durham Miners Association.

From the start The Bowes Museum struggled to attract interest and visitors. The Museum's location in thinly populated south west Durham was a problem and, seemingly, the county's small middle-class could not provide the scale of core support it needed to prosper. As early as 1922, Curator Sydney Harrison was complaining in a three page report to the Governors that lack of resources left him in a 'difficult position' (Harrison, 1922). Harrison's proposed solutions of the Museum's problems expose the contradictions which gave rise to them. The curator suggested on one hand that the museum should build up a strong local collection and open a Local Room '...to attract the local people and gradually build up civic patriotism for the collection', but on the other he argues that the elite art journals the 'Burlington' and the 'Connoisseur' be placed in a prominent position in the library in order to attract 'people of Knowledge' and that the museum should encourage 'artistic, antiquarian and literary societies to visit the museum. [Since] The members forming such societies are people who really care for Museums . . . To be advertised and quoted by members of learned societies would carry much more weight than the thousands of trippers who visit the museum during the Summer' (Harrison, 1922).

This was Bowes problem; like its founders, it represented the values and art of the elite but failed to value, and at best took a condescending view of, the potent folk culture that everywhere surrounded it. In this sense, since its inception Bowes had been moving against the grain of history. Following a further financial crisis in 1956, Bowes joined many of the other palaces of the great coal owners in the care of Durham County Council [dominated by the Labour Party and the National Union of Miners (NUM)]. Significantly, within two years of the acquisition, Frank Atkinson was appointed Director of Durham Museum Service. Atkinson was influenced by and associated with the Open Air Museum Movement which was dedicated to the celebration and preservation of folk culture. During the 1960s Atkinson's interests and intentions soon became apparent as he began to fill every available space in Josephine and John Bowes palace of European culture with industrial equipment salvaged from what he saw as the region's 'imploding' industrial economy. This vision was prophetic, but it made explicit the differences in cultural values that existed in the County and raised the associated question of the role of Bowes and the representation of North East England.

Atkinson's solution to these problems eventually resulted in the birth of the Beamish Open Air Museum, but its gestation during the 1960s and its consolidation during the 1970s and 1980s serves as an object example of the way in which the complexities associated with cultural representation can articulate the politics of heritage, for in North East England tensions associated with the formation of Beamish and the role of Bowes still rumble on. Nevertheless, during the 1970s the Beamish museum came to symbolise and mediate the strong current of traditional sentiment which had always resisted the 'imposition' of cultural forms and values associated with the paternalism of the past and ran counter to the 'top-down' ideology of economic and social modernization perpetuated by economic planners and sections of Labour's regional political elite.

Atkinson's tentative first steps to create Beamish were built upon his policy of popular participation, and it is clear that it found strong support in the post-war context of rising regional and class consciousness. This was most clearly reflected in 1970 when a consortium of local authorities agreed to start up and fund Beamish as the first regional Open Air Museum in the country. Their aim was to create 'an open air museum for the purpose of studying, collecting, preserving and exhibiting buildings, machinery, objects and information illustrating the development of industry and the way of life in the North of England' (Beamish, 2003). John Gall explained that the supporters of the museum were 'folk who could see that the great industrial past of the people of the region really was a far greater story than Bowes'. People, he said, who were fascinated that in Frank Atkinson they had 'a leader of society who cared about their history' (Howard, 1999).

It was this undercurrent of traditionalism and parochialism in County Durham, the very attitudes that so frustrated the modernizers, that made the creation of Beamish possible and it is the same current of consciousness, it is claimed, that rejects or at best resents Bowes. It is an irony that Bowes is seen as a purveyor of culture in the French Revolutionary tradition. According to this point of view Josephine and John Bowes' plan was to democratise European art and culture by allowing public access to it, just as the art and culture of revolutionary Europe was taken from the seats of power, the chateaux and the monasteries, and given over to the gaze of the common people. A large part of the frustration that is felt appears to turn upon the perception that this concept is not understood by many local politicians in Durham, particularly those representing 'socialist' mining constituencies in central and east Durham. To them Bowes is alienating because they cannot break from a mentality which associates art with aristocracy and a distorted view of social class.

Conclusion

The history of the Beamish and Bowes museums, it seems, reflects a much wider debate which seeks to establish what the subject of heritage should be, but it also presents a graphic demonstration of how the issues of social class and regional identity can inform it. In the context of County Durham at least, neither Beamish nor Bowes can dislocate themselves from the history of social class, the myths of popular memory or a politics embedded in the peculiar social structure of the region. Indeed in North East England, it could be argued that the politics of mining heritage tourism is very largely a product of an historic struggle to archive or assert cultural hegemony. In this sense at least, the Bowes Museum stands for many as a symbol of the economic and cultural paternalism from which generations of working class 'northeasterners' struggled to free themselves, just as Beamish represents the most recent institutional expression of popular cultural autonomy and identity.

References

Beamish (2003) *Beamish: The Experience of a Lifetime.* Beamish: Jarrold.

Bennett, T. (1995) *The Birth of the Museum.* London: Routledge.

Beynon, H. and Austrin, T. (1979) *Global Outpost: the Working Class Experience of Big Business in the North East of England.* ESRC Discussion Paper, Department of Sociology, Durham University.

Beynon, H. and Austrin, T. (1994) *Masters and Servants: Class and Patronage in the Making of a Labour Organization.* London: Rivers Oram Press.

Bowes Museum (2009) *History of the Bowes Museum.* Available online at: http//www.thebowesmuseum.org.uk/about/history (accessed 14 August 2009).

Bruner, E. (2005) *Culture on Tour: Ethnographies of Travel.* Chicago: University of Chicago Press.

Bulmer, M. (ed) (1978) *Mining and Social Change.* London: Croom Helm.

Byrne, D. (1996) 'Immigrants and the formation of the Northestern industrial working class', *North East Labour History*, 30.

Byrne, D. (2001) 'The reconstruction of Newcastle: Planning since 1945', in R. Colls, and B. Lancaster (eds) (2001) *Newcastle upon Tyne: A Modern History.* Chichester, West Sussex: Phillimore & Co, pp. 341–360.

Colls, R. (1977) *The Colliers Rant: Song and Culture in the Industrial Village.* London: Croom Helm.

Colls, R. and Lancaster, W. (1992) *Geordies: Roots of Regionalism.* Edinburgh: Edinburgh University Press.

Colls R. and Lancaster, W. (2001) *Newcastle upon Tyne: A Modern History.* Chichester, Phillimore.

Dobson, S. (1970) *Larn Yersel Geordie.* Newcastle: Frank Graham.

Durham County Council (1951) *Durham Development Plan.* Durham: Durham County Council.

Graham, B., Ashworth, G. and Tunbridge, J. (2000) *A Geography of Heritage.* London: Arnold.

Hannam, K. and Knox, D. (2010) *Understanding Tourism: A Critical Introduction.* London: Sage.

Harrison, S.E. (1922) *To the Trustees of the Bowes Museum.* 1 September.

House, J.W. (1954) *North Eastern England : Population Movements and the Landscape Since the Early 19th Century.* Durham: University of Durham.

Howard, S. (1999) *Interview with John Gall.* Beamish Museum.

Howey, P. (1971) *The Geordie Cook Book.* Newcastle: Frank Graham.

Johnson, N. (1999) 'Framing the past: time, space and the politics of heritage tourism in Ireland', *Political Geography*, 18: 187–207.

Kane, S. (1995) 'Greedy parsons: goose pies, drunken servants: the correspondence of John Bowes, 1860–1880', In Lancaster, B. (ed.) *Northern Review Magazine, Volume 2.* Northumbria University: Centre for Northern Studies.

Killen, L. Handle, J. and Ross, C. (1962) *The Colliers' Rant.* London: Topic Records.

Lawson, J. (1932) *A Man's Life.* London: Hodder & Stoughton.

McManners, R. and Wales, G. (1997) *Tom McGuiness: The Art of an Underground Miner.* Durham: Gemini Productions.

Moore, R. (1974) *Pit-Men Preachers and Politics: The Effects of Methodism on a Mining Community.* Cambridge: Cambridge University Press.

Northern Echo (2005) Visitor milestone reached. *The Northern Echo* 20 January. Available online at: http//archive.thenorthernecho.co.uk/2005/1/20/25666.html (accessed 14 August 2009).

Potts, A., Challinor, R. and Byrne, D. (1994) 'Perspectives of T. Dan Smith', *North East Labour History*, 28: 7–26.

Scot Dodson, E. (1970) *Learn Yersel Geordie.* Newcastle: Frank Graham.

Urry, J. (1990) *The Tourist Gaze.* London: Sage.

Williamson, B. (1982) *Class, Culture and Community.* London: RKP.

3 Chinese mining heritage and tourism in the goldfields of the Pacific Rim

Keir Reeves, Fiona Wheeler, Jennifer Laing and Warwick Frost

Introduction

The nineteenth century gold rushes of the Pacific Rim were multicultural affairs. Though located in English speaking settler societies (the USA and the British colonies of Australia, Canada and New Zealand), the lure of gold attracted a diverse polyglot of miners. Most prominent amongst these were the Chinese. Somewhere between a quarter and half a million Chinese crossed the Pacific in search of gold. Today these gold rush regions contain a common Chinese mining heritage and there is increasing interest in developing this as tourism attractions and products. Our intention in this chapter is to outline these developments, particularly noting their differing levels of maturity and success and discussing common issues affecting Chinese mining heritage around the Pacific Rim.

To understand the patterns and potential of Chinese mining heritage as a tourism product, there is value in taking a comparative approach. Accordingly this chapter takes in the four major gold regions of California (USA), Victoria (Australia), British Columbia (Canada) and Otago (New Zealand). By utilizing a comparative methodology, a range of differing patterns, experiences and options are observed within the context of a shared Chinese and mining heritage. At the core of this chapter is the assertion that a comparative analysis of distinct cultural landscapes enables tourism researchers to better understand the tourism potential of combining cultural history and the built environment. It is only by grounding this analysis in the field of heritage studies and applying it to tourism sites that sustainable strategies for sustainable heritage tourism can emerge (Rojek and Urry, 1997).

Three myths

In undertaking this comparative analysis, three conceptual qualifications must first be made. Indeed, these qualifications essentially deal with three myths of heritage tourism that have major implications for any tourism developments. As these myths have been examined in detail elsewhere, it is our intention to only briefly discuss them here.

The first myth is that heritage only attracts tourists with a close personal connection and accordingly the main market for Chinese heritage tourism products is Chinese tourists. Such a view might seem highly attractive, given the recent massive growth in Chinese outbound tourism. However, research in Australia has shown that currently Chinese tourists have little interest in heritage and tend to visit major cities rather than gold rush regions. Rather than the Chinese, the market for Chinese heritage is the incidental heritage tourist, generally interested in a wide range of tourist experiences, including but not limited to history (Frost *et al.*, 2007; Frost *et al.*, 2009).

An enduring, and in terms of gauging the relationship between heritage and tourism perhaps the most important, second myth is that heritage sites make successful tourism products.

As McKercher and Ho have observed this is not necessarily the case. This means that while sites may have high levels of heritage significance it does not necessarily follow that they will make for attractive tourist destinations or indeed provide an economic return even with significant spending on tourism infrastructure (McKercher and Ho, 2006).

A third myth is that history is factual and unchanging, 'the recording of the past as accurately as possible' (Timothy and Boyd, 2003: 4). Accordingly, visitor interpretation at heritage attractions should strive for authenticity by being factual and avoid being selective (Timothy and Boyd, 2003: 237). However, historians are subjective and their interpretations of the past change over time. This creates a challenge for heritage tourism interpreters in trying to decide what version of history to include (Chapter 9 by Frost in this volume). In recent years, the historiography of the Chinese on the Pacific Rim has undergone a major shift, moving from presenting the Chinese as sojourners and the passive victims of racism to highlighting their dynamic and enduring relationships with the European majority (Chan, 2001; Chan, 2000; Fitzgerald, 2007; Frost, 2002; McKeown, 2001; National Parks Service, 2009; Mountford and Reeves 2009; Reeves, 2004; Wilton, 2004).

Goldfields landscapes

A problem of current international scholarship is the tendency to interpret the built environment as an end in itself: a product of specific technologies, or materials or even social attitudes (Buggeln, 2001: 245; Lydon, 1999: 20, xiii; Prown and Haltman, 2000). What is so often missing from this approach is the human agency that created them and interpretation strategies for understanding them in the present day (Lydon, 1999: 1–24). In order to people these cultural landscapes and bring them to life for present day visitors it is necessary to weave human stories and experience of landscape together in order to provide a richer historical context. Cultural landscape analysis provides one way of identifying different historical layers of meaning in the remnant mining landscapes and interpreting them for their heritage tourism potential.

While the focus of this chapter concentrates on Chinese goldfields' heritage, the landscape of goldfields towns of the later nineteenth and early twentieth century provide excellent case studies through which the problems and possibilities of cultural landscape can be explored (Reeves and Nichols, 2007). They are difficult to gauge in terms of heritage tourism values as these areas typically consist of small, urbanized centres scattered between earth workings left behind by mining. Here the challenge is to innovatively and effectively interpret goldfields areas. This can involve considering how cultural and natural values interact in goldfields areas where remnant vegetation as well as strands of replanting are used to mask mine workings. Often there are typical series of buildings that exist in goldfields areas that relate to the community life of mining and the residential patterns that evolved in various stages of mining. It is worth noting that there are now well-defined trails or routes through goldfields areas that are central to tourism and which can be expressed as a linked narrative of sites rather than as a series of isolated mine workings or buildings. Such a linking process better demonstrates many of the themes of significance more completely and convincingly than through treating individual buildings and mine sites in isolation.

Heritage tourism sites and attractions

This section of the chapter will examine Chinese mining heritage tourism, linked to the nineteenth century gold rushes, in the United States, Canada, New Zealand and Australia.

It provides an overview of some of the key sites or destinations that are actively promoting Chinese mining heritage to visitors, and a comparison of the nature and scope of tourist experience available. While there are a number of Chinese heritage sites in gold rush countries much of what remains is relatively small-scale and not heavily promoted to tourists. These sites generally lack tourist infrastructure, particularly the provision of high-quality thematic interpretation. They also tend to be located in remote places and lack a critical mass of assets related to Chinese heritage within the one locale. The sites with the most potential for tourism development are found in Victoria, particularly around the Bendigo region, which will be discussed in more detail in the next section of this chapter.

United States

Although California is well known for being the focal point for the gold rushes in the United States, it has few heritage sites left to commemorate this era, with even fewer pertaining to Chinese mining heritage. There are a scattering of small and relatively isolated sites, with a number under the auspices of California State Parks at Weaverville, Coloma and Bodie (Frost, 2006). Weaverville's joss house was built in 1874 to replace the previous one that had been burnt (California State Parks, 2009). It is the oldest Chinese temple still being used as a place of worship in California. This makes it problematic for development as a tourist attraction, although it is open to the public for most of the week. The township of Coloma and the surrounding valley are better known today for their food and wine offerings, but there are two restored buildings, formerly the Man Lee and Wah Hop stores, in the Marshall Gold Discovery State Historic Park (Leung and Lee, no date). The ghost town of Bodie has been left in a state of 'arrested decay' since being declared a National Historic Site and a State Historic Park in 1962. The 'Chinatown' on the outskirts of Bodie, was one of the largest in California, but now contains only three buildings that remain from the gold rushes, all of which are 'badly deteriorated' (De Lyser, 1999). Despite its authentic veneer, Bodie is essentially a town 'wrapped in a mythic blanket' (Frost, 2006), woven in part from romantic notions of the 'Wild West' and cinematic qualities (De Lyser, 1999), rather than its associations with the Chinese.

The state of Idaho and, in particular the US Forest Service, has displayed a growing interest in developing heritage tourism experiences around Chinese mining sites (National Park Service, 2009; Wegars, 2004). For example, the Payette National Forest in Idaho has a number of sites linked to the Chinese miners who came to Warren between 1870 and the 1890s (National Park Service, 2009). Relics of the Chinese community include the Ah Toy Garden, Celadon Slope Garden and Chi-Sandra Garden, Old China Trail, the Chinese Cemetery and the Chinese Mining Camp Archaeological Site. Interpretation programs are offered by the US Forest Service, including self-guided interpretive tours, using brochures and interpretive signage (National Park Service, 2009) and a reconstruction of a semi-subterranean Chinese dwelling (Wegars, 2004). The town of Lewiston has a nineteenth century temple, the assemblage of which is now exhibited permanently at the Lewis-Clark Center for Arts & History (Wegars, 2004).

Canada

Canadian Heritage tourism attractions include 'regional historic gems' such as the towns of Kingston and Perth (Boyd, 2002), battlefields sites, assets associated with particular cultures or ethnic groupings (Boyd, 2002), such as the Mennonite culture in Ontario (Boyd, 2002), native Canadian culture (Li, 2000) and French Acadian culture, particularly on Prince Edward

Island (MacDonald and Jolliffe, 2003). There is also tourism development associated with gold mining history, mainly centered on the Klondike (Pretes, 2002; Smith, 2003). The Chinese story on the goldfields struggles to be heard; mainly the consequence of the paucity of built heritage that survives to the present day. The historic mining town of Barkerville on the Caribou in British Columbia contains a Chinatown, including a Chinese museum (Abraham and Wegars, 2005). Like Bodie in California, it was a ghost town by the 1950s, when revived by being declared a national park. As a tourist attraction, it suffers from its location, which is isolated and about 750 kilometres from Vancouver. Essentially Chinese heritage tourism in British Columbia is centered on the vibrant Chinatowns in the port cities of Vancouver and Victoria.

New Zealand

According to Huang, Fountain and Perkins (2008), there is a growing interest in redeveloping Chinese heritage sites in New Zealand. Most of the remaining Chinese heritage associated with the gold rushes is located in Central Otago in the South Island, with proposals for a Chinese heritage trail through the region (Huang *et al.*, 2008). While the region is largely known for its skiing and adventure tourism, mostly in and around Queenstown, there appear to be some attempts to balance this with limited development of nearby Chinese mining sites, chiefly at Arrowtown and Lawrence (Frost, 2005). The Arrowtown dig occurred in the 1980s and resulted in the construction of a number of replica Chinese buildings, which are mostly fairly simple, although in good condition. The Arrowtown interpretation trail and signage require a revamp, to link with new discoveries about the Chinese in New Zealand and personal stories or narratives, to provide interest and heighten the engagement of tourists. The small Chinatown at Lawrence, just outside the town's boundary, was excavated even more recently (2004). While this excavation has been used to promote Lawrence to visitors, there are neither visible remains, nor interpretation at the site. Funding pursuant to the Chinese Poll Tax Heritage Trust, established by the New Zealand Government, has been granted to the Lawrence museum (Chinese Poll Tax Heritage Trust, 2008).

Elsewhere in the region, Chinese heritage is a low-key element of the tourism experience. In the 1980s, the Chinatown at Cromwell was included in the Otago Goldfields Park, but was later destroyed by the construction of a hydro-electric dam. Kearsley (1990) notes the town is now pursuing a role in tourism, but unfortunately without the benefit of its former Chinese heritage, being based mainly on wine and recreation. Bonspiel Station at Moa Creek contains a number of stone huts built by Chinese miners in the 1860s, which now serve as rough and rustic visitor accommodation, without power or running water (Otago Central Rail Trail, 2009). There are also some Chinese sites in the Macrae's Flat area, such as a Chinese miner's mud-brick house and a camp site (Barnett, 2008). Once the existing mine site is decommissioned, there are plans to develop the area as a Heritage and Art Park, 'that combines history, art, designed landscape, the natural environment, ecology, technology, and traditional and alternative agriculture into a self-sustaining destination for tourism purposes' (Barnett, 2008: 27).

Australia

The gold rush sites associated with the Chinese are largely in regional Victoria. The main exception is Robe in South Australia, which was the place where many Chinese landed in Australia from China, after Victoria passed a head tax on them in 1855. Many Chinese circumvented this tax by walking to Victoria from Robe, a distance of 600 kilometres. There is a 'Gold Trail' monument to the Chinese near the Customs House, where the Chinese miners were processed upon their arrival (Golden Trail, 2009), and attempts have been made to develop the walk as an attraction, with visitors walking in the 'footsteps' of the miners.

Most of the Chinese mining heritage in Victoria is located in Bendigo, Beechworth, Ararat and Ballarat, with some remnants scattered across the Castlemaine Diggings National Heritage Park. The Ararat goldfields were discovered by the Chinese in 1857, during their cross-country walk from Robe to Victoria. Ararat has the Gum San Chinese Heritage Centre, which seeks to bring to life the history of the Chinese miners in the region, and link this with the influence of their culture on Victoria. Many of the exhibits are interactive and aim to involve visitors in a fun and participatory way. The museum is housed in a recreation of a traditional Chinese two-storey building and set in Chinese-style gardens.

Beechworth is actively working to highlight the Chinese connection with the town's history to broaden their touristic offering and range of experiences available to visitors. Within the Buckland Cemetery is a cairn and plaque commemorating the anniversary of the Buckland Riot in 1857, where Chinese miners demonstrated against restrictive and discriminatory practices on the goldfields. The Beechworth Cemetery has two burning towers, an altar and over 2,000 Chinese graves. The Burke Museum has the Chinese Community Collection, encompassing ceramics, photographs, weapons, ceremonial armour and banners, and hosts exhibitions that are connected with the Chinese story in Beechworth. Both these sites however, lack broad tourist appeal and are more likely to be sought out by heritage enthusiasts. The two sites that are currently in various stages of development and have the greatest potential for attracting tourists are the Chinese Gardens and the Beechworth Chinese Cultural Centre. The Gardens are being developed on a site where Chinese villages and market gardens existed until the twentieth century. There is some interpretation on site. The Cultural Centre is housed in a heritage listed building that was the office of the Chinese Protectorate, the only remaining office in Victoria. The Centre will house artifacts dating back to the Chinese gold rush, including items from the Beechworth Chinese Camp.

Ballarat is chiefly known to tourists for its successful heritage attraction Sovereign Hill (to be further discussed in Chapter 10), with its recreations of 1850s gold rush buildings, including a Chinese joss house, Chinese village and Chinese Protectorate Office (Evans, 1991; Frost, 2005). The Joss House acts as a portal to the village, with an introductory video for visitors. Interpretation is available in the huts forming the village, with computerized touch screens and audio recreations of the Chinese story in Ballarat. Visitors must take part in a tour, rather than visiting this site independently (Frost *et al.*, 2007). The attraction is visited by growing numbers of Chinese tour groups (Zhang and Murphy, 2009). It has however been criticized for placing an emphasis on 'infotainment' in the past, at the expense of the delivery of 'take-home messages' via interpretation (Ham and Weiler, 2004). The museum has been working on strategies to develop and strengthen their interpretive product (Ham and Weiler, 2004) but does not have a repository of authentic sites at its disposal such as are found in Bendigo or elsewhere in the Victorian goldfields.

The Castlemaine Diggings National Heritage Park was established in 2002, and contains various Chinese mining sites, none of which have yet been developed for visitors. There is a question mark as to whether they have any potential as tourism products, with little or no interpretation currently available (Frost, 2005). At Herons Reef, there is a Chinese village containing the ruins of stone huts and ceremonial ovens. There is also a Chinese hut at Vaughan, which was recently excavated by archaeologists. A limited Chinese presence remains in Vaughan, comprising the Chinese cemetery and a few small Chinese headstones, along with the Chinese section in the official Vaughan cemetery. The latter is, however, almost invisible with no signage available and unmarked gravesites (Frost *et al.*, 2007).

The Chinese heritage sites at Bendigo, in contrast with those discussed above, are more extensive and authentic, with greater potential for development as part of a Chinese Heritage *product*, which may increase visitation (McKercher and Ho, 2006). The next section of this chapter will discuss the issues involved with this process, including the need for an integrated experience, perhaps linked with thematic interpretation and tourist transport, as well as a marketing strategy that re-brands the precinct, in order to raise awareness of its unique qualities and attractions.

Bendigo heritage tourist sites: elements in a Chinese historical landscape

For the purposes of this chapter the collection of Chinese heritage sites include the Golden Dragon Museum precinct and also sites around Emu Point (and the Bendigo Creek to the north of the city) that we term the Ironbark Camp Precinct. This recognizes that Ironbark Camp was the name of the largest and most enduring Chinese Protectorate in Bendigo. The precinct contains four main sites: the Chinese Joss House, the Chinese brick-kiln, the White Hills Cemetery and the Chinese Diggings Historic Reserve. This precinct is not as well developed for visitors as the Golden Dragon Precinct, which is why it is considered separately. It is located approximately 1.5 kilometres north of the city and currently has very few visitor facilities. The Ironbark Camp Precinct is close to Lake Weeroona, Bendigo Tramway Museum and Depot and the currently undeveloped Bendigo Gasworks site. It is also situated at the northern terminus of the Bendigo Tramway. While this area is rich in European and Chinese heritage significance, it is also in need of interpretation and infrastructure to be an attractive tourist destination.

Golden Dragon Museum

Bendigo's main tourism product related to Chinese heritage is the Golden Dragon Museum, though there are a number of other associated heritage products nearby. The Golden Dragon Museum celebrates the contribution of the Bendigo Chinese Community to the gold mining, cultural and commercial life of Bendigo from the Gold Rush to the present day. The Golden Dragon Museum describes itself as the 'the hub of Chinese cultural activity in Australia'. Its significance and influence extend well beyond Bendigo. Its location in Bridge Street is in an area historically known as the Chinese Precinct. It was developed by the Bendigo Chinese Association as a way to house and preserve the Association's collection of materials of historical and cultural significance to the Chinese community. In exhibiting the collection to the public, and also communicating through publications, research and public programs, the museum also hopes to promote goodwill and understanding between Chinese Australians and the broader Australian population through the sharing of less commonly presented voices of Australian history and Australia's present. As well as presenting a historical narrative of the Bendigo Chinese community, the museum allows visitors to experience Chinese arts, crafts and artifacts first hand and provides an opportunity to meet with visiting artisans and tradespeople.

Chinese Joss House

The Chinese Joss House (or temple as Joss is a corruption of the Portuguese the Deus used to describe Chinese temples) is a visible marker and a prominent symbol of the cultural and religious differences between the Chinese and European communities in Bendigo (Chamberlain, 1999). The Joss House was erected in the 1860s in order that Chinese sojourners and settlers would be able to practise their culture and religion. The Joss House was, and is still an important site for community gathering and worship and as such is an important record of the Chinese presence and experience on the goldfields. The continuation into the present day of both the physical building of the Joss House and its cultural use, demonstrates how the Chinese were able to preserve their culture in a European society where religions other than Christianity were not viewed in a favourable light. The fact that the Chinese Joss House in Bendigo is the only remaining temple of its kind in Victoria gives it considerable heritage significance.

The Chinese Joss House is in an isolated place, with limited car-parking and visitor facilities, although there is a tram stop nearby to board the Bendigo Talking Tram. Until, recently the Joss House was operated by the National Trust with limited opening hours. It has now been transferred to The Bendigo Trust, which has commenced some basic site works on its fencing and grounds. Works include restoration of a picket fence, provision of lighting around the building; and the development of a pathway with interpretive signage between Joss House and the Emu Point tram stop. This site signage should be integrated with interpretation across the Ironbark Camp Precinct, as well as the Golden Dragon Precinct and associated heritage sites across Bendigo.

PepperGreen Farm/kiln Site

In 2005, participants in a Community Jobs program run by Bendigo Access Employment, during re-fencing work, uncovered the remains of a Chinese brick-kiln. This was at PepperGreen Farm, about 200 metres south of the Chinese Joss House. The kiln was constructed in 1859 by A'Fok, Fok Sing and Co. A large conical structure, a great deal of it remains buried under the ground. Its distinctive bricks were used to build a large wall at the corner of Forest and Rowan streets near the Golden Dragon Precinct. The kiln is believed to be the only one of its type still in existence outside of China. It is currently not accessible to visitors. It was only partially excavated in 2005 by Heritage Victoria and the potential remains for a more extensive dig, in order to reveal its secrets and be placed in a situation where it can become a tourism product.

The kiln was abandoned probably in the 1880s and the surrounding area taken up by Chinese for market gardens. PepperGreen Farm is littered with artifacts from this period, including pottery, bricks and European bottles. There are a number of mature peppercorn trees along the banks of the nearby Bendigo Creek which probably date from this period. PepperGreen Farm is leased by Access Employment who is supportive of new development in order to better present the kiln as a tourist attraction. Because there is no interpretation on the site to mark the presence of the kiln it is not currently a tourism destination despite its high heritage significance.

White Hills Cemetery

The physical presence of the Chinese community in Bendigo is manifested in the large Chinese presence in the Bendigo White Hills Cemetery. The Cemetery is thought to be one of the largest sites of Chinese burials in Australia. In accordance with traditional Chinese beliefs, the graves are marked by small footstones and in 1877 a funeral tower was built so that offering could be burnt. The size and very existence of the Cemetery, brings to light multiple facets of the Chinese goldfields' narrative but like PepperGreen interpretation is minimal. According to Bendigo's Golden Dragon Museum, it was 'the ardent wish of all Chinese' to be buried near their ancestral grounds. That necessarily meant having one's remains returned for burial in China. This supports the representation of the Chinese as sojourners who integrated little with the wider Australian community. However, the large and highly visible Chinese section of Bendigo's White Hills Cemetery suggests another story of Chinese life on the Victorian goldfields in which the Chinese became settler Australians. In fact a small, yet significant, minority of Chinese gold seekers did not return to China. Their history is visibly demonstrated at the White Hills Cemetery and suggests an enduring presence of the Chinese population in Victoria. Moreover, it is a history that adds another layer to the story of the development of the Australian community and the Australian identity.

In many respects, the Chinese burial sections of the cemeteries, most notably White Hills, stand alongside the Chinese temple at White Hills as the key cultural and historical record of the Chinese on the Bendigo goldfields during the nineteenth century. However as alluded earlier in this chapter the cemeteries, like other more subtle relict heritage sites, require heritage interpretation as well as being integrated into a tourism plan for the City of Greater Bendigo.

Conclusion

At the heart of the Chinese gold seeking tourism experience both in a localized regional setting such as Bending or in a trans-Pacific gold seeking context lies a cultural exchange that has been obscured by the passage of time. Today, questions of treatment of the Chinese and their role in the respective nineteenth century gold rushes have triggered a dialogue central to considerations of national identity and how to interpret and promote Chinese heritage tourism sites. In this examination of the Chinese heritage throughout the Pacific Rim the authors have examined sites that may help to explain the complexity not only of Chinese–European relations in the past, but also of interpreting and experiencing them as tourism destinations in the present day. On a broader level using a cultural landscape methodological approach to interpret tourism sites enables a multiplicity of viewpoints to be expressed in areas rich in cultural (and sometimes natural) heritage values (Lennon, 1999). To ignore the cultural landscape approach is to make the task of bringing the various historical themes to the fore and developing them for their tourism potential. It would be remiss to forego opportunities to enhance the significance of heritage resources when narrow and single discipline approaches are applied to Chinese mining heritage tourism sites throughout the Pacific Rim.

In this chapter we have discussed a number of Chinese mining heritage sites situated throughout the Pacific Rim with considerable tourism potential. Our argument has been that while each region has its own distinctive Chinese themed sites it is in Australia, particularly regional Victoria, that the majority of remnant Chinese mining landscapes can be viewed (this is particularly the case in Bendigo and its surrounding settlements). Moreover, this heritage can be equally understood in terms of its local heritage tourism significance and also the broader international historical linkages that connect sites across the entire sweep of the Pacific Rim. Or, more simply, regional and global perspectives are both necessary to understand the interpretive narratives that make visitation of individual sites more rewarding.

A key question that emerges for tourist operators and cultural tourism practitioners is how do sites with mining heritage tourism values present the stories and remnant landscapes of Chinese sites to tourists? Cultural landscapes provide one way of understanding the complexity of the sites. This is particularly the case with the nuanced and sometimes difficult Chinese-European encounter that occurred in the diggings. Reading the assemblage of sites as a cultural landscape enables more integrated tourism interpretation strategies as Chinese heritage is treated as a composite whole rather than a series of peripatetic elements of little, or limited, visitor appeal. Finally, our contention has been that in order to address the three myths briefly outlined in the formative section of this chapter, then it is necessary to develop engaging tourism experiences that reflect the more complex nature of the mid-nineteenth century goldfields society. We have argued that this is best achieved by developing historically informed regional heritage tourism strategies and understanding them in an international context.

Acknowledgements

The authors acknowledge the support of Kathryn Mackenzie and the marketing team at the City of Greater Bendigo, Anita Jack and the Bendigo Golden Dragon Museum and David Bannear of Heritage Victoria.

References

Abraham, T. and Wegars, P. (2005) 'Respecting the Dead: Chinese Cemeteries and Burial Practices in the Interior Pacific Northwest', in Fawn Chung, S. and Wegars, P. (eds), *Chinese American Death Rituals: Respecting the Ancestors*. Lanham MD: Rowman Altamira, 147–73.

Barnett, R. (2008) 'Gold and the Gift: Theory and Design in a Mine-Reclamation Project', in Berger, A. (ed), *Designing the Reclaimed Landscape*. Abingdon, Ox; New York: Taylor and Francis, 26–35.

Boyd, S. (2002) 'Cultural and Heritage Tourism in Canada: Opportunities, Principles and Challenges', *Tourism and Hospitality Research*, 3(3): 211–33.

Buggeln, G.T. (2001) 'American Artifacts: Essays in Material Culture', *Winterthur Portfolio*, 36(4): 245.

California State Parks (2009) *Weaverville Joss House State Historic Park*. Available online at: http://www.parks.ca.gov/?page_id=457 (accessed 28 May 2009).

Chamberlain, J. (1999) *Chinese Gods* (6th edn). Subang Jaya: Pelanduk Publications.

Chan, H. (2001). 'Becoming Australian but Remaining Chinese: The Future of the Down under Chinese Past', in Ann Curthoys, Lan-hung Nora Chiang, and Henry Chan (ed), *The Overseas Chinese in Australasia: History, Settlement and Interactions: Proceedings*, (Taipei: Interdisciplinary Group for Australasian Studies (IGAS) and Centre for the Study of the Chinese Southern Diaspora, 2001).

Chan, S. (2000) 'A People of Exceptional Character: Ethnic Diversity, Nativism and Racism in the California Gold Rush', *California History*, 79(2): 44–85.

Chinese Poll Tax Heritage Trust (2008) *Annual Report 2007–2008*. Available online at: http://www.dia.govt.nz/pubforms.nsf/URL/CHPTAnnualReport08.pdf/$file/CHPTAnnualReport08.pdf (accessed 28 May 2009).

De Lyser, D. (1999) 'Authenticity on the Ground: Engaging the Past in a California Ghost Town'. *Annals of the Association of American Geographers*, 89(4): 611–32.

Evans, M. (1991) 'Historical Interpretation at Sovereign Hill', *Australian Historical Studies*, 24(96): 142–52.

Fitzgerald, J. (2007) *Big White Lie: Chinese Australians in White Australia*, Sydney: UNSW Press.

Frost, W. (2002) 'Migrants and Technology Transfer: Chinese Farming in Australia, 1850–1920', *Australian Economic History Review*, 42(2): 113–31.

Frost, W. (2005) 'Making an Edgier Interpretation of the Gold Rushes: Contrasting perspectives from Australia and New Zealand', *International Journal of Heritage Studies*, 11(3): 235–50.

Frost, W. (2006) 'Heritage Tourism at Eureka Stockade in an International Context', in A. Mayne (ed.), *Eureka 1854–2004: Reappraising an Australian Legend*. Fremantle: Fremantle Arts Centre Press.

Frost, W., Laing, J., Wheeler, F., Reeves, K. and Weiler, B. (2007) *Bendigo Chinese Heritage Precinct: Marketing and Interpretation Plan*, Melbourne: Monash University Tourism Research Unit.

Frost, W., Reeves, K., Laing, J. and Wheeler, F. (2009) 'Villages, Vineyards and Chinese Dragons: Constructing the Heritage of Ethnic Diasporas', *Tourism, Culture & Communication*, 9(1/2): 107–14.

Ham, S.H. and Weiler, B. (2004) 'Diffusion and Adoption of Thematic Interpretation at an Interpretive Historic Site', *Annals of Leisure Research*, 7(1): 1–18.

Huang, H., Fountain, J. and Perkins, H. (2008) 'The (Re)development of New Zealand's Chinese Heritage Sites and the Potential for the Chinese Visitor Market. Working Paper', *New Zealand Tourism and Hospitality Research Conference Proceedings*, Hanmer Springs, December 3–5.

Kearsley, G.W. (1990) 'Tourism Development and Users' Perceptions of Wilderness in Southern New Zealand', *Australian Geographer*, 21(2): 127–40.

Lennon, J., Egloff, B., Davey, A. and Taylor, K. (1999). *Conserving the Cultural Value of Natural Areas: A Discussion Paper*. Canberra: University of Canberra.

Leung, P.C.Y. and Lee, C. (no date) *Welcome to the Wah Hop Store*, Coloma: Gold Discovery Park Association.

Li, Y. (2000) 'Ethnic Tourism: A Canadian Experience', *Annals of Tourism Research*, 27(1): 115–31.

Lydon, J. (1999) *'Many Inventions': the Chinese in the Rocks, Sydney 1890–1930*, Clayton: Department of History, Monash University.

MacDonald, R. and Jolliffe, L. (2003) 'Cultural Rural Tourism: Evidence from Canada', *Annals of Tourism Research*, 30(2): 307–22.

McKeown, A. (2001) *Chinese Migrant Networks and Cultural Change: Peru, Chicago, Hawaii, 1900–1936*, Chicago: The University of Chicago Press.

McKercher, B. and Ho, P. (2006) 'Assessing the Tourism Potential of Smaller Cultural and Heritage Attractions', *Journal of Sustainable Tourism*, 14(5): 473–88.

Mountford, B. and Reeves, K. (2009) 'Reworking the Tailings: New Gold Histories and the Cultural Landscape', in McClisky *et al.*(ed), *ReOrienting Whiteness*, Sydney: University of Sydney Press (forthcoming).

National Parks Service, US Department of the Interior (2009) *Asian Reflections on the American Landscape: Identifying and Interpreting Asian Heritage, Washington DC: National Centre for Cultural Resources.*

Otago Central Rail Trail (2009) Accommodation – Bonspiel Station and Old Moa Creek Hotel. Available online at: http://www.centralotagorailtrail.co.nz/accommodation_advertising/moa_creek/index.html (accessed 28 May 2009).

Pretes, M. (2002) 'Touring Mines and Mining Tourists', *Annals of Tourism Research*, 29(2): 439–56.

Prown J.D. and Haltman, K.(eds) (2000) *American Artifacts: Essays in Material Culture*. East Lansing: Michigan State University Press.

Reeves, K. (2004) 'A Songster, a Sketcher and the Chinese on Central Victoria's Mount Alexander Diggings: Case Studies in Cultural Complexity during the Second Half of the Nineteenth Century', *Journal of Australian Colonial History*, 6(1): 175–93.

Reeves, K. and Nichols, D. (2007) 'Deeper Leads: A Cultural Landscape of the Victorian Gold Fields', in K. Reeves and D. Nichols(ed), *Deeper Leads: New Approaches to Victorian Goldfields History*, Ballarat: Ballarat Heritage Services, 7–9.

Rojek, C. and Urry, J. (1997) *Touring Cultures: Transformations of Travel and Theory*, London: Routledge.

Smith, S.L.J. (2003) 'A Vision for the Canadian Tourism Industry', *Tourism Management*, 24: 123–33.

Timothy, D.J. and Boyd, S.W. (2003) *Heritage Tourism*, Harlow: Prentice Hall.

Wegars, P. (2004) 'Heritage Tourism of Chinese Archaeological and Historical Sites in Idaho', *The SAA Archaeological Record*, November: 18–21.

Wilton, J. (2004) *Golden Threads: The Chinese in Regional New South Wales, 1850–1950*, New England Regional Art Museum, Armidale.

Zhang, Y. and Murphy, P. (2009) 'Supply-chain Considerations in Marketing Underdeveloped Regional Destinations: A Case Study of Chinese Tourism to the Goldfields Region of Victoria', *Tourism Management*, 30(2): 278–87.

4 Pilgrimage in heritage tourism

Finding meaning and identity in the industrial past

Mary Beth Gouthro and Catherine Palmer

Introduction

Industrial heritage tourism is a well-established niche within the heritage tourism sector. Depending upon the social and economic history of the country being visited, tourists can experience a diverse range of attractions depicting aspects of the industrial past. This chapter focuses on one particular type of industrial heritage, that of coal mining. Specifically, it is concerned with exploring the phenomenon of mining tourism rather than management issues associated with operating mines as tourist attractions. This perspective is important because it can shine a light on the ways in which visitors think about and experience this form of tourism, as Pretes (2002) asks 'why are tourists interested in mines?'

In responding to this deceptively simple question this chapter discusses the results of research into meaning making at two industrial heritage museums that display the coal mining history of their respective regions (Gouthro, 2008). One museum is in the UK, the big Pit National Coal Museum, Wales (a UNESCO World Heritage Site) and the other is located in Canada, the Cape Breton Miners' Museum, Nova Scotia. The ethnographically informed study of these sites sought to explore the meaning and significance of the sites for the visitors through an analysis of visitor behaviour. The research investigated the ways, in which meaning-making linked to mining heritage might influence how individuals come to understand their place in the wider world; come to understand who they are, their identity if you like. Such a focus is less well documented in the literature than those aspects dealing with the evolution of mines into tourist attractions and the management issues associated with this process.

Within the tourism literature the metamorphosis of coalmines into tourism attractions has been examined from varying perspectives and in some depth (Cole, 2004; Edwards and Llurdés i Coit, 1996; Light and Prentice, 1994; Prentice *et al.*, 1998; Pretes, 2002; Rudd and Davis, 1998; Wanhill, 2000). The development of former mining sites into tourist attractions is discussed by authors such as Cole (2004), Edwards and Llurdés i Coit (1996) and Wanhill (2000). Whilst other studies have focused on the socio-demographic factors linked to the types of visitors attracted to such sites (Light and Prentice, 1994), and on motivation and experience linked to socio-economic factors (Prentice *et al.*, 1998). The relationship between particular mining sites and the local community has likewise been explored. For example, Pretes' (2002) narrative study of tourist–local interactions at the silver mines in Potosí, Bolivia; and Rudd and Davis' (1998) research into the way in which a Utah corporation promotes copper mine tourism to alleviate public concern for the environment. In this research, Rudd and Davis show how copper is positioned as 'a gift of nature' brought to life through the activity of mining to alleviate concerns that some visitors may associate mining as a polluting activity.

The features of mining attractions are unique in that they are not typical of other tourist attractions such as theme parks. Former coal mines turned museums are themselves former production spaces transformed into places for tourism consumption (Richards, 1996). In the case of the Big Pit Mining Museum in Wales, it was felt that while the scars of the industrial revolution may not appeal to tourists, the landscape, society and culture of the mining communities should be documented and conserved before they were forgotten (Wanhill, 2000). The uniqueness of former mining sites turned attractions is that they tell the history of a time when exploiting a region's resources was commonplace (Edwards and Llurdés i Coit, 1996). An exception to this is the Potosí silver mines in Bolivia (Pretes, 2002) where active mining is combined with tourism through the provision of underground tours. Even though the Potosí silver mines are still in operation, like the visitors to Welsh mines (Wanhill, 2000), those drawn to Potosí come to hear the story of struggle, decline, hardship and ultimate success through tourism of an industry and its people.

The hardship and danger associated with mining provide an insight into why tourists visit coal mines. However, such factors do not in themselves explain what visitors get out of the experience of touring the industrial landscape of coal mining. The research discussed here sought to delve a little deeper into the phenomenon of mining tourism by drawing from the insights to be gained from the anthropology of tourism; a subdiscipline that among other things seeks to understand what the experience of tourism can tell us about the human condition. For the purposes of this discussion a key interest is, as previously stated, the link between visits to mining heritage sites and issues of identity and belonging. A key argument being that visits to such sites can be seen as a form of pilgrimage, not to the locations and relics of organized religion but to those associated with an activity that communicates something about the human condition, something about what it means to be human and this is the sense in which we are applying the concept of identity. Before discussing the two case studies, it is necessary to explore why the concept of pilgrimage is instructive in this instance.

From pilgrimage to tourism

The word pilgrimage has long been associated with the ancient practice of many different world religions where long arduous journeys were undertaken to enable the faithful to worship at significant shrines and temples. This image is most often conjured up from what is known about medieval pilgrimage, a common social practice of movement dating back to the fifth century (Digance, 2006). At the height of medieval pilgrimage in Europe, half a million pilgrims made the journey to the Christian shrine of Saint James de Compostela in Spain. By the late twentieth century, that number had risen to between four and five million visitors (Murray and Graham, 1997), leading Digance (2003: 144) to argue that pilgrimage is realising 'a marked resurgence' and is as popular now as it has been in the past. The motivating factors providing the rationale for people to engage in a pilgrimage journey are many and varied, particularly in the context of modern preoccupations and concerns where religion may not always be the prime motivator.

The anthropologists Eade and Sallnow (1991: 3) argue that pilgrimage is as much about historical and cultural behaviours and meanings as it is about pious intentions. In Frey's (2004) study of the Camino de Santiago, the pilgrims are often on the road '. . . for a host of cultural, spiritual, athletic, and personal reasons'. Similarly, for Morinis (1992: 4–5), pilgrimage is motivated by the pursuit of embodied ideals, a '. . . journey undertaken by a person in quest of a place or a state that he or she believes to embody a valued ideal', an ideal which one cannot achieve at home. Some of the intrinsic rewards of engaging in modern pilgrimage are not unlike those found in traditional, religious applications where people search for identity or spiritual rebirth (Osterreith, 1997). These examples reflect the fact that religion may not always feature as a prime motivator for pilgrimage.

Indeed, pilgrimage also has a close relationship with the roots and growth of tourism (Graburn, 1977), and could even be considered as an ancient form of tourism as it draws upon traditions grounded in varying religious beliefs. For anthropologists Turner and Turner (1978:240), a modern characteristic of contemporary pilgrimage is that it is indeed '. . . blended with tourism, and involves a major journey, usually by modern means of transportation, to a national or international shrine'. In anthropological discussions of tourism, pilgrimage features as a dominant motivator for visits to many different sites (Graburn, 1989; Badone and Roseman, 2004). Also as Morgan and Pritchard (2005: 41) point out certain tourism places are deemed sacred because they are charged with personal and social significance and visits to such places serve '. . . to shape the images we have of ourselves'.

Our understanding of the concept of pilgrimage has therefore undergone significant change in response to engagement with tourism and leisure pursuits (Badone and Roseman, 2004) such that the concept is also applied in a secular sense (Reader and Walter, 1993). For example, Aitchison *et al.* (2000) show how druids have for many years worshipped at the English tourist attraction of Stonehenge, using it as both a meeting place and a centre for ritual practice. Stonehenge also holds a similar appeal for New Age travellers, who make annual visits to the site in order to celebrate the summer solstice. They tend to gravitate to heritage sites such as this because of their association with the sacred beliefs of prehistoric and indigenous peoples (Digance, 2003). This is not unlike the pull of the sacred Aborigine landmark, Uluru Rock, which is seen by many people as 'Australia's geographic and spiritual or emotional centre' (McGrath, 1991).

Equally relevant are secular examples of pilgrimages made to more unassuming sites: Elvis Presley's *Graceland* in Tennessee (Rigby, 2001), motorcycling pilgrimages to Washington by Vietnam veterans (Dubisch, 2004) and pilgrimages to conventions celebrating the American television programme *Star Trek* (Porter, 2004). Similarly, in his study of the ancient English village of Walsingham, referred to on signs and posters as England's Nazareth, Coleman (2004) points out that it is the shrines, the narrow lanes and the pubs of this tiny village that comprise some of the components consumed through pilgrimage: '. . . pilgrimage does not merely involve an intensified form of prayer; it also involves for the pilgrim (and even for the tourist) complex and varying forms of engagement with the physical environment . . .' (Coleman, 2004: 53). Such forms of engagement have been identified at other, non-religious sites such as Hever Castle, Chartwell and Battle Abbey in the UK (Palmer, 2005). As with places of sacred worship such as religious shrines, heritage sites also have the ability to possess heightened meaning for their visitors (Palmer, 2003).

These examples illustrate that religious pilgrims coexist with tourist pilgrims at the same site and that tourist pilgrims may invest non-religious places and attractions with meaning and emotions more usually attributed to those of religious pilgrims. So, although the discourse underpinning pilgrimage is dominated by religious connotations, it is clear that 'a pilgrim' is not always a traveller seeking to commune with a religious deity. Likewise, it is not only tourist attractions that can be seen as akin to pilgrimage sites, experiences can also function in this way. For example, Franklin (2003:177) notes how those who watch the Royal Tournament in the UK, an event he describes as a '. . . ritual display of national military capability, frequently do so in order to reaffirm or perhaps celebrate national pride and identity'. Hence, to say that a site has spiritual significance does not mean that site is linked to a particular religious faith. It is a description increasingly applied to tourist sites and even to experiences that enable the visitor to explore deeper meaning about the self and the self's relationship with the wider world.

Making meaning at mining museums

The findings gathered for the discussion derived from two different mining museums. One being the Big Pit National Mining Museum of Wales which lies at the north-eastern tip of the South Wales coalfield and falls within the Blaenavon[1] UNESCO World Heritage Site. Mining started on the site as early as 1812, and at the time supplying coal and ironstone to the nearby Blaenafon Ironworks (National Museum Wales, 2005). The Big Pit Museum is made up of both above ground and underground features of a former mining site. The above ground features include various elements that comprise a winding house, saw mill and mortar mill, tram circuit, pithead, fan house, pithead baths and exhibitions, original canteen, explosives magazine, blacksmith's shop and simulated mining galleries displaying modern mining techniques. The Big Pit Museum's underground tour drops visitors 300 feet to the pit bottom in a shaft cage. From there visitors can accompany a guide to see the inner workings of the mine and its machinery, and learn more about the work of miners underground.

The second site is the Cape Breton Miners' Museum and is located in Glace Bay on Cape Breton Island, Nova Scotia, Canada. The museum opened in 1967 as a result of a community effort to preserve the story of the region's industrial past. The museum is made up of a number of similar elements as well. The museum complex comprises a main exhibition area and the entry way to Ocean Deeps Colliery, the underground tour. The site also has a small miner's village on site which displays replicas of period buildings and possessions from the mid 1850s onwards. The Ocean Deeps Colliery is a purposely dug coal mine which is accessed within the museum and offers underground tours to visitors. A drift mine by design, it has a gradual slope dropping 1000 feet into the earth and leads to a coal face where retired miners take visitors on a tour led by their first hand knowledge of mining experience (Newton, 1992). In contrast to the Big Pit Museum where visitors to the underground workings descend the shaft in an open cage lift, visitors to the Ocean Deeps Colliery at the Cape Breton Miners' Museum walk down a gradual slope to enter the mine.

Because of their respective geographies, both regions experienced the growth of coal mining industries due in part to their geological assets (coal) and ideal location for shipping (on or close to water ports enabling global trade and shipping of minerals). These assets made each region a prime candidate for industrial growth. Each region also had convenient access to iron ore (ironstone) and limestone, thus making the production of iron and steel as sister industries to coal mining. These respective industrial heritage museums depict a similar tale of the growth, suffering, hardship and triumph encountered during the rise of fall of the industry and its people.

Nevertheless, taking into consideration the features of these museums, the industrial heritage site is not – by definition – a religious site. Yet, for some visitors the characteristics of the activity associated with the site enable them to contemplate their own past, their own familial connections. This is the sense in which the concept of pilgrimage is employed here. However, it is important to note that not all visitors will connect with the site they visit; some will describe their experience in terms of a 'good day out'. Hence use of the pilgrimage concept is in relation to those visitors who expressed a particular affinity with the site.

The excerpts included here are interpreted in terms of how the visitors expressed their own thoughts and feelings when asked to discuss their experience of the site. The data collected draws from excerpts taken from visitors in social groupings; that of family and friends at both sites. Thus visitors become pilgrims of industrial heritage, because they find meaning from the site by making the journey with others. It is the interplay of memories, the sharing of familial stories among friends and family that illustrate what the site 'means' to people. In essence, identity is reinforced among groups of visitors by making *pilgrimage* together to the heritage site. It is within this dynamic of the collective group that meaning and identity in terms of an individual's place in the world is constructed and reinforced.

The following excerpt from a conversation with an elderly lady visiting the Cape Breton Miners' Museum with her children and grandchildren illustrates the site's ability to trigger feelings and memories linking her to her own family's connection to the mining industry. Although she left the area over 60 years ago, the firsthand accounts given by the miner guides enable her to make links to her own past, her own sense of self identity and that of her family:

> *Elderly Lady:* I think it was fabulous (underground tour and the museum). I'm from the area, my dad was a miner, he pulled out one of the guys that was mentioned earlier, the fires in the Phalen. This place is not promoted enough. It is a real diamond in the rough.

This particular visitor brought her family to the museum because it is an important part of her past, her identity. She is sharing this history, this familial identification with mining, with her companions and in so doing outlining how they should understand and interpret the mine. In effect, her family are all located within the context of the 'industrial past' on display. In the following excerpt from a conversation with a family also visiting the Cape Breton Miners' Museum the family's link with mining is likewise identified and in addition a clear reference is made to the characteristics of mining: the hardship and danger experienced on a daily basis by the miners:

> *Family of 5 from Ontario:* My father's grandfather was a miner. Sheldon (miner guide) did a great job of bringing the danger to life, hard to imagine unless you go underground yourself. I'll be interested to hear what my children (teenagers) think. They were not scared because she (youngest aged nine) knows her parents wouldn't take them underground to anything that was unsafe. I was less taken with the exhibit area, it is standard. But Sheldon is an actual coal miner that speaks from firsthand experience . . .

In this instance, the father is speaking on behalf of the whole family and, collectively, they have the opportunity to reflect upon a time long since passed, and to learn more about 'how their ancestors' lived and survived. He also comments on the miner guide Sheldon's involvement in this process of meaning making by bringing to light firsthand accounts of what it was like for his father's grandfather to work as a coal miner. As this account illustrates an element of storytelling emerges as a key factor in helping visitors to link the mining experience to their own lives and in so doing locate themselves in the present through what has occurred in the past. In effect, the father brought *his* family to this site to learn more about whom *they* are and he is curious to know what his children think about their visit. The excerpt that follows is also indicative of this practice:

> *Father and two daughters:* Wishie (tour guide) had great stories, this is part of our heritage, I wanted to bring my girls and show them, they're old enough to understand and remember. This is our history, my family goes way back, right up until my father's generation. *Daughters:* It was dark, dirty and dangerous!

What is being created and reinforced in these cases is a sense of identity in relation to the family; this is my family, my family history being passed down through the generations. Storytelling is key here as it is one of the ways in which the links between the family and the mining heritage are established. Through the process of storytelling, they come to understand more about whom they are and where they came from.

The following accounts taken from the Big Pit in Blaenavon, South Wales illustrate not just the familial links created by the site and the history of the mining region but also the site's ability to trigger a sense of national identity:

> *Man:* This is part of me; my family came from this region so it is part of me.
>
> *Woman:* Going down underground was really interesting. You can hear about it, you can read about it – but there is no replacing seeing it for yourself and how they worked . . . it was outstanding.
>
> *Young Welsh couple:* We are both Welsh, proud of our heritage. My grandfather was a miner. It is really interesting to be underground. I am proud to be Welsh, (lifts up shirt sleeve to display tattoo of red dragon).

Here, a direct link is again made through having family working in the area. In the second passage, the respondent contextualizes his identity further, as he makes a strong connection with his identity as a Welshman. For some visitors it is not necessarily the presence of a former coal mining industry per se that brings them to an industrial heritage museum. Rather, it is a link to the industrial past more generally, and in this case the exploitation of the earth through coal mining, which at one time was the major employer in Wales. For the following visitors, the death of their father in the North Wales slate quarry and the relationship between him and heavy industry resonates at the Big Pit:

> *Couple (brother and sister) from Scotland, he was 70, she was 61:* Our father died in a quarry, he was the quarry man, the manager . . . he had two accidents, he said that if there was ever a third, he wouldn't get out. There was a blast, he was hit from behind suffered a fractured skull and punctured lung . . . he didn't make it. He couldn't get away quick enough . . . children need to see this, to see how hard it was . . . our grandfather was a miner. And our dad a quarry man . . . I don't think men would go down the mines today . . . would they? Not that they shouldn't, mind you. Would be good for them. They had no choice then . . . did they . . .

This brother and sister have made a pilgrimage to the Big Pit to remember the untimely death of their father in an industrial accident. The visit not only enables them to remember their family it also enables them to identify with and articulate the sibling bond between them through an encounter with the industrial past.

In another example, a Welsh couple brought American visitors to the Big Pit. In bringing them to the museum, they show their American guests more about the mining history of their region. These two couples met through their children, who are now married. In essence, the family connection to the industrial past is thus extended and furthermore shared with 'new' social members of their family. The Welsh lady shares a story about her personal connections to the industry and the area, and reminisces about times past:

> *Welsh lady with American visitors:* My father was a miner, he went underground with his father, and his brothers were miners. But one of them had an accident on the coalface . . . my uncle and his father, they were miners in the valley where I lived, and he was one of the homeless men on the lifesaving, he had to go if there were any accidents around – there was a [roof] fall – something fell on his pelvis and fractured his spine – and he was a paraplegic for the rest of his life, yet he was telling the best way to remove the coal to get him out, he was, actually, in all his pain, explaining how to get him out. So you know, so much is mingled with coal mining, just the word coal.

In the accounts described above, visitors to these sites are engaging in 'tourist' pursuits yet they come to these sites for particular, frequently very personal reasons. Hence the history associated with industrial heritage sites resonates with some of its visitors, making the account meaningful *because* it is shared with family and friends. Visiting the site together also nurtures a collective sense of 'belonging' to a way of life that existed in the 'industrial' past. The link with pilgrimage here is illustrated by Bauman's (1996) application of the pilgrimage metaphor to identity where he argues that 'identity-building' occurs because it is done *together.* Reflecting upon the similarities between the journeys made by religious pilgrims and those of the modern day pilgrim: the stroller, the vagabond, the tourist, and the player Bauman argues that pilgrims walked:

> . . . in a kind of world in which one can tell life as a continuous story, a 'sense-making' story, such a story as makes each event the effect of the event before . . . each age a station on the road pointing towards fulfilment. The world of pilgrims – of identity-*builders* . . . must be a kind of world in which footprints are engraved . . . so that the trace and the record of past travels are kept and preserved. (1996: 23 original emphasis)

The above examples are 'pilgrimages' not just of identity building but of identity reinforcing as the visitors share personal connections to the site, the history, with friends and family members. Coal mining leaves tangible, physical footprints in the ground, footprints that provide the key ingredient from which stories of the self are constructed and told to others. This is not unlike a 'pilgrimage' to Australia's Uluru Rock to connect with its spiritual and emotional significance, so too does pilgrimage to industrial heritage possess similar appeal to visitors. It matters not then if a pilgrimage is sacred or secular in nature; a person is able to make 'a spiritual step forward' (Turner and Turner, 1978) in recognising something they deem personal and significant for their sense of self, their identity. As such, an industrial heritage site can, in its own right, be considered a site or destination of pilgrimage.

Furthermore, for the following couple, coming to the Big Pit Museum has through repeated visits become a ritual in which meaning is created and reinforced. Not unlike paying homage to sites of spiritual significance such as Stonehenge, the ritual of visiting the same site reinforces the identity building capacity of the site. As Franklin (2003) asserts, a feature of ritual in tourism is the repeated nature of visiting some sites more than once, as the following excerpts from the Big Pit Museum illustrate:

> *Couple from Forest of Dean:* We've been a couple of times, we keep coming back. We like to take our time, take things in and not to rush. So we haven't seen it all yet, but we check things out. As far as history is concerned, Napoleonic wars, what's left behind is very general, with this it is very real.

> *Couple from Abergavenny, Wales:* We've been here loads of times . . . we like to come back and see the reactions of people . . . that is why we come back . . . basically that is why we come back. The bangs and so on . . . I find it a very moving experience. I like to hear what the miners have to say, their stories.

Through repeat visits, each of these couples find new meaning each time they visit. These meanings are contextualized by listening to the different stories of the miner guides, from observing the reactions of other visitors and through their own contemplation of the whole experience.

In both industrial heritage museums: the Big Pit Museum and the Cape Breton Miners' Museum, the ritual of storytelling emerges naturally among visitors. This becomes evident from what they have to say, and in the process they are telling stories amongst each other. This causes these visitors to then reflect on their own experience of consuming the site, thus impacting on the meaning they draw from the site. In effect, the visitors are encouraging other people who they know to visit the Big Pit Museum, and construct their own meaning from the industrial heritage displayed:

> *Couple from Merthyr Tydfil:* Merthyr Tydfil was the centre of the Industrial Revolution . . . it is very important to keep heritage sites as this – you have to pass it on. You had a real sense of community then – your neighbours were your best friends. It's important to pass on that history. My grandparents lived through this. It is important to know what they lived through. You have to know your history.

Such suggestions by visitors illustrate the passing on through storytelling of a past deemed to be in need of preservation. The occurrence of such recollections was consistently evident in both sites – the Big Pit Museum and the Cape Breton Miners' Museum. The story of the past is important, and by passing it down, layers of meaning can be added by other visitors. In this way the process of identity building continues through the act of pilgrimage connecting individuals to significant stories from their past.

Conclusion

Here we have argued that meaning making from the industrial past can be likened to a form of pilgrimage, not in the religious sense where the pilgrim searches for a higher truth but in terms of what Bauman (1996) refers to when he describes pilgrimage as a form of identity building. A site's 'meaning' then for visitors to the Big Pit and to Cape Breton is in relation to identity, the visit as a way of locating oneself in the world, primarily but not exclusively in terms of felt familial ties. Industrial heritage museums such as those discussed here offer the opportunity for visitors to imagine, remember, forget, hope, discover, or feel sad as a result of their experience. It is everyday behaviour such as visiting museums that is important for understanding the significance of the social context in which the industrial heritage is displayed. As Shotter (1993) argues, there is deep-rooted significance in how modern society constructs meaning in the everyday.

Both pilgrimage and tourism then, provide spaces and experiences that offer individuals the potential to construct meaning and negotiate identity (Badone, 2004:181). Hence tourists can engage in spiritual behaviour in alternative settings such as those provided by heritage tourism. Indeed, Cohen (1979:190) seemed to be offering an alternate concept for spirituality when he proposed that tourists develop 'elective' spiritual centres that are '. . . external to the mainstream of his/her native society and culture'. Badone and Rosman (2004) likewise argue that the once rigid dichotomies between pilgrimage and tourism do not hold the same currency in postmodern forms of travel and tourism. So, although these mining museums are not religious per se, they do have the ability to offer a secular form of spirituality for the visitors.

Eade and Sallnow (1991) also acknowledge the theoretical limitations of harbouring a narrow view of the diverse meanings related to sacred 'things'. They contest the view that religious symbols such as shrines possess meanings that are largely predetermined since a variety of perceptions and interpretations may be brought by the pilgrim/visitor and employed to impose their own meaning on the site being visited. Such a view highlights the power of the visitor in constructing his/ her own understanding of what is important about a particular experience. As Digance (2006:36) states, 'the twenty-first century calls for a newer and more flexible definition of pilgrimage, namely "undertaking a journey that is redolent with meaning" '.

The research discussed here has interesting implications for the managers of heritage sites generally and specifically in relation to coal mines turned tourist attractions where the ongoing running costs are often prohibitive (Alfrey and Putnam, 1992; Cole, 2004; Wanhill, 2000). For example, heritage sites can become as important to their visitors and have as much meaning for them as religious shrines do for their visitor/pilgrims. Acts of pilgrimage are often exercised in similar social contexts, in that they are practiced with others: family, friends, and members of a (religious) community. As the visitor accounts illustrate, people come to these sites to remember a family member, an industrial past, or a previous way of life and the experience is usually shared with other people as family or friends. This shared experience is made meaningful through the power of storytelling, what marketers might refer to as 'word of mouth'. In managing heritage attractions, managers must keep the needs of the collective groups in mind. However, although some mining museums already offer special archival services to visitors if required, not all do and more can be done to match industrial records with those of family members to create rich and detailed accounts of the significance of the coal mines beyond the colliery gates, accounts that can be shared with future visitors. Strengthening the ties between these mining museums and their visitors will help to preserve the financial sustainability of these attractions.

Note

1 UNESCO adopts the name Blaenavon as opposed to the Welsh name Blaenafon.

References

Aitchison, C., MacLeod, N. and Shaw, S. (2000) *Leisure and Tourism Landscapes: Social and Cultural Geographies.* London: Routledge.

Alfrey, J. and Putnam, T. (1992) *The Industrial Heritage: Managing Resources and Uses.* London: Routledge.

Badone, E. (2004) 'Crossing Boundaries; Exploring the Bordrrlands of Ethnography, Tourism, and Pilgrimage', in E. Badone and S. Roseman (eds) *Intersecting Journeys: The Anthropology of Pilgrimage and Tourism*, Chicago: University of Illinois Press.

Badone, E. and Roseman, S. (2004) 'Approaches to the Anthropology of Pilgrimage and Tourism', in E. Badone and S. Roseman (eds) *Intersecting Journeys: The Anthropology of Pilgrimage and Tourism*, Chicago: University of Illinois Press.

Bauman, Z. (1996) 'From Pilgrim to Tourist – or a Short History of Identity', in S. Hall, and P. duGay (eds) *Questions of Cultural Identity*, London: Sage.

Cohen, E. (1979) 'A Phenomenology of Tourist Experiences', *Sociology*, 13(2):179–201.

Cole, D. (2004) 'Exploring the Sustainability of Mining Heritage Tourism', *Journal of Sustainable Tourism*, 12(6): 480–94.

Coleman, S. (2004) 'Pilgrimage to "England's Nazareth": Landscapes of Myth and Memory at Walsingham', in E. Badone and S. Roseman (eds) *Intersecting Journeys: The Anthropology of Pilgrimage and Tourism*, Chicago: University of Illinois Press.

Digance, J. (2003) 'Pilgrimage at Contested Sites', *Annals of Tourism Research*, 30(1): 143–59.

Digance, J. (2006) 'Religious and Secular Pilgrimage: Journeys Redolent with Meaning', in D. Timothy and D. Olsen (eds) *Tourism, Religion and Spiritual Journeys*, Abingdon: Routledge.

Dubisch, J. (2004) '"Heartland of America": Memory, Motion and the (Re)construction of History on a Motorcycle Pilgrimage', in S. Coleman and J. Eade (eds) *Reframing Pilgrimage: Cultures in Motion*, London: Routledge.

Eade, J. and Sallnow, M. (1991) 'Introduction', in J. Eade and M. Sallnow (eds) *Contesting the Sacred The Anthropology of Christian Pilgrimage*, Routledge: London: Routledge.

Eade, J. and Sallnow, M. (1991b) *Contesting the Sacred the Anthropology of Christian Pilgrimage*, London: Routledge.

Edwards, J. and Llurdés i Coit, J. (1996) 'Mines and Quarries: Industrial Heritage Tourism', *Annals of Tourism Research*, 23(2): 341–63.

Franklin, A. (2003) *An Introduction to Tourism*, London: Sage.

Frey, N. (2004) 'Stories of the Return: Pilgrimage and its Aftermaths', in E. Badone and S. Roseman (eds) *Intersecting Journeys: The Anthropology of Pilgrimage and Tourism*, Chicago: University of Illinois Press.

Gouthro, M.B. (2008) 'Consuming Heritage, Experiencing Identity: Finding Meaning through Interpretations of Industrial Heritage Tourism', Unpublished PhD Thesis, University of Brighton, UK.

Graburn, N. (1977) 'Tourism: The Sacred Journey', in V. Smith (ed) *Hosts and Guests: The Anthropology of Tourism*, Oxford: Basil Blackwell.

Graburn, N. (1989) 'Tourism: The Sacred Journey', in V. Smith (ed) *Hosts and Guests: The Anthropology of Tourism*, Philadelphia: University of Pennsylvania Press.

Light, D. and Prentice, R. (1994) 'Who Consumes the Heritage Product? Implications for European Heritage Tourism', in G. Ashworth and P. Larkham (eds) *Building a New Heritage: Tourism, Culture and Identity in the New Europe*, London: Routledge.

McGrath, A. (1991) 'Travels to a Distant Past: The Mythology of the Outback', *Australian Cultural History*, 10: 1113–24.

Morgan, N. and Pritchard, A. (2005) 'On Souvenirs and Metonymy: Narratives of Memory, Metaphor and Materiality', *Tourist Studies*, 5(1): 29–53.

Morinis, E. (1992) 'Introduction: The Territory of the Anthropology of Pilgrimage', in E. Morinis (ed) *Sacred Journeys: The Anthropology of Pilgrimage*, Westport, Conn: Greenwood Press, pp. 1–27.

Murray, M. and Graham, B. (1997) 'Exploring the Dialectics of Route-based Tourism: The Camino de Santiago', *Tourism Management*, 18: 513–34.

National Museum Wales (2005) *Big Pit: National Coal Museum A Guide*. Cardiff: National Museum Wales.

Newton, D. (1992) *Where Coal is King: The Story of the Cape Breton Miners' Museum*, Glace Bay, Nova Scotia: Cape Breton Miners Foundation.

Osterreith, A. (1997) 'Pilgrimage, Travel and Existential Quest', in R. Stoddard and A. Morinis (eds) *Sacred Places, Sacred Spaces: The Geography of Pilgrimages, Geoscience and Man*, Baton Rouge: Department of Geography and Anthropology, Louisiana State University, 34: 25–39.

Palmer, C. (2003) 'Touring Churchill's England: Rituals of Kinship and Belonging', *Annals of Tourism Research*, 30(2): 426–45.

Palmer, C. (2005) 'An Ethnography of Englishness: Experiencing Identity through Tourism', *Annals of Tourism Research* 32(1): 7–27.

Porter, J. (2004) 'Pilgrimage and the IDIC Ethic: Exploring Star Trek Convention Attendance as Pilgrimage', in E. Badone and S. Roseman (eds) *Intersecting Journeys: The Anthropology of Pilgrimage and Tourism*, Chicago: University of Illinois Press.

Prentice, R., Witt, S. and Hamer, C. (1998) 'Tourism as Experience: The Case of Heritage Parks', *Annals of Tourism Research*, 25(1): 1–24.

Pretes, M. (2002) 'Touring Mines and Mining Tourists', *Annals of Tourism Research*, 29(2): 439–56.

Reader, I. and Walter, T. (eds) (1993) *Pilgrimage in Popular Culture*. Basingstoke: MacMillan Press.

Richards, G. (1996) 'The Scope and Significance of Cultural Tourism', in G. Richards (ed) *Cultural Tourism in Europe*, Wallingford: CAB International.

Rigby, M. (2001) 'Graceland: A Sacred Place in a Secular World?', in C. Cusack and P. Oldmeadow (eds) *The End of Religions? Religion in an Age of Globalization Sydney Studies in Religion 4*, Sydney: The University of Sydney.

Rudd, M. and Davis, J. (1998) 'Industrial Heritage Tourism at the Bingham Canyon Copper Mine', *Journal of Travel Research*, 36: 85–89.

Shotter, J. (1993) *The Cultural Politics of Everyday Life.* Milton Keynes: Open University Press.

Turner, V. and Turner, E. (1978) *Image and Pilgrimage in Christian Culture.* New York: Columbia University Press.

Wanhill, S. (2000) 'Mines – A Tourist Attraction: Coal Mining in Industrial South Wales', *Journal of Travel Research*, 39: 60–69.

5 The branding of Beaconsfield

How a mining disaster put the Tasmanian town on the tourist map

Leanne White

Introduction

This chapter will examine how the 2006 Beaconsfield Mine rock fall in northern Tasmania forever changed the close-knit community and put the little-known destination on the tourist map. More than two weeks of unplanned national and international media coverage raised the awareness of the town and enhanced its prospects for future tourism. The towering head frame of the mine became a key signifier for the cautious rescue mission which came to be known as 'The Great Escape'. After such prolonged media exposure, many visitors were curious to actually see the Beaconsfield mine for themselves.

The mine disaster was a global news story with the miraculous rescue of Todd Russell and Brant Webb, and the tragic death of co-worker Larry Knight. Russell and Webb were trapped for 14 days in a small wire cage 925 metres below the surface. They survived underground in a pocket of air that measured just two by four metres. The website of their management company proclaims, 'The story of the Beaconsfield miners has a place in history as one of the greatest stories of Australian mateship, courage and survival ever told.' (Twenty-Two Management, 2009) The newsworthiness of the story was further heightened when leading Australian journalist Richard Carleton died at the location having suffered a stroke just seconds after asking a question while standing in a media pack. Apart from the media coverage, a book, a musical, a movie and songs were also written about the dramatic event and subsequent rescue mission.

The Beaconsfield mine collapse was clearly not the first time that a mining accident or incident had the capacity to dramatically change a town. Mining disasters that have affected communities in just Canada alone include: the Drummond Colliery disaster in 1873; the 1891, 1956 and 1958 (fire, explosion and earthquake) incidents in Springhill; the Fernie or Cold Creek accident in 1902; the Hillcrest disaster in 1914; the Moose River incident in 1936; and the Westray disaster in 1992.

After the Beaconsfield mining tragedy 53 miners were made redundant as the mine was closed for more than a year until it was safe to re-open. The Australian government allocated \$8.3 million to assist the mine and the community in its rebuilding efforts. The Beaconsfield Community Fund was established to support economic, social and cultural projects to assist the community to grow and prosper. The town was supported financially by government at all levels – local, state and federal. The main tourist attraction – the Beaconsfield Mine and Heritage Centre – received more than \$1 million to upgrade the site.

Beaconsfield with a population of around 1500 residents is located in Tasmania's picturesque Tamar Valley, 40 kilometres northwest of Launceston. Gold mining has been taking place in the town since the 1840s and more than 50 mining companies operated in the town at the peak of the boom. The place where the town is situated was originally known as

Cabbage Tree Hill, after that it was known as Brandy Creek. In 1879, the town was named Beaconsfield in honour of the Prime Minister of Great Britain – Benjamin Disraeli – also the Earl of Beaconsfield. By 1881, Beaconsfield was producing more gold than any other town in Tasmania. Beaconsfield also holds the inimitable honour of being the first town in Australia to fluoridate its water supply.

Beaconsfield's tourist attractions and striking architectural features were brought into sharp focus during the media coverage of the mine collapse. The town experienced a significant tourism boom directly after the event. Beaconsfield's Mayor Barry Easther explained, 'They're coming to see where this disaster occurred and Council are very pleased' (Australian Broadcasting Corporation, 2006a). This chapter explores the tangible and intangible benefits that the town now enjoys as a result of the highly intensive (some might say intrusive) media coverage.

Tragedy and eventual triumph – Anzac Day 2006

The Beaconsfield Mine rock fall (earthquake) measured 2.2 on the Richter scale. The small earthquake or seismic shift was felt at 9.26 pm on the evening of April 25 and resulted in miners being trapped 925 metres underground. Seventeen miners were trapped, 14 escaped soon after, leaving three down the mine – Russell, Webb and Knight. The mine disaster was a global news story with the miraculous rescue of Todd Russell and Brant Webb, and the tragic death of co-worker Larry Knight. Todd Russell was born in Beaconsfield in 1972; Brant Webb was born in the Victorian town of Sale in 1968, while Larry Knight was the eldest of the three being born in Launceston in 1962.

Anzac Day is a respected and solemn day of the year. It is the public holiday when Australians and New Zealanders pay tribute to legendary battles and fallen war heroes. Consequently, the irony that this mine event occurred on Anzac Day and 'heroes' Russell, Webb and Knight came to the attention of the public was an aspect that was emphasized in the media coverage. While the political genesis of Australia can be traced to the events of 1901, some historians have argued that it was the Anzac (Australian and New Zealand Army Corps) landing at Gallipoli on 25 April 1915 that marked the emergence of Australia, the nation. While Gallipoli was nothing less than Australia's greatest military defeat, the campaign also marked Australia's loss of innocence. For Australia, Gallipoli represented the first foray into the world arena and was both a watershed moment in the national consciousness and a 'coming of age' for the newly formed nation.

The three trapped miners were thought to have died. The body of Larry Knight was found on 27 April. Little hope was held for Todd Russell and Brant Webb but they were found on 30 April with the aid of heat cameras. The rescue of the trapped miners was a long and arduous process as it involved cutting and carefully blasting through 36 metres of solid rock with sophisticated low-vibration equipment. A 5.4 tonne $6 million piece of rock-cutting equipment known as a raise bore was immediately brought in from a zinc mine in Tasmania. The raise bore was used to drill through solid rock during the 14 day rescue mission. Around 70 low-impact explosives were also detonated in the final stages as rescuers edged closer to the trapped men.

A 12 metre long and 90 millimetre wide plastic pipe was also put in place so that Russell and Webb could receive food, drink and other items including dry clothes, medical supplies, letters from loved ones, toiletries, a torch, MP3 players loaded with music, and an all-important digital camera. Photographs and vision of the surrounding rock surface enabled rescuers to have a clearer understanding about the complex nature of the mission. From the photographs that the men took, they could see that a large pile of rocks sat directly above the telehandler vehicle in which the men were trapped. Some of the rocks above them had fallen into the cage and the men were partly buried. Thus, digging through rock to approach the trapped men from the side or below provided the best chance of success. A video-link that was established with the trapped men allowed rescue workers to remain in constant contact. Despite their dire situation, Russell and Webb remained in good spirits. Dean Mackrell, one of the 10-member Beaconsfield Mine rescue squad (of which Todd Russell is also a member) said, 'they were always laughing and joking' (Rose, 2006). Russell and Webb survived by drinking groundwater. Before food was made available to them, they also had a muesli bar to share. After an extremely careful rescue mission with the world watching and waiting, the two men finally came to the surface on May 9.

To announce their rescue the church bell, which had not been used since the end of World War II, rang out with such force that it broke. Fire sirens were also heard around the small town and the noise encouraged the townspeople to gather at the mine. The much-awaited news images that were immediately beamed around the globe showed Russell and Webb walking from a lift, punching the air in victory, clocking off from their shift and embracing their families. John Howard, Australia's Prime Minister at the time said, 'they're tough blokes' and added, 'Overall it has been a wonderful demonstration of Australian mateship and perseverance' (Australian Broadcasting Corporation, 2006b).

More than 200 journalists, photographers and television production crew came to the tiny town to cover the story. The situation was nothing short of a media frenzy or media circus. As the rescue mission took longer than originally expected, journalists worked around the clock to out-scoop their media rivals. Police investigated a report that a journalist offered rescue workers a large amount of money to smuggle a video camera down into the mine so that their work could be filmed (Phillips, 2006). The 'event' reported by the media was almost entirely played out within a 300 metre radius of the mine (Easther, 2006).

The timing of the Beaconsfield disaster was slightly fortuitous for the main television networks in Australia. As it turned out, a number of outside broadcast vans were on their way to Port Arthur in Tasmania to broadcast from the site on the tenth anniversary of what has become known as 'The Port Arthur Massacre'. On 28 April 1996, a gunman named Martin Bryant killed 35 people at the Port Arthur Historic Site in southern Tasmania. The unprecedented tragic event resulted in tighter gun laws in Australia. Bryant remains in prison to this day and Tasmanians generally do not like to discuss either him or the massacre.

In order to cope with the voracious appetite of the assembled media, the Tasmania Police media unit, the Department of Premier and Cabinet media unit, the Beaconsfield Mine media consultant and the local council worked closely together. Key media spokespeople included: Mine Manager Matthew Gill, Australian Workers Union National Secretary Bill Shorten, District Police Inspector Paul Reynolds, and West Tamar Mayor Barry Easther. To keep new dimensions of the story alive throughout the two-week rescue mission psychologists, paramedics, explosive experts and even the Minister of the Beaconsfield Uniting Church – Frances Seen, were strategically fed to the hungry media pack. About 250 people from Tasmania and what Tasmanians refer to as the 'mainland' were involved in the rescue effort (Rose, 2006). Mining specialists from as far away as Cornwall in the United Kingdom were also involved.

An inquest was conducted into the rock fall and in February 2009, Coroner Rod Chandler found that the Beaconsfield Gold Mine had an inadequate ground support system. While the Coroner's findings were damaging for Beaconsfield Gold and Workplace Standards Tasmania, they stopped short of actually attributing blame to any person or company for Larry Knight's death (Clark, 2009). Brant Webb told the inquest that Beaconsfield Gold

Mine management had failed to provide a safe workplace. He said, 'I'm still bitter – they are supposed to supply a safe workplace' (Australian Associated Press, 2008). Ironically, Webb had flipped a coin with Larry Knight to decide who was going to use what equipment that night. The decision would prove a fatal one for Knight. After Todd Russell gave his evidence he told the assembled media, 'That's the final chapter mate . . . It's all over – it's done and dusted' (Australian Associated Press, 2008).

Global media celebrities Todd Russell and Brant Webb

After the significant global media coverage and their subsequent miraculous rescue, Todd Russell and Brant Webb became global media celebrities for a limited time. They were also described as 'Australia's new folk heroes' (Gibson and Cubby, 2006). As *Sixty Minutes* reporter Charles Wooley explained, 'This was the great Australian story with something for everyone – mateship, heroism, good luck and good old fashioned workers' solidarity.' (Wooley, 2006). In order to cope with the thousands of requests for information suddenly placed upon them by a curious media, the miners employed the services of manager Sean Anderson as their agent. The company negotiated the most lucrative deal in Australian television history with the Nine Network ($2.6 million) for the program *The Great Escape* – a two-hour documentary. The promotion for the program boldly declared, 'The interview the world's been waiting for'. The program aired on Sunday 21 May across Australia during prime viewing time – 8.30 pm (Nine Network, 2006).

The negotiations for the deal effectively began when former Channel Nine Chief Executive Officer Eddie McGuire asked Todd Russell on national television what it would take to get an exclusive interview with the men. Knowing the value of his story Russell replied 'Tell me how big your chequebook is and we'll talk.' Nicola Goc and Jason Bainbridge argue that it was the first time in Australia that a member of the public had been so brazen about their monetary value on national television (Goc and Bainbridge, 2008). For the program, Channel Nine doubled its usual advertising rates and charged advertisers $100,000 for a 30 second commercial. As a result, the money spent securing the exclusive deal with the miners was more than recouped with the broadcast of this program alone.

The interview was conducted by long-standing journalist Tracy Grimshaw and was watched by almost 2.8 million viewers. *The Great Escape* was the highest rating program of 2006 (Enker and Newsome, 2006). Australian Workers' Union National Secretary Bill Shorten had dubbed the rescue 'the great escape'. As a result, television news editors used the soundtrack of the 1963 movie as the theme song of the rescue mission (Phillips, 2006). The eventual emergence of the trapped miners was a carefully orchestrated marketing moment as Russell and Webb handed out business cards entitled 'The Great Escape' to the waiting crowd. The message on the card read, 'To all who have helped and supported us and our families, we cannot wait to shake your hand . . .' (Gibson and Cubby, 2006). Further 'exclusive' interviews were also arranged with Dianne Sawyer on the American Broadcasting Corporation's *Good Morning America* and with the US *Primetime* program. Todd, Brant and their families were flown to New York and given the 'royal treatment'. As is often the case when a big story breaks in Australia, the Australian media spent significant air time and print space reporting on what the rest of the world made of 'The Great Escape'.

Australia's long-running magazine *The Bulletin* produced a souvenir issue of the events at Beaconsfield and named the issue 'The Great Escape'. *The Bulletin* and the Nine Network are owned by the same organisation – Publishing and Broadcasting Limited (PBL). Consequently, it is no surprise that the printed version of the 'exclusive' story went to the magazine.

Advertising the special edition the editors announced, 'Because of its historical importance and the enormous emotional impact it has had on the nation, *The Bulletin* has printed a special souvenir edition . . .' (Magshop, 2006). The magazine's national affairs editor Tony Wright was also the author of a book published within six months of the rescue entitled *Bad Ground: Inside the Beaconsfield Rescue* (2006). Some of the proceeds of the book sales were donated to the Russell Webb Legacy – established by Todd and Brant to assist the youth of Beaconsfield and surrounding areas. 'Exclusive' interviews with the trapped miners and their families were also carried in other PBL publications the *Australian Women's Weekly* and *Woman's Day*.

A number of songs were also written about the well-publicized event and subsequent rescue mission. The Foo Fighters wrote a piece of music called *Ballad of the Beaconsfield Miners* after it was discovered that Brant Webb was a fan of the band. The miners and their families were also invited to concerts with the highly appreciative band. A single *321 Hours* was also recorded by the Adelaide band Unitopia. It was also reported that the two trapped miners sang the Kenny Rogers song *The Gambler* to keep their spirits up as it was the only song to which they both knew the words.

To further capitalize on the event, a tribute CD of 40 songs entitled *Solid Gold Tassie Rock* was created. Some of Brant Webb's selection included: *Solid Rock* by Goanna, *Down Under* by Men at Work, *The Only Way is Up* by Yazz, *Living on a Prayer* by Bon Jovi, *Comfortably Numb* by Pink Floyd, *Blue Sky Mine* by Midnight Oil, *I Feel the Earth Move* by Carole King, and *Long Way to the Top* by AC/DC. Todd Russell's picks included: *Rescue Me* by Pat Benatar, *We Gotta Get Out of this Place* by The Angels, *SOS* by Abba, *I Want to Break Free* by Queen, *Help* by The Beatles, *The Air that I Breath* by The Hollies, *Hold On, I'm Coming* by Sam Cooke and *We Will Rock You* by Queen.

A musical was also written about the event. Entitled *Beaconsfield: A Musical in A Flat Minor*, it later became *Beaconsfield: The Musical*, then finally *Beaconsfield: A Musical in No Particular Key* (as a reminder of the original dry title). Creator of the musical Dan Ilic presents a behind-the-scenes examination of the intense media coverage of the mine disaster. When *Beaconsfield: A Musical in A Flat Minor* was released, there was much public outrage. Family and friends of Larry Knight were incensed by the title with Larry Knight's father-in-law Phil Stevenson referring to Dan Ilic as a 'sick bastard' (Burgess and Spencer, 2008). One song *Carleton Cardiac* deals with the sudden death of veteran *60 Minutes* reporter Richard Carleton who suffered a heart attack moments after he had asked a question about previous safety concerns being ignored. Reviewer Cameron Woodhead claims that the musical is a 'tasteless, insensitive and riotously funny skewering of press excess' which portrays the 'unedifying scramble' of the media to exploit the disaster. It includes 'merciless lampoons' (Woodhead, 2009) of Eddie McGuire and Tracy Grimshaw (Channel Nine), Naomi Robson, Melissa Doyle and David Koche – known colloquially as Mel and Kochie (Channel Seven) and former union boss, now Labor politician Bill Shorten.

In May 2009, it was announced that the story of the trapped miners and the relentless media coverage which followed the event would be made into a movie for television by Southern Star Endemol (Lees, 2009). It is expected that Eric Bana will play the role of Brant Webb and Todd Russell may be played by actor Shane Jacobsen. Other casting decisions will be made for actors to play the role of Eddie McGuire, Tracy Grimshaw, Matthew Gill and Bill Shorten. Webb and Russell will apparently advise the creators of the telemovie about some of the personal and technical aspects to ensure that the movie remains relatively accurate.

The names Todd Russell and Brant Webb (usually referred to in that order) became household names in the weeks and months following the Anzac Day 2006 rock fall. While many media outlets frantically fought to cash in on any news remotely connected with the men and the event, an enterprising person registered the domain names toddrussell.com.au and brantwebb.com.au in the hope that he might be able to on-sell the domain names for a profit. The same person also registered nathanrees.com.au (Premier of New South Wales) and then attempted unsuccessfully to sell the name to him for 'tens of thousands of dollars' (Australian Broadcasting Corporation, 2009a).

In June 2009, Brant Webb announced his nomination for pre-selection to possibly become a candidate for the Tasmanian seat of Bass at the forthcoming election. Accompanying Webb at the announcement were Todd Russell, Bill Shorten and the Tasmanian Premier David Bartlett. Webb claimed that he chose to enter politics because he wants to 'give something back to the community that has supported me' (Wright, 2006). He plans to campaign on a platform of occupational health and safety (Australian Associated Press, 2009). Todd Russell still lives in the area and works as a salesman for an explosives company.

Government assistance for small town

After the mining tragedy 53 miners were made redundant as the mine was closed for about a year before it was safe to re-open. The miners represent around half the town's workforce (Nettlefold, 2006). As such, the Australian government allocated $8.3 million to assist the mine and the community in its rebuilding efforts. The Beaconsfield Community Fund recognizes the economic significance of the Beaconsfield mine to the community. However, there was much animosity about the government handouts as more than $4.8 million of the so-called 'community funds' were given to the mine operators (despite the fact that they are a private company).

In November 2008, Prime Minister Kevin Rudd opened the $1.7 million Beaconsfield Mine and Heritage Centre (Figure 5.1). The Centre is an important part of tourism infrastructure for the town and was redeveloped using $1.28 million from the Community Fund. When Rudd officially opened the Centre he declared, 'Beaconsfield is proud of its history and optimistic about its future' and that 'Opening the Centre today symbolises one of the great traits of Australians – resilience.' (Rudd, 2008) Todd Russell and Brant Webb (who feature prominently in the new exhibition) also attended the official opening of the Heritage Centre. West Tamar Mayor Barry Easther explained that the Heritage Centre has had an influx of visitors from around Tasmania, from all over Australia and overseas. He said, 'It's a wonderful mine rescue display in our new building and people who visit there are very pleased indeed with what they see' (Australian Broadcasting Corporation, 2009b). Importantly for the local community, the museum was transferred from state government ownership to being owned locally. Beaconsfield is being supported financially by government at all levels – local, state and federal. At the local level, Barry Easther explained that the Beaconsfield residents are 'just good country folk' who 'go about their business day by day with no great fuss' (Easther, 2006).

A number of events were also held in the weeks and months after the rescue including: a Beaconsfield broadcast of the Channel Nine *Footy Show*, the Channel Seven *Sunrise* program broadcast from what they colloquially referred to as 'Beacy Central', a BBQ Breakfast at the Beaconsfield Primary School, a Premier's Community BBQ, and a Prime Minister's Civic Reception at Parliament House in Canberra. Apart from the obvious media interest, these events also aimed to raise funds for the local community. With the assistance of their management team, Russell and Webb also established a charitable trust so that they can give back to the community that supported them.

Figure 5.1 The Beaconsfield Mine and Heritage Centre.

Images of Beaconsfield

More than two weeks of unplanned national and international media coverage resulted in a new appreciation for the tiny town of Beaconsfield. The rescue made international headlines and received top billing on the *New York Times* and BBC websites (Gibson and Cubby, 2006).

After such prolonged media exposure, many were curious to actually see the Beaconsfield mine for themselves. The town experienced an obvious tourism boom directly after the event. Beaconsfield's Mayor Barry Easther explained, 'They're coming to see where this disaster occurred and Council are very pleased' (Australian Broadcasting Corporation, 2006a).

Beaconsfield's tourist attractions and striking architectural features were brought into sharp focus during the media coverage of the mine collapse. The towering head frame of the mine became a key signifier for the rescue mission. The A-frame mine head is indeed a powerful signifier for the town (Figure 5.2). Semiotics is the study of signs, codes and culture, and a methodology for reading 'soft data' such as visual representations of a town. The word 'semiotics' is derived from the Greek name for an interpreter of signs – 'semeiotikos' (Cobley and Jansz, 1999: 4). Semiotics is a useful tool for examining the sometimes multi-layered images of a town.

The world famous image of the mine shaft has been replicated in the form of a number of souvenirs that are available for purchase. A pewter Beaconsfield belt buckle is one of many made by Master Cast – a Tasmanian family business which manufactures cast metal products. Their product range includes key rings, bottle openers, belt buckles, medallions and car badges. Not to be outdone on the souvenir stakes, Noel Russell (father of Todd) produced his own fashion range including T-shirts and caps. He claimed, 'I thought if I don't do it, someone else will so I'd better go and sell a few of these' (Nettlefold, 2006).

Conclusion

The *Lonely Planet* guide to Tasmania informs readers that the 'Beaconsfield mine disaster of 2006 will be remembered for many reasons, not least for the media circus that surrounded the incident' and that the two-week rescue operation 'made global headlines' (Rawlings-Way *et al.*, 2008: 218). The *Lonely Planet* authors ungraciously describe Beaconsfield as a 'slightly shabby town' and a 'quiet rural backwater' (Rawlings-Way *et al.*, 2008: 217).

Figure 5.2 The head frame of the mine received global media coverage.

The tour company Evergreen Tours offers a trip that includes 'a visit to Beaconsfield – a town full of history and interesting old buildings'. When visiting the Heritage Museum, travellers can 'learn about the miraculous survival of miners Webb and Russell during April 2006' (Evergreen Tours, 2009). Other tourist attractions in the area include Platypus House and Seahorse World. In order to maximize visitation at these key tourist sites, operators have used cross-promotion strategies to offer visitors a 'Tamar Triple Pass'. Beaconsfield also boasts one of the world's first cable-stayed truss bridges in the Batman Bridge which was opened in 1968.

The events following the rock fall on Anzac day 2006 certainly united the community and raised awareness of the tiny town of Beaconsfield and enhanced prospects for future tourism. Journalist Jocelyn Nettlefold reported that with an increased number of curious visitors coming to the town 'tourism is shaping up to play a big role in Beaconsfield's future'. One tourist she interviewed explained, 'We thought we'd drop in at Beaconsfield because it's been put on the map now' (Nettlefold, 2006). Being put on the map is one thing but sometimes towns become famous for the wrong reasons. Local resident John Whitely explained that the mine collapse 'put Beaconsfield on the map for the wrong reasons' and added that it was now 'time to move on' (Rout, 2008).

References

Australian Associated Press (2008) 'Todd Russell, Brant Webb speak at Beaconsfield mine inquest', *Herald Sun*, 24 July. Available online at: http://www.heraldsun.com.au/news/national/safety-complaints-fell-on-deaf-ears/story-e6frf7l6-1111116999437 (accessed 6 September 2009).

Australian Associated Press (2009) 'Brant Webb Aims for Politics', *The Mercury*, 5 June. Available online at: http://www.themercury.com.au/article/2009/06/05/77855_print.html (accessed 27 August 2009).

Australian Broadcasting Corporation (2006a) 'Council Monitors Beaconsfield Mine Redundancies Impact', 13 June. Available online at: http://www.abc.net.au/news/stories/2006/06/13/1661756.htm (accessed 20 June 2008).

Australian Broadcasting Corporation (2006b) 'Russell, Webb Walk from Gold mine', 9 May. Available online at: http://www.abc.net.au/news/newsitems/200605/s1633713.htm (accessed 7 September 2009).

Australian Broadcasting Corporation (2009a) 'Cybersquatter Targets NSW Labor Party', 21 August. Available online at: http://www.abc.net.au/news/stories/2009/08/21/2662951.htm (accessed 6 September 2009).

Australian Broadcasting Corporation (2009b) 'Tourists Flock to Beaconsfield Mine Rescue display', 23 January. Available online at: http://www.abc.net.au/news/stories/2009/01/23/2473200.htm?site=news (accessed 6 September 2009).

Burgess, M. and Spencer, S. (2008) 'Outrage at Mine Tragedy Musical', *The Age*, 7 October. Available online at: http://www.theage.com.au/national/outrage-at-mine-tragedy-musical-20081006-4v1z.html (accessed 7 October 2008).

Clark, N. (2009) 'Beaconsfield Safety Slammed', *The Mercury*, 26 February. Available online at: http://www.themercury.com.au/article/2009/02/26/57761_tasmania-news.html (accessed 6 September 2009).

Cobley, P. and Jansz, L. (1999) *Introducing Semiotics*, Cambridge: Icon Books.

Easther, B. (2006) 'Plenary Address to the 2006 National General Assembly', (Australian Local Government Association), 30 November. Available online at: http://www.alga.asn.au/newsroom/speeches/2006/NGA2006/MayorBarryEasther2006.php (accessed 8 September 2009).

Enker, D. and Newsome, B. (2006) 'Defining Moments', *The Age*, 19 December. Available online at: http://www.theage.com.au/news/tv--radio/defining-moments/2006/12/19/1166290549568.html (accessed 9 September 2009).

Evergreen Tours (2009) 'Tasmanian Spectacular'. Available online at: http://www.evergreentours.com/destinations/TourItinerary.aspx?ID=11763 (accessed 12 September 2009).

Gibson, J. and Cubby, B. (2006) 'From Great Escape to Great Wake', *Sydney Morning Herald*, 9 May. Available online at: http://www.smh.com.au/news/national/tough-pairs-great-escape/2006/05/09/1146940478162.html (accessed 7 September 2009).

Goc, N. and Bainbridge, J. (2008) 'The Beaconsfield Mine Disaster and the Evolution of Chequebook Journalism', *Australian Journalism Review 30(1)*, Thirroul, NSW: Journalism Education Association of Australia, pp. 99–112.

Lees, P. (2009) 'Beaconsfield Mine Survivor Fields TV Deal', *The Examiner*, 18 May. Available online at: http://www.examiner.com.au/news/local/news/arts-culture-entertainment/beaconsfield-mine-survivor-fields-tv-deal/1515269.aspx (accessed 6 September 2009).

Magshop (2006) '*The Bulletin* – Tasmanian Mine Rescue Souvenir Edition'. Available online at: http://www.magshop.com.au/Bulletin_Tasmanian_Mine_Rescue_Souvenir_Edition.htm (accessed 7 September 2009).

Nettlefold, J. (2006) 'Beaconsfield Struggles after Mine Closure', *The 7.30 Report*, 30 October. Available online at: http://www.abc.net.au/7.30/content/2006/s1777153.htm (accessed 20 June 2008).

Nine Network (2006), *The Great Escape*, 21 May.

Phillips, R. (2006) 'The Australian Media and the Beaconsfield Mine Rescue', World Socialist website, 20 May. Available online at: http://www.wsws.org/articles/2006/may2006/mine-m20.shtml (accessed 10 September 2009).

Rawlings-Way, C., Worby, M. and Mocatta, G. (2008) *Lonely Planet Tasmania*, 5th edn, Footscray: Lonely Planet Publications.

Rose, D. (2006) 'World Unites for Mine Rescue', *The Australian*, 9 May. Available online at: http://www.theaustralian.news.com.au/0,20867,19080897-1702,00.html (accessed 27 May 2008).

Rout, M. (2008) 'Beaconsfield Wants to Move on', *The Australian*, 24 July. Available online at: http://www.theaustralian.news.com.au/story/0,25197,24068142-5006788,00.html (accessed 12 September 2009).

Rudd, K. (2008) 'Beaconsfield Mine Heritage Centre' (Media Release), 5 November. Available online at: http://www.pm.gov.au/node/5493 (accessed 12 September 2009).

Twenty-Two Management (2009) 'Todd Russell and Brant Webb'. Available online at: http://www.22.net.au/Clients/Beaconsfield-Miners.html (accessed 7 September 2009).

Woodhead, C. (2009) 'Beaconsfield: The Musical', The Age, 16 April. Available online at: http://www.theage.com.au/news/entertainment/arts/arts-reviews/beaconsfield-the-musical/2009/04/16/1239474978814.html (accessed 6 September 2009).

Wooley, C. (2006) 'The Great Escape', *Sixty Minutes*, Nine Network, 21 May.

Wright, T. (2006) *Bad Ground: Inside the Beaconsfield Mine Rescue*, Millers Point: Murdoch Books.

PART III

Transforming mines into heritage attractions

6 Mining the mining museum on New Zealand's North Island

Rich veins of dissent

Jane Legget

Introduction

This account discusses two museums that collect and interpret industrial heritage in different mining communities on New Zealand's North Island. The coal mining centre of Huntly is home to the Waikato Coalfields Museum, while gold mining on an industrial scale takes place at the bottom of the Coromandel Peninsula at Waihi, location of the Waihi Heritage Museum. Gold and coal feature strongly in the South Island's tourism scene, but why has this mining heritage achieved less profile in the North Island, especially since both Huntly and Waihi have rich related resources in their local museums and are located on major highways offering tourists easy access? This enquiry revealed both internal and external constraints on these two museums' ability to reach their potential to contribute effectively to their local economies through heritage tourism.

Cultural heritage tourism is still developing in New Zealand (Angus Associates, 2009; Colmar, 2003). The country remains better known for its scenery and nature-based and adventure tourism through Tourism New Zealand's successful *100% Pure New Zealand* promotional campaign. Tourism products and services drawing on indigenous Maori culture and outdoor pursuits have a long history (McClure, 2004), but museums, galleries, historic houses and heritage sites have only recently gained recognition as tourism assets (Warren and Taylor, 2001).

Prentice's heritage attractions typology classes mining-related tourism as 'attractions concerned with primary production' while museums are 'socio-cultural attractions' (2005: 248). New Zealand has agricultural and vineyard tourism sites aplenty, but heavier industries have been overlooked for major heritage investment, especially in the North Island. No heritage-led economic regeneration based on declining industries has occurred on the scale of Ironbridge Gorge (England), Big Pit (Wales), Lowell (Massachusetts, USA) or Le Creusot (France). New Zealand's industrial history remains in the landscape and in museums, but rarely heads the tourism menu.

Thornton (1982) and Smith (2001) surveyed New Zealand's industrial sites and Thomson (1981, 1991) noted several museums collecting industrial heritage material. More recently, the national on-line directory, *NZMuseums* (2008), categorised industrial heritage collections under 'technology' and 'transport'. Warren and Taylor (2001) reported 61 (3 per cent) industrial heritage tourism sites from a sample of 2266 operators, but their broad definition included workplace tours, potteries, open air museums and mills. They also found that 20 percent of 236 heritage tourist operators interviewed considered old mines and industrial sites attractive to a generalised notion of the heritage visitor. However, it is likely that some of these respondents were located near existing mining heritage which would thus positively influence their responses. It would be fair to say that Smith's observation (2001: 12) still holds true:

> The words 'industry' and 'heritage' do not readily trip off the lips of Kiwis and certainly not in the same sentence. Ours has been a predominantly agricultural society, not the Southern hemisphere's cradle of industrialization. Ask about the heritage of our industry and the usual response is just a quizzical look and a few fading folk memories.

Heritage legislation (e.g. Historic Places Act 1993 and Resource Management Act 1991) governs identification and regulation of recognised heritage sites and landscapes. The Register of Historic Places records industrial sites and their significance. The Institute of Professional Engineers (IPENZ) installs markers acknowledging engineering achievements including mining structures, while regional heritage trails (e.g. those produced through New Zealand Heritage Trails Foundation) and regional guides (Hayward *et al.*, 2008) often mention industrial monuments. Nonetheless, persistent imagery of sheep, dairying, fruit-growing, forestry and fishing eclipses industrial heritage with its associations of scarred landscapes, pollution, smells and noise (Bell and Lyall, 1995).

South and north

Colonial settlers discovered gold and coal on both Islands, but only the South Island has seriously developed this heritage for tourism. The South Island West Coast's coal-bearing seams produced both the Denniston Incline, and New Zealand's Labour movement. Coaltown Museum (established 1977) is being redeveloped within a coal heritage area incorporating the Denniston Incline, 'eighth wonder of the engineering world' (Smith, 2001: p. 138). The South Island's alluvial gold has generated tourism attractions in the north-west and central Otago. Gold-panning, abandoned settlements, exhausted goldfields, archaeological remains of Chinese encampments and heritage recreations such as Shantytown (established 1971) combine gold rush themes and impressive scenery.

The two North Island communities discussed here both depended directly and indirectly on hard rock mining, coal in Huntly and gold in Waihi. Their fortunes continually fluctuated but only in the twenty-first century has tourism been an option.

Methodology

South Island mining heritage has attracted the attention of tourism academics (Balcar and Pearce, 1996; Bell, 1997; Frost, 2005; Kirby, 1993), unlike the North Island. New Zealand's larger museums are increasingly studied, especially the indigenous issues (Butts, 2003; Cameron, 2000; McCarthy, 2007; Legget, 2006), but museum sector data has only recently become available (Museums Aotearoa 2007, 2008, 2009). The museums discussed here typify smaller community institutions: located in country towns, uncertain income, professionally isolated, unstructured collecting of community history, mostly reliant on volunteers and under-researched. Among the few to study them are Wirick (2007), Simpson (2000) and Bell (1997).

With little published on the museums (Lockwood, 2003; Ward, 2003), this investigation mainly used discussions with those managing them. Limited internal documentation was accessed at this stage. At the Waikato Coalfields Museums, the director was the principal informant, while at Waihi, two members of the recently re-activated museum committee were the principal informants. From these sources, a comparative SWOT (strengths/weaknesses/opportunities/threats) analysis provided a framework for eliciting understandings of issues influencing or impeding these museums' tourism possibilities.

Experiencing mining heritage on the North Island

How do visitors experience mining heritage in New Zealand? Smith (2001) mapped the accessible mining legacy: stamping batteries, tramways, mine heads, boilers, pump houses, furnaces, settling tanks and deserted communities. Visitor encounters include: viewing the man-made landscape from a distance, walking over it and inspecting tramways, stamping batteries, other relics in situ and traces of abandoned settlements, following gold miners' trails to exhausted mines, and controlled visits to working sites. On-site information varies from none through basic signposting and interpretive panels to guided walking tours with differing levels of walking packaged with gold-panning opportunities. Museums are rarely integral to the mix, although, depending on their location, they could serve as starting or finishing points for special tours, or at least feature in themed excursions, given the rich stories that they hold.

Waikato Coalfields Museum, Huntly

Huntly Mining and Cultural Trust operates Waikato Coalfields Museum (WCM) in a former mine manager's house in a pleasant garden (Figure 6.1). Huntly with a population of approximately 7,000 is the largest town within Waikato District Council (WDC), which oversees 44,000 people from the Hauraki Gulf westwards to the Tasman Sea, in an otherwise agricultural and fishing economy. State Highway One, New Zealand's main trunk road, takes approximately18,000 vehicles through Huntly daily, and many carrying tourists. Despite poor signposting 2,500 determined visitors found WCM in 2008. Situated in a residential area, WCM has modest annual visitation, despite being open daily for six hours.

Figure 6.1 Former mining manager's house.

WCM's collection strengths relate to coal mining and associated industries (power generation, coking and bricks) and general social history from the Huntly area. Its comprehensive documentary heritage is an important resource for genealogists. The wooden heritage building's domestic scale limits the amount of heavy machinery shown indoors, so larger exhibits are displayed outdoors, with a miner's cottage and other buildings. The upper floor exhibits material on home and community life found in similar museums elsewhere. Downstairs, permanent exhibitions cover coal mining and life underground; showcases, reconstructions, photographs and videos evoke memories and new information from local visitors. Changing exhibitions in a side room use WCM's resources, mostly with community input. Although mining tragedies are featured, the museum storyline avoids contentious history and politics, instead profiling past and present mine-owners rather than the miners' lived experiences (Day, 2007, 2009).

WCM's origins are typical in country towns. History-minded people established an association in 1973; a museum followed in 1980 (Thomson, 1991). A voluntary committee governed and operated the museum, funded through admission charges, memberships, donations and, latterly, Council support. Staffing varied with levels of Council funding. In 2001, WDC employed the first professional director. She inherited one part-time paid guide, soon hired a registrar and administrative assistance, and won grants for contract collections assistance. After initial enthusiasm for imaginative plans for a purpose-built museum, *Te Waro*, capable of raising Huntly's profile and attracting travellers off the State Highway (WCM, 2003; Great Leap Forward, 2007), WDC's attitude changed. '*Te Waro*' is the Maori word for coal. A bicultural coal-themed heritage attraction had unique appeal, but once full costs were appreciated, Council interest waned and relationships seriously broke down (Day, 2009). Despite strong support from key local employers, mining company Solid Energy, and Genesis Energy who operate Huntly's landmark Power Station, disagreements about location, capital investment and on-going operating expenses followed.

WCM introduced professional collections management (Ward, 2003), built networks with community organisations, iwi (indigenous Maori tribes) and sponsors, and pragmatically reframed *Te Waro* with a modular building design, but mayoral support shifted to a rival volunteer museum proposal in Raglan, a pretty seaside town. WDC distanced itself from the WCM, no longer employing staff directly. Instead WCM's Committee received a reduced operating grant, becoming the employer. By 2009, a sole-charge paid director manages the Museum with volunteers doing weekend duties. WCM remains committed to *Te Waro* but negativity from mayor, councillors and WDC staff hampers its timely realisation. Meanwhile the director quietly strengthens relationships with other, albeit less powerful, community interests, facilitating heritage registration of the carbo works, developing heritage trails, publishing local history and developing an informative website. The Waikato River, Rangiriri Battlefield and Woodlands, an historic house, are the only competing attractions nearby.

Waihi Arts Centre and Museum Association, Waihi

Waihi Arts Centre and Museum Association (WACMA) operate the Waihi Heritage Museum (Figure 6.2). The former Technical School fronts onto a main road through Waihi (pop. 4,500); near a section of State Highway Two designated the 'Pacific Coast Highway'. WACMA's former class rooms are conducive to museum and community use. Faced brick renders the heritage collections more secure and less vulnerable to fire. Visitor facilities include off-street parking and disability access.

WACMA's focus is the history of Waihi and its surroundings, specifically its gold mining heritage, with good archives. In 2008, visitation was 4,860 (including school groups).

Figure 6.2 Waihi Heritage Museum.

Since 1962, WACMA has held exhibitions of art works – mainly by local artists and craftspeople – and provided meeting venues for community groups such as the Camera and Pottery Clubs (Lockwood, 2003; Thomson, 1981, 1991). These changing exhibitions and community events were intended to knit WACMA firmly into Waihi's cultural life. Unlike WCM, Waihi is part of a regional gold mining story, potentially competing against other attractions sharing the theme in, arguably, more attractive surroundings: Thames and Coromandel Township further north on the main Coromandel Peninsula which is itself a destination.

Current museum displays are somewhat dated: traditional showcase displays, mining dioramas (both scale models and 'life size'), basic interpretive labels and limited audio-visual. A unique three-dimensional glass model of the main mining excavations illustrating multiple levels of underground tunnelling, enables locals and researchers to pinpoint extraction zones many metres below Waihi's buildings and landmarks, and offers rich interpretive potential. Equally memorable are preserved severed thumbs, required as evidence for accident insurance claims.

WACMA's recent history has been troubled. A 1974 bequest created a false sense of security; interest payments covered basic operations for many years. However, the Committee started spending the principal, ultimately putting WACMA at risk (Gorbey, 2007), compounding the situation by divorcing itself from wider community concerns. Declining to participate in *Vision Waihi* (2004), an initiative to guide Waihi's future, should mining cease, was unfortunate. Hauraki District Council (HDC – pop. 17,900) intervened and WACMA's new committee now engages with this project, which will open a $20NZ million Waihi Gold Discovery Centre in 2010.

Strengths, weaknesses, opportunities and threats

Comparing strengths, weaknesses, opportunities and threats which influence these two institutions' ability to exploit their heritage resources shows that WCM and WACMA have much in common. They interpret industries where luck, hardship and tenacity created a sense of community, seasoned by danger, disaster and strikes. Given their similar origins in arts and history organisations, WCM, with one qualified manager and supportive coal and power companies, might appear better placed to develop mining history and grasp tourism opportunities for Huntly. Having a single employee and weekend volunteers means WCM is a 'small' museum, while WACMA's wholly volunteer venture is a 'micro' museum (Museums Aotearoa, 2007, 2008, 2009).

Strengths and weaknesses

The museums' relative strengths and weaknesses (Table 6.1) are influenced by internal factors, mainly structural, relating to physical, financial and human resources. Other factors include intangibles: local knowledge, museum know-how and relationships. Both museums' strengths are their underexploited heritage resources – objects, oral histories, archives – capable of generating engaging narratives for exhibitions, educational programmes and events. Their capacity for managing these enterprises and resources differ.

Most New Zealand museums face uncertainty around operating revenue. A weak committee prevented WCM from capitalising on its collection strengths (Table 6.1). Losing 10 percent from its 2009–2010 Council grant increased pressure on WCM's director to make an effective case for WDC annual support despite continuing Council ambivalence (Day, 2009). Evaporating hopes of imminent investment in modern exhibitions at *Te Waro* mean maintaining outmoded displays in WCM's unsatisfactory premises. The risk of 'sole-charge curator burnout' increases when workload expectations are unrealistic and volunteer assistance is unpredictable.

WACMA's positive responses to HDC's review secured three-year funding, enabling implementation of achievable plans and alignment with Vision Waihi (Daly and Verrall, 2009). WACMA may resolve its dated display issues through collaboration with the Goldfields Discovery Centre (Vision Waihi, 2004; Gorbey, 2007). WACMA's new committee faced its failings, acknowledged their limitations, and inaugurated a four-day opening regime (Gorbey, 2007). Now run on a business-like basis, WACMA generates income from research services to genealogists, while still charging its exhibition visitors. Gold and coal mining required good records for pay, health, safety, insurance and other business needs: rich seams for researchers and other 'questing knowledge seekers' (Gorbey, 2007) at both museums.

Although WCM is professionally managed, its committee members have been unsuccessful advocates with WDC and the wider community. Heritage structures impose restraints, as does physical location, hence WCM's plans for *Te Waro* (WCM, 2003). Acrimonious stakeholder relationships have held back both museums, but recruiting enterprising new committee members has already turned WACMA around (Daly and Verrall, 2009). WCM has been less fortunate in its committee's calibre. Key differences concern staffing, governance and, especially, stakeholder relationships (Table 6.1).

Opportunities and threats

External factors present both opportunities and threats to WCM and WACMA (Table 6.2). Tourism's growing importance to New Zealand is conducive to enterprising museum

Table 6.1 Comparing the strengths and weaknesses of Waikato Coalfields Museum (left) and Waihi Arts Centre and Museum Association (right)

	Waikato Coalfields Museum	*Waihi Arts Centre and Museum Association*
STRENGTHS	Rich documentary heritage	Rich documentary heritage
	Coal mining and Waikato local history objects	Gold mining and Waihi local history objects
	Well-catalogued collections – accessible	Revitalised committee with realistic ambitions
	Sound educational offerings and expertise	Active engagement with stakeholders/ rebuilding relationships
	One qualified full-time staff member	Council representation on Committee
	Fitful engagement with stakeholders	Gold industry support – educational partnerships
	Active rebuilding relationships	Engagement with wider museum community
	Council representation on Committee	Proactive – applying for grants
	Coal, energy and brick industry support – educational partnerships	Success in securing Council funding support
	Engagement with wider NZ museum community	Responding positively to local heritage tourism developments
	Focused museum collecting	Changing exhibitions in arts centre
	Changing community history exhibitions	Focused museum collecting
	Supporting local heritage campaigns	Attractive heritage building
	Producing heritage trails and publications	Good stories
	Interest and support of local iwi (tribes)	Web presence
	Open seven days	Good space for community exhibitions
	Good stories	Meeting space available for hire
	Supporting local volunteer-run museums	Emerging business model for research service fees
	Te Waro development plans with strong concept and realistic ambitions	Off-street parking
	Openness to partnerships	Gold theme appeal
		Openness to partnerships
WEAKNESSES	Diminishing operational funding from Council	Operational funding not secure beyond three years
	Arm's length relationship with Council	Dependence on energy, skills and enthusiasm of key committee members
	Calibre of Committee members	Uncertain income
	Dependence on sole staff member's capacities	Legacy of poor image within the community
	Local perception that little changes	Large physical scale of gold mining – not the normal tourist image of gold rushes etc. (but more likely to be initially of interest to NZers)
	Dated displays	Local perception that little changes
	Poorly sign-posted	Dated museum displays
	Location – not visible from State Highway	No full-time staff, no qualified staff
	Limited success in securing Council funding support	Only open Thursday to Sunday
	Few potential partners	Recent history of internal dissent and distrust
	Legacy of difficult relations with Council	
	Over-reliant on limited volunteer pool	
	Wooden heritage building – fire risk	
	Limited scope for expansion on present site	
	Low awareness of museum among locals	
	Upper floor inaccessible to visitors with disabilities	
	Minimal visitor facilities	

initiatives. WCM's first director recognised that a coal-themed museum could divert passing tourists into Huntly, opting for a site beside Puketirini, despite its distance from State Highway One. This landscaped former open-cast mine with deep water lake, walking and bike trails, already caters for recreational activities. Partnerships with rail preservation groups are possible on an existing railway. Huntly's old railway station has been relocated there to house *Te Waro*'s facilities for both recreational and museum visitors. As 'good corporate citizens' actively encouraging WCM's plans, Solid Energy made the rehabilitated land available with WDC. In the current energy economy, Solid Energy and Genesis Power remain supportive of WCM, but reluctant to act independently of WDC.

Waihi's 'Heart of Gold' branding creates a positive environment for WACMA (Table 6.2). Having relocated a landmark pumping engine house, replicated an iconic poppethead, installed interpretive panels along the main street and operating an interim Gold-Mining

Table 6.2 Comparing the opportunities and threats of Waikato Coalfields Museum (left) and Waihi Arts Centre and Museum Association (right)

	Waikato Coalfields Museum	*Waihi Arts Centre and Museum Association*
OPPORTUNITIES	Coal and energy companies' corporate citizenship	Town commitment to its gold mining
	Industry recognition of *Te Waro's* innovative interpretation potential	Waihi's Heart of Gold branding
	Town located on State Highway 1, between Auckland and Hamilton	Gold mining companies' corporate citizenship
	National recognition from the Coal Council	Council investment recognising heritage value
	Te Waro site in Puketirini with outdoor recreation and rail access	New Gold Discovery Centre (2010)
	Director's creative ideas and relationship building skills	Track record of collaboration
	Supportive visitor information centre	Accepted partner in Vision Waihi
	Positive outlook for coal and energy industries and employment	Range of possible partners
	Popularity of genealogy research	Recognised potential to tell interesting stories in new ways
	Attractive web presence	Popularity of genealogy research
	Popular novel set locally (Challinor, 2005)	Popularity and promotion of other goldfields heritage sites
		Emerging tourism infrastructure
		Town location on State Highway 2 (Pacific Coast Highway)
		Location near Waihi Beach
		Anniversaries in 2011/2012
THREATS	Town image not associated with tourism	Promotion of Coromandel/Thames – competing gold mining heritage experiences, different appeal
	Disbandment of Tourism Waikato	Potentially saturated market
	Few potential tourism partners	Exhausted gold mining – lose supportive companies
	Lack of accessible related sites for packaging	Review of LGA 2002 – risk of core services restriction
	Review of LGA 2002 – risk of core services restriction	Review of minerals on conservation estate – increased mining may spoil landscape and deter tourists
	Council's annual operating grant reduced	Antipathy to polluting industries
	Council support lacking for *Te Waro* development	Tourist expectations increasingly sophisticated
	Local politics and interests	New Gold Discovery Centre (2010)
	Low Council interest in Huntly's tourism potential	
	Low Council interest in iwi	
	Antipathy to polluting industries	
	Clash with tourism's green clean image	
	Minimal tourism infrastructure	
	External demands on sole charge director	
	Raglan museum proposal	
	Lower education levels in community	
	Limited visibility of directional signage	
	Tourist expectations increasingly sophisticated	

Discovery and Information Centre, Waihi extended the theming to street furniture, bronze sculptures and banners celebrating gold-mining heritage. An annual gold festival, a mini-bus tour and a perimeter walkway around the opencast Martha goldmine are all recent ventures. Educational tours of the commercial goldmine, nearby heritage sites (e.g. Victoria Battery, the Karangahake Gorge trails) and a heritage railway between Waihi and Paeroa encouraged WACMA to identify complementary interests. The Department of Conservation (government agency responsible for Karangahake Gorge) and the Historic Places Trust, which manages the Thames School of Mines and published a Waihi and Paeroa goldfields heritage trail, are possible partners (Table 6.2).

Near the Pacific Coast Highway, and only 10 kilometres from Waihi Beach, a popular seaside destination, Waihi can attract 'passing trade' from travelling tourists and offer 'wet day activity' for beach-goers. WACMA can catch up on opportunities missed earlier and feature in goldfields daytrips or weekend packages, *and* be a history resource. Anniversary celebrations are already planned for Waihi's 1911 Strike and WACMA's own jubilee in 2012, both occasions for demonstrating the Museum's local contributions.

These opportunities are tempered by external constraints on tourism development and risks to the museums' viability (Table 6.2). Legislation amending the Local Government Act 2002 may restrict Councils' spending to 'core services' (roads, rubbish, street lighting and libraries). Reinstating the 1990's 'user pays' paradigm (Kelsey, 1997; Scott, 2001) bodes ill for museums – WDC and HDC would be prohibited from providing financial assistance, leaving the museums unable to meet visitors' increasingly sophisticated expectations.

Huntly takes little overt pride in its mining heritage (Table 6.2). There is little for tourists other than a visitor information bureau and views of the power station from a riverside park with picnic area. No iconic pithead structures, no prominent signposts welcoming visitors to 'coal country' (or similar). Although supplying its agricultural hinterland, Huntly still needs coal mining for employment. WCM's director suggested that class is another factor inhibiting the Museum's development (Day, 2009). Decisions for this blue-collar town may still be made by rural landowners (Day, 2009; Wedgwood, 2009).

Some factors may be both strengths and weaknesses or present both opportunities and threats, or alternate, depending on the circumstances. This holds true for both WCM and WACMA, but effective stakeholder relationships appear to be vital for engaging with heritage tourism.

Discussion

This study aimed to explore why the heritage of two industries, coal and gold mining, had achieved less tourism profile in the North Island than in the South Island. It centred on WCM and WACMA in the belief that these community heritage resources would prove useful starting points. The principal revelations from the SWOT analysis are these financially insecure operations' vulnerability to changing community dynamics and their relative lack of power and credibility to fulfil their heritage stewardship missions, let alone a tourism role, when other forces, mostly external but sometimes internal, seem ranged against them.

Options and issues

No definitive answers were anticipated, but a range of issues worth further investigation surfaced. Paired factors or options relate to aspects of these museums' circumstances. Some are oppositional, some complementary and some both, and all are inter-dependent. These

Table 6.3 Museum situations with options that can contribute to success in the community

Museum situation	*Options*
Positioning in wider community	Leading or following
Operational focus	Heritage resource or visitor attraction
Primary function	Preservation or tourism offering
Scope of operation	'Doing it all' or playing to strengths
Mode of operation	Stand-alone or community collaborator
Level of operation	Professional or volunteer

factors (Table 6.3), not all necessarily 'either/or' options in the short or longer term, are discussed next.

How do the museums position themselves in their communities – as *leaders* or *followers*? Resource dependency issues determine the roles which the case museums can realistically adopt in their communities. What is the best approach for small industrial heritage museums wanting to play an active role in developing local industrial heritage tourism? WCM had the drive to initiate a significant community project, but shouldering responsibility for leadership while facing indifference and occasionally active hostility of key stakeholders, especially the Council, proved a struggle (Day, 2009). WACMA is an emerging partner in the larger Waihi Gold Discovery Centre project, following rather than leading, although this need not preclude it taking a leadership role at a later date.

In common with many smaller institutions, both museums juggled multiple foci and functions. Are they operating as *heritage resources* or *visitor attractions?* WACMA appears to be prioritising its role as a heritage research resource, serving the 'questing knowledge seeker' (Gorbey, 2007), supplemented by art exhibitions for the community and some local heritage displays. WCM remains ambitious for *Te Waro* as a major visitor attraction, but keeps operating its current museum meantime. How do these small museums determine exactly what business they are in? Should their primary focus be *preservation* or *tourism offering?* Can such museums balance their mission to preserve industrial heritage (their own collections and other local sites) and their wish to contribute to the local economy as tourism offerings? This can result in 'scope creep', where museums try to *'do it all'*, and fail to *'play to their strengths'*, while reluctance to make hard decisions threatens institutional viability. The temptation to stretch scarce resources too thinly can disappoint visitors with unadventurous displays and low-grade facilities and inadvertently allow heritage material to deteriorate.

These museums each belong to a larger story. By default, WCM chose to 'go it alone' in trying to establish a purpose-built museum complex at Puketirini without natural partners in either the tourism or heritage fields, especially any sharing the coal theme. Council antipathy to tourism's economic or social value discouraged potential community partnerships. Waihi, however, wholeheartedly embraced its goldfields heritage, with WACMA belatedly joining its 'hub and spoke' model for tourism development, identifying its special contribution. WACMA's archives and collections become recognised resources for all partners, while serving a distinct niche of 'learning tourists', 'questing knowledge seekers' and educational groups. Whether operating in *stand-alone* or *community collaborator* mode, both organisations have limited human resources. Availability and calibre of both workforce and governance influence operational success. Paid professionals bringing knowledge, skills and experience raise standards of visitor experiences and heritage practice, and, when adequately resourced, manage heritage attractions effectively, delivering community value. Wholly volunteer-run museums require energetic team-players welcoming different ideas and potential partners, while replenishing themselves over time through attracting recruits.

Professionals and *volunteers* can make their museums a focus for pride in local achievements, create strong 'place identity' and provide volunteers with a sense of purpose. Clarity of roles and responsibilities for management and governance is critical where professionals and volunteers work together.

WCM has the vision and imagination for an exciting heritage-led visitor attraction, but its diminishing resources necessarily narrow its ambitions. Nonetheless, as the only professionally staffed museum between Auckland and Hamilton, it risks overextending itself. Maintaining current level of operations, pursuing a new development, championing heritage preservation locally and advocating for support in a difficult environment places heavy demands on a sole-charge director. Outside intervention has guided WACMA to recognise and play to its strengths; in doing so its volunteers have re-focused and re-engaged with Waihi community, observing the miners' trade union motto: 'unity is strength'.

Gold and coal

Are these heritages competing directly? Geography and geology shape any response to this question. WCM is a lone outpost of coal-mining heritage, with great stories to tell but a local government that chooses not to acknowledge or capitalise on them. WACMA is part of a bigger, better known gold story within the Waihi district and the greater Coromandel peninsula, and is cutting out its own niche within it.

Gold-seeking undoubtedly benefits from the sense of quest, romance and hard-won value. It has attracted historians (Eldred-Grigg, 2008; Downey, 2002; Salmon, 1996) and writers of both children's and adult fiction (Tremain, 2003). New Zealand's coal history is less well-documented (Cook, 2006; Jones, 2003) but recently, however, two novelists have mined it. Pattrick's popular works (2003, 2004, 2006) set in the South Island's West Coast around the Denniston Incline are best sellers in the local Coaltown Museum (2009), while Challinor's novel (2005) set during the 1951 Strike, where miners supported Auckland's waterfront workers in the Waterfront Lock Out, may yet put Huntly on the map. The hard rock mining of Huntly and Waihi has associations with heavy toil, dirt and danger less prominent in South Island areas where alluvial gold offer visitors hands-on gold-panning experiences (Larson, 2007).

For both museums, their visitors probably represent McKercher's 'purposeful cultural tourists' (2002).These tourists visited museums more than most, also chose to visit lesser known sites and were more likely to absorb themselves in local culture. However, they comprised only 12 per cent of McKercher's diverse sample of cultural tourists, an audience still under-developed in New Zealand. Both museums are fishing in a small audience pool. Waihi plans to entice tourists to take State Highway Two (the Pacific Coast Highway) (Daly and Verrall, 2009) rather than State Highway One, potentially diverting WCM's market from Huntly.

North and south

North Island gold mining heritage's main advantage over coal mining lies in belonging to a larger cluster of tourism activities within an established tourism destination, the greater Coromandel Peninsula (Nolan, 1977; Moore and Ritchie, 1996). Waihi maximised its opencast mine's accessibility to stimulate tourism, contrasting with and complementing the goldfields trails of the Coromandel bush and the Karangahake Gorge. It benefits in the same way that clusters of both gold and coal heritage attractions increase tourists' choices in the South Island's West Coast. WCM has industry recognition as the national coal museum, acknowledging the importance and value of its heritage holdings. However, as a solo operation, lacking a local network of either coal heritage attractions or active coalition of industrial heritage sites, WCM is doubly isolated – professionally and physically. Travel costs stretch slim budgets, inhibiting collegial visits and professional exchange at the South Island attractions.

In both islands, vocal 'green' advocacy against renewed Government manoeuvres encouraging mineral extraction, may present coal and gold heritage attractions with additional challenges, potentially prompting allegiance of mining heritage attractions in both North and South. Whatever happens, acknowledging stakeholder legitimacy (e.g. Legget, 2001) and engaging with different interests groups provide the oxygen necessary for survival.

Relationships – the canary in the mining museum

The health of key community relationships determined the two museums' recent histories. The Council's wavering support for WCM played out in public meetings and the local media, garnering unwelcome press coverage. Without the express support of the Council, WCM became ineligible for capital development grants from Government and certain foundations, thus impeding *Te Waro's* progress. Local expectations of an exciting development were raised, apparently in vain, potentially undermining public confidence in the Museum. WCM was left to progress a more modest venture as funding allows, while its outmoded exhibitions perpetuate a static image, risking dissipating staff and volunteer morale and energy. Prospects of alienating the Council inhibit more rounded presentation of history, covering hardships, working class culture, politics, unionism and disasters (Day, 2007, 2009). Unsympathetic relationships can induce timidity; adhererence to the 'authorised heritage discourse' producing one-dimensional visitor experiences (Smith, 2006; Wedgwood, 2009). A vicious circle leaves the Museum unable to attract 'new blood' and strengthen community ties.

In Waihi, relationships have been equally crucial. By dismissing Vision Waihi (2004), WACMA became sidelined. It's depleted bequest exacerbated internal dissent around lack of transparency and accountability (Daly and Verrall, 2009). Sympathetic Council intervention provided a second chance and a rational direction (Gorbey, 2007). Realistic reappraisal of capabilities identified a role complementing the Waihi Gold Discovery Centre initiative, inventorying stakeholders' expectations and defining achievable projects aligned with specific stakeholders' interests and support. Successful recruitment of a new committee brought a fresh outlook and vital re-engagement with Waihi community, building the alliances essential for WACMA to contribute to Waihi's new tourism-based economy. The new Centre, due to open in 2010, may initially overshadow WACMA, but this will allow WACMA time to consolidate its specialist target audiences of educational groups, research visitors, family historians and McKercher's 'purposeful cultural tourists' seeking a deeper level of experience.

Conclusion

This investigation revealed multiple challenges facing two North Island museums seeking to add value to their communities through celebrating their respective mining histories. Site location, fluctuating workforce energy, uncertain finances and inexperienced committees all contributed, with breakdowns in key relationships overriding these. WCM's coal-themed visitor attraction stalled through changing mayoral allegiances, while WACMA only belatedly recognised that its viability depended on active collaboration with community-led economic regeneration initiatives. Bell's observation (1997: 149–50) about Ross, a remote gold-mining village on the South Island's West Coast tourist route, applied equally to Huntly and Waihi: 'The people who run the local historic society work hard to insist that, even if it doesn't look a fascinating colourful place to people passing through today, well, it used to be.' In Huntly that hard work continues, still to convince the local authority that its heritage matters, while in Waihi, the whole community has put gold at the centre of its branding strategy, turning itself into a gold-themed destination. Both museums are now alert to their changing environments and better able to respond as circumstances alter.

The study suggests several fruitful directions for future research: visitor studies comparing the relative attractiveness of coal and gold heritage in the North and South Islands; class politics in the valuing (or not) of mining heritage where power has traditionally been held by land-owning interests; the applicability of ecomuseum models in the greater Coromandel peninsula. Writing about Europe, Cole (2004: 481) suggests that 'the coal mining industry's late emergence within the area of industrial tourism may be linked with its association with harsh working conditions'. In New Zealand industrial heritage tourism is still emerging, especially in the North Island, but there are rich seams to be mined.

Acknowledgements

The author would like to thank Ian Day, Director, and Linda Wigley, former Director, of the Waikato Coalfields Museum, and Berys Daly and Ian Verrall of the Waihi Arts Centre and Museum Association for sharing their experiences.

References

Angus and Associates (2009) *Demand for Cultural Tourism Update June 2008. Report prepared for the Ministry of Culture and Heritage.* Wellington: Ministry of Culture and Heritage.

Balcar and Pearce (1996) 'Heritage Tourism on the West Caost of New Zealand', *Tourism Management*, 17(3): 203–12.

Bell, C. and Lyall, J. (1995) *Putting our Town on the Map: Local Claims to Fame in New Zealand.* Auckland: Harper Collins.

Bell, C. (1997) 'The 'Real' New Zealand: Rural Mythologies Perpetuated and Commodified', *The Social Science Journal*, 34(2): 145–58.

Butts, D. (2003) *Maori and Museums: the Politics of Indigenous Recognition.* Unpublished thesis, Massey University, Palmerston North.

Cameron, F.R. (2000) *Shaping Maori Identities and Histories: Collecting and Exhibiting Maori Material Culture at The Auckland and Canterbury Museums from the 1850s to the 1920's.* Unpublished thesis, Massey University, Palmerston North.

Challinor, D. (2005) *Union Belle.* Auckland: Harper Collins.

Coaltown Museum (2009) Personal Communication, 20 September, 2009.

Cole, D. (2004) 'Exploring the Sustainability of Mining Heritage Tourism', *Journal of Sustainable Tourism*, 12(6): 480–94.

Colmar B. (2003) *Demand for Cultural Tourism – Final Research Report.* Wellington: Tourism New Zealand.

Cook, D. (2006) *Lake of Coal – the Disappearing of a Mining Township.* Nelson: Craig Potton Press.

Daly, B and Verrall, I. (2009) Personal Communications, Waihi Arts Centre and Museum Association. August and September.

Day, I. (2007) *The Exploded Watch and Tales of the Carbo: An Examination of the Poetics of Heritage and Display in a North Waikato Coalmining Museum*, Unpublished.

Day, I. (2009) Personal Communications. Waikato Coaldfields Museum. August and September.

Downey, J.F. (2002) *Goldmines of the Hauraki District.* Christchurch: Cadsonbury Publications.

Eldred-Grigg, S. (2008) *Diggers, Hatters and Whores: The Story of the New Zealand Gold Rushes.* Auckland: Random House.

Frost, W. (2005) 'Making an Edgier Interpretation of the Gold Rushes: Contrasting Perspectives from Australia and New Zealand', *International Journal of Heritage Studies*, 11(3): 235–50.

Gorbey, K (2007) *A Review of Waihi Arts Centre and Museum Association (Incorporated) for Hauraki District Council.* Unpublished.

Great Leap Forward (2007) *Waikato Coalfields Museum Development Feasibility study and economic benefit overview.* Unpublished.

Hayward, B., Murdoch, G. and Cameron, E. (2008) *A Field Guide to Auckland Exploring the Region's Natural and Historic Heritage.* Auckland: Godwit.

Jones, G.M.E. (2003) *When Coal was King.* Tauranga: G.Jones.

Kelsey, J. (1997) *The New Zealand Experiment: A World Model for Structural Adjustment?* Auckland: Auckland University Press/Bridget Williams Books.

Kirby, V.G. (1993) 'Landscape, Heritage and Identity: Stories from the West Coast', in Hall and MacArthur (eds) *Heritage Management in Australia and New Zealand: Visitor Management, Interpretation and Marketing:* 119–29. Melbourne: Oxford University Press.

Larsen, M (2007) *Kiwi Gold: A Practical Guide to Finding Gold in New Zealand.* Penguin: Auckland.

Legget, J. (2001) *The Waikato Coalfields Museum and its Stakeholders: A Report to the Director of the Waikato Coalfields Museum on a Methodological Case Study Investigating Museum Stakeholders and their Expectations Prepared by Jane Legget, April 2001.* Auckland: Unpublished.

Legget, J. (2006) *Mapping What Matters in New Zealand Museums: Stakeholder Perspectives on Museum Performance and Accountability.* Unpublished thesis, Massey University, Albany, Auckland.

Lockwood, D. (ed) (2003) *Pukewa Waihi.* Waihi: Don Lockwood.

McCarthy, C. (2007) *Exhibiting Maori – A History of Colonial Cultures of Display.* Oxford: Berg.

McClure, M. (2004) *The Wonder Country: Making New Zealand Tourism.* Auckland: Auckland University Press.

McKercher, B. (2002) 'Towards a Classification of Cultural Tourists', *International Journal of Tourism Research*, 4: 29–38.

Moore, P. and Ritchie, N. (1996) *Coromandel Gold: A Guide to the Historic Goldfields of Coromandel Peninsula.* Palmerston North: Dunmore Press.

Museums Aotearoa (2007) *New Zealand Museum Sector Web Survey. Prepared for Museums Aotearoa by the New Zealand Tourism Research Institute, Auckland University of Technology March 2007.* Wellington: Museums Aotearoa.

Museums Aotearoa (2008) *New Zealand Museums Barometer No. 2: 2007. Special focus: Staffing and Labour. Prepared for Museums Aotearoa by the New Zealand Tourism Research Institute, Auckland University of Technology March 2008.* Wellington: Museums Aotearoa.

Museums Aotearoa (2009) *New Zealand Museums Barometer No. 3: 2009. Special focus: Collections. Prepared for Museums Aotearoa by the New Zealand Tourism Research Institute, Auckland University of Technology June 2009.* Wellington: Museums Aotearoa.

Nolan, T. (1977) *Historic Gold Trails of the Coromandel.* Wellington: A.H. & A.W. Reed.

Pattrick, J. (2003) *The Denniston Rose.* Auckland: Black Swan.

Pattrick, J. (2004) *Heart of Coal.* Auckland: Black Swan.

Pattrick, J. (2006) *The Illustrated Denniston Rose and Heart of Coal.* Auckland: Black Swan.

Prentice, R. (2005) 'Heritage: A Key Sector in the "New" Tourism', in Corsane, G. (ed) *Heritage, Museums and Galleries: An Introductory Reader.* (243–56). Routledge: Abingdon.

Salmon, H.M. (1996) *A History of Gold Mining in New Zealand.* Christchurch: Cadsonbury Publications.

Scott, G. (2001) *Public Management in New Zealand: Lessons and Management.* Wellington: New Zealand Business Roundtable.

Simpson, K.G.M. (2000) 'Customer Satisfaction and Behavioural Intentions in a Rural Community Museum Environment', *Journal of Quality Assurance in Hospitality and Tourism*, 1(3): 1–28.

Smith, L. (2006) *Uses of Heritage.* London and New York: Routledge.

Smith, N (2001) *Heritage of Industry: Discovering New Zealand's Industrial Heritage.* Auckland: Reed.

Thomson, K.W. (1981) *Art Galleries and Museums of New Zealand.* Wellington: Reed.

Thomson, K.W. (1991) *A Guide to Art Galleries and Museums of New Zealand.* Auckland: Reed.

Thornton, G. G. (1982) *New Zealand's Industrial Heritage.* Wellington: A.H. & A.W. Reed.

Tremain, R. (2003) *The Colour*. London: Chatto and Windus.

Vision Waihi (2004) *Waihi's Golden Legacy: A 2020 Vision for the Future of Waihi. Summary of the Ideas and Projects from the Waihi Community Consultative Committee.* Unpublished.

Waikato Coalfields Museum (2003) *National Coal Mining Museum Business Plan*. Unpublished.

Ward, S. (2003) 'Focus on Photographs – Cataloguing the Collections at Waikato Coalfields Museum', *Te Ara – Journal of Museums Aotearoa*, 28(2): 47–48.

Warren, J.A.N. and Taylor, C.N. (2001) *Developing Heritage Tourism in New Zealand.* Wellington: CRESA.

Wedgwood, T. (2009) 'History in Two Dimensions or Three? Working Class Response to History', *International Journal of Heritage Studies*, 15(4): 277–97.

Wirick, E.L. (2007) *Remembering and Belonging: Colonial Settlers in New Zealand Museums.* Unpublished thesis, Massey University, Albany, Auckland.

7 Transforming working mines into tourist attractions

Conceptual and practical considerations

Elspeth Frew

Introduction

When tourists visit operational mines, the managers of such sites face the challenge of managing two sometimes conflicting operations, namely an operational mine and a tourist attraction. Indeed, the development of tourism at an operational mine can create conflicts due to incompatible objectives, namely the need for the mine to produce effectively its core product and, at the same time, allowing visitors onto the site. For example, the tourism department may want to bring large groups of tourists to the site and this may conflict with the production department which may want to use more space to increase production. On the other hand, the need to develop site visits as a public relations opportunity may override concerns regarding the associated financial and developmental costs. This chapter considers the concept of industrial tourism at contemporary mines and the key factors from a supply side perspective in the conversion of an operational mine into an attraction, including safety issues and allowing access to visitors of all ages and abilities. The chapter considers the need for adequate and appropriate interpretation to ensure tourists have fulfilling experiences with supporting facilities and services. The chapter considers illustrative case studies of industrial tourism at two Australian operational mines, namely PowerWorks in Victoria and the Super Pit in Kalgoorlie, Western Australia.

Conceptual background

Industrial heritage tourism forms a distinctive, but under-researched, subset of the wider field of heritage tourism (Edwards and Llurdés i Coit, 1996) whereby it is concerned with the 'remnants of the *past* that are associated with primary (extractive) or secondary (manufacturing) economic histories, and it is typically associated with the *past* of the working class' (Timothy, 2007: xiii). Since many aspects of contemporary mining production are strongly linked to the history and culture of a local region, particularly when the site has existed for many years, and is a major local employer, then such operational sites can also be described as industrial heritage tourism despite being currently operational. At these operational sites, the industrial processes may have 'emerged from earlier periods' (Edwards and Llurdés i Coit, 1996: 342) but they remain in use now, so the concept of industrial heritage tourism can be extended to include visits by tourists to contemporary, operational industrial sites. Industrial heritage has been classified into four types of attractions: productive, processing, transport and socio-cultural, with mine and quarry heritage attractions being most commonly concentrated in the 'productive attraction group', since they are linked to an immovable geological formation (Edwards and Llurdés i Coit, 1996). Since the contemporary mines are currently operational there is no need to discuss whether the industrial heritage tourism replaces the mining industry (Cole, 2004; Edwards and Llurdés i Coit, 1996; Vargas-Sánchez *et al.*, 2009; Ruiz-Ballesteros and Hernandez-Ramirez, 2007) since operational mines continue to make a valuable contribution to the local culture and economy.

Little research has been undertaken into the tourism potential of mines, quarries and other kinds of industrial installations, perhaps because mining areas are often 'peripheral to accepted and well-known economic and tourist circuits' (Edwards and Llurdés i Coit, 1996: 342). However, in more recent times there has emerged a growing appreciation of the physical beauty of mines and mining landscapes, alongside recognition of 'their historical and cultural significance' (Cole, 2004: 484). Indeed, one of the appeals of mines may be that they are traditionally located in peripheral locations 'in the realm of the rural idyll' (Cole, 2004: 484). Referring to a mine in Belgium, Jansen-Verbeke (1999: 71) noted that very slowly an awareness of the 'cultural value' of this industrial heritage and an appreciation of its distinctive landscape, 'has penetrated into the public psyche – and into the consideration of some authorities'. In addition to the physical attributes of mining sites, consideration should also be given to the intangible industrial culture and the related 'collective memories of hundreds of thousands of "ordinary" people' (Leung and Soyez, 2009: 61). Thus, mining sites need to be analysed within the context of the scenery that surrounds them, and should recognise 'less famous and rejected past, protagonized by lower social classes' (Edwards and Llurdés i Coit, 1996: 360). Pretes (2002) considered operational silver mine tours in the town of Potosí in Bolivia, South America and found that the mine tours are offered and narrated by the local indigenous people, with many of them being the direct descendants of those who laboured in the mines from the sixteenth century onwards. Thus, Potosí is an example of a mining community that has successfully developed industrial heritage tourism largely centered on its currently operational mines and, the indigenous miners.

Around the world there are many examples of tourists visiting operational sites described here as industrial tourism. The industrial tourism phenomenon has existed for centuries with 'visits to French vineyards and chocolate factories, Greek and Maltese lace makers, and Dutch cheese factories and flower markets' (Stevens, 1988: 43) being the forerunners of industrial tourism. Thus, industrial tourism occurs when a site is open to the public with or without prior booking, with facilities provided for the access of tourists, where the core business is the production of goods and/or services not intended solely for visitors (Frew, 2008). The industries involved in industrial tourism can produce goods and/or services, they can be involved in the processing of raw materials or semi-processed inputs, they can have a large or small number of employees, and the processes can be automated and/or they can be labour intensive. The organisation can have public or private sector ownership and it can be profit or non-profit oriented. This emphasises the potential size of the industrial tourism sector (Frew and Shaw, 1999). If no tourist ever visited a site producing goods and/or services the site would still exist, in principle, since the core activity is not to cater for tourists. At any time the industrial tourism attraction could return to the core function, namely, the production of non-tourism goods and/or services. In other words, at an industrial tourism attraction, tourism is not required for the organisation to exist, as the core function is not tourism but the site contains facilities to handle tourists (Frew, 2008). Industrial tourism can be regarded as an extension of traditional factory and plant tours organised for customers, sales representatives, and suppliers. Organisations with experience in handling these types of tours may find it relatively easy to adapt the tours for the general public (Frew and Shaw, 1995). However, while all producers *could* engage in industrial tourism, not all necessarily should. For example, if viewing the production process is likely to alienate customers, then some organisations may be well-advised to forego inviting visitors (Frew and Shaw, 1999). Table 7.1 lists terms used by authors to describe aspects of industrial tourism.

Table 7.1 Terms used to describe visits to operational industrial sites

Agricultural tourism	'an enterprise that produces and/or processes plants or animals and which also strives to attract visitors to enjoy the agricultural attributes of the operation and its site, and/or to purchase agricultural products produced or obtained by the enterprise' (Cox and Fox 1991: 18)
Farm stays	'guests can observe and occasionally participate in farm activities' (Pearce 1990: 338)
Wine tourism	'visitation to vineyards, wineries, wine festivals and wine shows for grape wine tasting and/or experiencing the attributes of a grape wine region' (Macionis 1996: 269)
Work displays	'examples of work displays include guided tours of banks, the telephone company, industrial plants' (MacCannell 1976: 36)
Work places	'to observe and understand the operations of other people's workplaces, and to shop for the products or mementos of the company' (McBoyle 1994: 517)
Factory tourism	'visits to working factories to see industry in action, mainly manufacturing industry' (Swarbrooke 1995: 51)
Sideline tourism	'organised activities mounted to appeal to tourists, but as sidelines, peripheral to the primary functions of non-tourism-oriented activities' (Kelly and Dixon 1991: 21)

Practical considerations

There are many management issues associated with developing an operational mine as a tourist attraction. The financial costs involved in allowing tourists to visit an operational site means that some industrial site managers may choose not to open their doors to the public. Some managers may give the possibility of theft, tampering or industrial espionage as the reason why they do not want to allow visitors on-site (Frew and Shaw, 1996). However, the public relations benefits of opening to the public may overcome this reluctance, particularly when the site tour has the potential to inform the public about the positive aspects of mining and, help visitors become 're-educated' about mining companies being 'responsible environmentally aware producers of vital commodities' (Wheatley, 1989: 26). This public relations outcome is particularly important if the mine has traditionally left a 'long-lasting, permanent and highly visible imprint on the earth's surface' (Edwards and Llurdés i Coit, 1996: 342). Some mine tours show visitors the effects of planned rehabilitiation, how operational scars have healed, and perhaps how the landscape has improved since mining has been completed. Accordingly, some mine managers may view conducting visitors around the site mine as a valuable community activity to enhance the company's public image and that of the mining industry (Wheatley, 1989).

Some operational mining sites may have been originally developed with visitor access in mind but the majority were not designed for visitors, so there are logistic and safety problems to overcome, such as the need to:

- install safe walkways and/or viewing galleries to allow the visitor to view the whole process from a safe vantage point, protecting the visitor from any dangerous processes;
- provide toilet facilities, reception area and, car and coach parking;
- provision of visitor safety equipment where necessary, such as hard-hats, overalls, safety glasses, and boots (which also has the added benefit of increasing the authenticity of the

tourist experience as it confirms that the visitors are entering an area which is dangerous or risky) (Frew and Shaw, 1996);

- equip with signage and interpretation to assist visitors in understanding the site's significance;
- provide ancillary services such as all weather access roads and internal footpaths (Mallam, 1989); and
- provide access for disabled and poor mobility visitors, with reliable, up-to-date information about the physical obstacles they might face before arriving on site and ensuring the exhibits and interpretive signage to allow such visitors to engage with them (Poria *et al.*, 2009).

More ambitious visitor programs involve providing refreshment facilities, shops, and exhibitions (Carter, 1991). To achieve this level of development a site requires capital investment, staffing and management (Mallam, 1989). Tools for heritage site management can fall into three categories namely, developments, controls and communication (Carter and Grimwade, 1997) (Table 7.2). This chapter will now consider in particular one of the more important tools of communication, namely interpretation.

Interpretation is an 'opportunity for non-commercial experiences to provide visitors with cultural insights' (Carr, 2004: 434) and is an important component of a visit as it enables visitors to 'learn informally about, and appreciate, the place they are visiting' (Light, 1995: 145). As such, interpretation can enhance the value of the attraction such as an operational mine and can help promote 'sustainable visitor behaviour on and off the site' (Pearce *et al.*, 1998: 266). Thus, interpretation can be viewed as any activity which seeks to give visitors information about the place they are visiting and can contribute to the management of tourist attractions in several important ways. For example, tourists can be educated about the nature of the host region and culture, and inform them of the consequences of their behaviour, thus encouraging them to behave appropriately. Further, interpretation can enhance the quality of visitor experiences and add value to tourism products and encourage visitors to stay longer at the attraction (Pearce *et al.*, 1998). McArthur and Hall (1996: 88) note that interpretation in Australia and New Zealand is often focused on 'sites or artifacts of "high culture" (explicitly European), or "high nature", nature's monuments, the aesthetically pleasing (to the "trained" or "cultured" European eye) or, the picturesque'. However, visitors to operational mines have the opportunity

Table 7.2 Tools for site management

Management Tool	*Action*	*Example*
Development	Facilities	Toilets, cafeteria
	Access	Board walks
	Technological enhancement	Structural reinforcement
Controls	Barriers	Glass cases, rails
	Laws and guidelines	Burra charter, heritage legislation
	Surveillance/enforcement	Security cameras, patrol staff
Communication	Education	School programmes
	Extension	Conferences and workshops
	Interpretation	Role play, re-enactment

Source: Carter and Grimwade, 1997, p. 48.

Table 7.3 Principles for designing effective interpretation

Vary the interpretative experience	Variation can be offered in: The degree of physical activity from walking around the site to being seated while listening to a guide. The number of visitors taking part in the experience, from sitting alone in an exhibition, to being part of a large audience. The sophistication of the technology used in the interpretation, from a pen and paper to record visitor experiences, to using an interpretative computer to design buildings. The five physical senses, such as touching and smelling aspects of the exhibition. The media used, from signs to audiovisual displays.
Provide personal connections for visitors	By connecting to the visitor's own experience via their nationality, regional location, age, gender or occupation.
Practice participation	Whereby visitors are encouraged to participate, interact and engage with some aspect of the exhibit.
Organise orientation	To orientate the visitor as to what is available at the attraction and allow them to find these areas of interest. Provision of maps, have sound architectural design, provide arrows to follow for special exhibits, floor signs and marked trails.
Provide clear content	Allows visitors to know how to behave at the attraction. For example, to avoid touching nature fauna or, to dress appropriately.
Allow for alternate audiences	To allow different types of visitors to follow the interpretation which encourages a greater appreciation of the site recognising that visitors vary greatly in age, educational background and experience.

Source: Pearce, Morrison, Routledge (1998).

to experience 'real work' or, engage in 'work watching' which provides an opportunity to understand how ordinary working life functions (Carter, 1991: 10). Six principles for designing effective interpretation have been identified by Pearce *et al.* (1998) and are described in Table 7.3. The following case studies describe the tourism activities at two operational mines in Australia namely, PowerWorks in Victoria and the Super Pit in Western Australia. Each will then be discussed in the context of tools used for management, including interpretation.

Case studies

PowerWorks, La Trobe Valley, Victoria, Australia

PowerWorks is located in the town of Morwell which is 160 kilometres south-east of Melbourne, the capital city of the state of Victoria, Australia. PowerWorks is described as 'Victoria's leading industrial tourism attraction' which showcases the science of electricity generation and the role played by the Latrobe Valley in generating 85 per cent of Victoria's electricity needs (Tourism Excellence, 2009). A major part of the La Trobe Valley's history revolves around the brown coal resource which is one of the largest in the world and is a major industry in the region (The Age, 2008). The visitor centre and tours were established in 1995 to promote Victoria's power and coal mining industry and to educate visitors about the brown coal and the energy industry (Walker, 1997). The brown coal field has coal lying as little as 7 metres below the surface and the open cut mine is currently 5 km long, more than 4 km wide and up to 260 m deep, with the capacity to supply brown coal continuously until 2065. The open cut mine digs up thousands of tonnes of coal a day to supply coal to three coal-fired power stations in the region, namely Yallourn, Hazelwood and Loy Yang power stations. The PowerWorks visitor centre is named the Energy Technology Visitors Centre and contains multi-media and interactive displays, working models and large scale equipment in the surrounding gardens. The centre also highlights the importance of visitors learning about a wide-range of alternative, renewable energies and how to reduce their carbon footprint (Tourism Excellence, 2009). Tours are offered once daily during the weekdays by coach around the open cut mine and visitors are offered a 'near hands on' experience of large machinery and dredgers. This is followed by a guided walk through one of the three power stations (Victorian Tourism Awards, 2009). In addition, PowerWorks provides educational programs and tours to suit school children (from pre-schoolers to university students) and can focus on areas such as energy, electricity, the environment, geology, mine rehabilitation, brown coal chemistry, wetlands management and alternative energies (PowerWorks, 2009). The site has been described as an 'exceptional, practical and interactive attraction and loads of fun' (Herald-Sun, 1998).

The sheer size of the operation at the open cut mine and associated power stations is highlighted when the dredgers are described as being '12 storeys high and 200 metres long . . . used to excavate the coal . . . can dig more than 4000 tonnes of coal an hour and could bury a house in minutes' (The Age, 1995). The size of the mine is also noted by Kuiper (2002) who states, 'You may have caught a glimpse of this giant hole from the Princes Freeway, but nothing short of the tour prepares you for just how big it really is', with the dredges which extract the coal being described as being as big as football fields (Kuiper, 2002). Similarly, the PowerWorks (2009) website states:

> 'Whilst in the mine, you will gain an understanding of the sheer size and scale of mining. Boilers, turbines, generators and all of the ancillary equipment are on full display while you walk through an operating power station'.

Thus, the size and importance of the site is highlighted to visitors and is incorporated into educating the visitors about power generation.

PowerWorks is currently dealing with the public's growing concern about coal burning electricity generation due to its heavy production of greenhouse gases. Hazelwood Power Station, one of the three power stations offered as part of the tour, has been described as Australia's most greenhouse-intensive power station. It was built in the 1960s and was due to be decommissioned in 2005 but controversially had its lifespan extended until 2031. It is estimated to emit 18 million tonnes of carbon dioxide a year (Morton, 2009). In September 2009, the environment group *Friends of the Earth* organised a major rally at the Hazelwood Power Station described as a 'peaceful mass civil disobedience' to highlight the impact of coal-burning electricity generation on global warming. One of the organisers described the rally as 'something everyone can get involved in who cares about climate and does want to see a transition away from coal' (ABC, 2009). Given this current controversy, it is understandable that the mining and electricity companies are keen to encourage a greater understanding about energy production and renewable energies via the daily site tours.

The Super Pit, Kalgoorlie, Western Australia

The Super Pit is the biggest open pit gold mine in Australia and is located in the city of Kalgoorlie-Boulder, which is about 600 kilometres east of Perth, the capital city of the state of Western Australia. The site was named the Super Pit following the amalgamation in 1989 of ten smaller pits (Super Pit, 2009). When the gold mining is completed in 2017 the Super Pit will span an area 4 km in length, 1.5 km in width and 650 m deep (Shackleton *et al.*, 2003). The Super Pit allows a private tour company named *Finders Keepers Gold Prospecting Adventures* to run twice daily tours of the mine, with adults being charged $60 per tour. The tour involves viewing an induction video at the visitor centre, listening to the tour company guide provide commentary into and around the Super Pit, and then visitors receive a certificate of completion. In addition, on the third Sunday of every month the Super Pit mine runs its own tours four times a day originating from the mine retail shop, while the local Sunday market is being staged. Visitors can purchase tickets a week before the tour for $5 per adult or, are free on the day of the tour (Super Pit, 2009). To be allowed to participate in the tour all participants are expected to wear enclosed shoes, long sleeved shirt and long trousers and must have a zero blood alcohol level (Finders Keepers, 2009). The tour operator website states:

> Feel what it's like to be a miner and throw on the provided hi-visibility vest, hard hat and a pair of safety glasses . . . View firsthand the mining and milling operations from the safety of the . . . four wheel drive safari truck . . . Be amazed by the size of the machinery (CAT 793 haul trucks) and watch a Komatsu PC 8000 face shovel load a haul truck from the internal lookout, Harvey Hut.
>
> (Finders Keepers, 2009)

Similar to the tour at PowerWorks, the Super Pit also provides interesting facts and figures about the mine such as:

> Although they look like toy trucks from the visitor lookout, the workhorses of the Super Pit are 34 giant Caterpillar 793 mining trucks costing $AU4 million each. Each truck weighs 166 tonne and has a payload of 225 tonne. It costs about $AU26 thousand just to replace a single tyre.
>
> (Super Pit, 2009)

Similarly, on the tour company web page, visitors are asked thought provoking questions such as:

> 'Did you know that, up to 800,000 ounces of gold a year is produced from the Fimiston Open Pit (Super Pit) and the underground operation Mount Charlotte? That's a lot of Gold!'.
>
> (Finders Keepers, 2009)

Thus, the Super Pit mine has outsourced the daily mine tours to a private operator and, allows the private company to market the tours, take bookings, train their guides, maintain the tour vehicle and operate the tour. This has allowed the pit to concentrate on its core operation, namely the mining of gold and they simply need to provide a once a month tour, maintain an internal lookout point and a video at the very basic visitor centre. The partnership with the private tour operator may reflect that the mine is due to close in 8 years time (in 2017) and so any investment in tourist infrastructure has been kept to a minimum.

Similar to the situation at PowerWorks, the Super Pit is also facing controversy when the company applied to extend the size and capacity of the mine and they were then accused by residents of enlarging the mine to within 200 metres of locals' homes and roads.

Environmental campaigners and Kalgoorlie-Boulder residents said such an expansion would make it dangerous for people to walk into a public street in the town as it would be within 150 metres of rock blasting. The local indigenous Ninga Mia community said the impact of the blasting from the mine had created unsuitable levels of dust and noise, which was making them sick (Towie, 2008). In addition, local residents have complained about damage to their homes such as cracking due to the impact of mine blasting (ABC News, 2008). The Super Pit has a strong public relations department which produces regular company newsletters, posts regular media releases, maintains the Super Pit website and engages in a range of community activities, including having staff members volunteer in the community. These activities, along with the regular public tours of the mine, may help improve the company's public image and so are worth pursuing.

Discussion

Mines and quarries as industrial heritage attractions would seem to have greatest chance of success if they are part of a wider range of tourism attractions in the locality (Edwards and Llurdés i Coit, 1996) and indeed, both PowerWorks and the Super Pit are promoted as part of the regional tourism offerings, with PowerWorks being promoted by the local tourism organisation *Gippsland Tourism* as an interesting attraction, particularly for tourists enroute to the nearby Tarra-Bulga National Park (The Age, 1995). Similarly, the Super Pit is promoted on the official tourism website for Western Australia as such:

> 'Kalgoorlie is still an active mining town and . . . Super Pit Mine . . . is mind-blowing . . . This massive hole in the Earth is as deep as Uluru is high with about the same circumference'.
>
> (Western Australia, 2009)

Both mine sites compared favourably with the findings of Cole (2004: 489) in that both sites provide surface tours, educational rooms or facilities, 'hands-on' activities with exhibits and videos/films/tapes explaining exhibits. PowerWorks does not currently operate a retail shop on site but the Super Pit has recently opened a shop in the nearby town with a range of related merchandise, a 'wealth of information about the Super Pit' and, features a model of the final 2017 pit design (Super Pit, 2009). The Super Pit retail shop recently installed a live 'Super Pit Cam' which feeds live video footage from five different locations inside the Super Pit, back to the shop (KCGM, 2008). As a result, shop visitors can remotely experience the mine which may be enough to fulfil their curiosity or, it may whet their appetite and encourage them to take a tour. Access for disabled and limited mobility visitors is not mentioned by either tour. Given that both visits involve bus tours, access for wheelchair bound visitors and people who have difficulty climbing stairs may be limited. Burnett and Baker (2001) note that few consumer groups have greater potential and are more ignored than the disabled so this is a weakness for both tours, particularly when the tourism industry is becoming more considerate of disabled visitors, not only for business reasons, but for corporate citizenship in helping to create a more inclusive society (Shaw and Coles, 2004). Similar to the mine tour in Bolvia, the Super Pit provides safety equipment for each person in the group, namely safety helmets, safety glasses and high visibility vests and, allows the tourist to interact directly with a knowledgeable guide, rather than public relations personnel (Pretes, 2002). At PowerWorks tour participants are expected to wear covered shoes and the tours are conducted by the education officer.

A visitor's centre on a mining site can be a strong magnet to visitors of all kinds (Jansen-Verbeke, 1999) and lively communication about coal mines and the production process and about the coal miners and their families has the potential to attract tourists (Edwards and Llurdés i Coit, 1996). Both the sites examined used interesting facts and figures to ensure that the communication is not purely educational, but includes entertaining information as well ('infotainment') (Edwards and Llurdés i Coit, 1996). Both the mine sites compared favourably with the findings of Frew and Shaw (1996) who determined that in Australia industrial tourism sites were typically open for at least six hours per day, every day of the year and about a third offering guided tours. Both sites charge an entrance fee to visitors, with adults to PowerWorks paying $20, and $60 for the private tour at the Super Pit; $5 or free for the once a month tour operated by the mine. However, since the frequency of tours operated by the mines is minimal and, the entrance prices are relatively low, this would suggest that such tours are offered for public relations reasons rather than for commercial benefit.

There is a variety of interpretative experiences offered at the operational mines as both tours involve travelling on a bus, walking around a visitor centre and watching a video. Thus, at least three of the five physical senses are utilised namely, sight, hearing and touch. There is evidence of 'personal connections' at PowerWorks when an explanation is given as to how the electricity generated by the power stations runs an individual's home or business. There is 'physical participation' at the Super Pit when the visitors are kitted out with safety equipment. Visitors are orientated at both sites through the provision of a visitor centre and induction video before the tour begins. There is 'clear content' provided, since all visitors know that they have to wear protective clothing and footwear before the tour commences. The displays and activities at both sites 'allowed for alternative audiences', particularly as the PowerWorks tour can be tailored for pre-school, primary, secondary and tertiary students. Both sites allow tours of open cut mines and both are controversial in that one generates power via burning coal in a heavy carbon producing manner and the other has a negative physical impact on the surrounding community. Both sites offer very controlled bus tours which ensures safety but also ensures that there is limited opportunity for sabotage. Both tours show examples of a range of interpretation and use all of the interpretation principles. Both tours utilise particular tools for site management namely, the development of facilities, allowing access, using technology to enhance the visit, demonstrating concern for visitor safety and, using education and interpretation as communication.

Conclusion

Developing industrial tourism by allowing tourists to visit operational mining sites has implications for mining managers as decisions have to be made regarding the facilities and services to be offered. There has been a move in the United States to replace some plant tours with visits to company museums and visitor centres where people do not have the opportunity to view the actual production process. However, to separate the visitors from the production process may eliminate the 'real life' element in the visit which may make the experience less fulfilling (Frew and Shaw, 1996). For managers who are considering opening their door to the public, consideration should be given to the type of interpretation used. Light (1995) believes investment in interpretation is money well spent, since visitors make use of it and that interpretative media have differing roles and can be targeted at different visitor groups.

Industrial tourism is derived from a position of subordination to the non-tourism activities of the organisation (Frew, 2008) and the preceding discussion focused on the supply-side of industrial tourism, namely, the creation and management of industrial tourism products at operational mines. There is the opportunity to capitalise on the public's interest in visiting operational mine sites as such a scheme can be a useful public relations opportunity, particularly when the site is currently experiencing controversy and associated attention from the general public. Industrial tourism of this type has been described as an oddity as it is 'essentially recreation in the workplace, where work and leisure meet' (Green, 1994: 16). Oglethorpe (1987: 271) points out that although society remains 'perpetually and totally dependent upon industrial activity' fewer people now work in industry so people may become interested in visiting industrial sites as they are unaccustomed to the industrial environment and are eager to discover more about the operations. As such, mining managers should not treat visitors as an 'irksome afterthought that leaves the visitor feeling as if they've intruded' but should use the opportunity to educate and entertain the visitors to create a positive image in their minds of the site (Wheately, 1989: 24). In addition, operational mines have a role to play in developing regional tourism as they can help visitors appreciate why the local area developed. This suggests that industrial tourism can help in the development of cultural tourism, where the culture visited signifies a 'whole range of human creations, customs, historic structures and activities' (Derrett and Welch, 2008: 75).

References

The Age (2008) 'Experience the Latrobe Valley: Advertising Feature', *The Age*, 14 June: 26.

The Age (1995) 'Tours Give Bird's-eye View of Operations', *The Age*, 7 August: 22.

Australian Broadcasting Corporation (ABC) News (2008) 'Kalgoorlie Residents Disappointed with Super Pit Expansion Kalgoorlie Super Pit', *Australian Broadcasting Corporation (ABC) News*, 26 June.

Australian Broadcasting Corporation (ABC) News (2009) 'Rally Organisers Defend Civil Disobedience Call', *Australian Broadcasting Corporation (ABC) News*, 4 August.

Burnett, J.J. and Baker, H.B. (2001) 'Assessing the Travel-related Behaviors of the Mobility-disabled Consumer', *Journal of Travel Research*, 40: 4–11.

Carr, A. (2004) 'Mountain Places, Cultural Spaces: The Interpretation of Culturally Significant Landscapes', *Journal of Sustainable Tourism*, 12: 432 59.

Carter, B. and Grimwade, G. (1997) 'Balancing Use and Preservation in Cultural Heritage Management', *International Journal of Heritage Studies*, 3: 45–53.

Carter, J. (1991) 'Watching Work go by', *Environmental Interpretation*, 6(4): 10–11.

Cole, D. (2004) 'Exploring the Sustainability of Mining Heritage Tourism', *Journal of Sustainable Tourism*, 12(6): 480–90.

Cox, L.J. and Fox, M. (1991) 'Agriculturally Based Leisure Attractions', *Journal of Tourism Studies*, 2(2): 18–27.

Derrett, R. and Welch, J.St.V. (2008) '40 Sheds and 40 Kilometers: Agricultural Sheds as Heritage Tourism Opportunities', in B. Prideaux, D.J. Timothy and K. Chon (eds) *Cultural and Heritage Tourism in Asia and the Pacific*, Abingdon, Oxon: Routledge, pp. 73–83.

Edwards, J.A. and Llurdés i Coit, J.C. (1996) 'Mines and Quarries: Industrial Heritage Tourism', *Annals of Tourism Research*, 23: 341–63.

Finders Keepers (2009) Available online at: http://www.finderskeepersgold.com/Tours/SuperPitTour/tabid/93/language/en-AU/Default.aspx (accessed 31 July 2009).

Frew, E.A. (2008) 'Industrial Tourism Theory and Implemented Strategies', in A. G. Woodside (ed) *Advances in Culture, Tourism and Hospitality Research*, Bingley, Emerald Group Publishing Limited, 2: 27–42.

Frew, E.A. and Shaw, R.N. (1995) 'Industrial Tourism: The Experience and the Motivation to Visit', in R. N. Shaw (ed.) *Proceedings of the 1995 National Tourism and Hospitality Conference*, Canberra: Bureau of Tourism Research, pp. 88–104.

Frew, E.A. and Shaw, R.N. (1996) 'Profiling the Industrial Tourism Phenomenon', in G. Prosser (ed) *Tourism and Hospitality Research: Australian and International Perspectives. Proceedings of the*

Australian Tourism and Hospitality Research Conference, Canberra: Bureau of Tourism Research, pp. 467–79.

Frew, E.A. and Shaw, R.N. (1999) 'Industrial Tourism Attractions: A Conceptual and Empirical Identification', in J. Molloy and J. Davies (eds) *Tourism and Hospitality: Delighting the Senses, 1999, Part 1. Proceedings of the Ninth Australian Tourism and Hospitality Research Conference*, Canberra: Bureau of Tourism Research, pp. 211–18.

Green, S. (1994) 'Industry Tourism', *Environmental Interpretation*, 9(3): 16–17.

Herald-Sun (1998) 'Hit the Road for Kids' Fun', *Herald Sun*, 6 January: 27.

Jansen-Verbeke, M. (1999) 'Industrial Heritage: A Nexus for Sustainable Tourism Development', *Tourism Geographies*, 1(1): 70–85.

Kalgoorlie Consolidated Gold Mines (KCGM) (2008) *KCGM Super Pit Shop goes Interactive: KCGM media release 19 September.* Available online at: http//www.superpit.com.au (accessed 31 July 2009).

Kelly, I. and Dixon, W. (1991) 'Sideline Tourism', *Journal of Tourism Studies*, 2(1): 21–28.

Kuiper, W. (2002) 'Our Broad Brown Land'. *Herald Sun*, 30 August.

Leung, M.W.H. and Soyez, D. (2009) 'Industrial Heritage: Valorising the Spatial – Temporal Dynamics of Another Hong Kong Story', *International Journal of Heritage Studies*, 15(1): 57–75.

Light, D. (1995) 'Visitors' Use of Interpretative Media at Heritage Sites', *Leisure Studies*, 14: 132–49.

McArthur, S. and Hall, C.M. (1996) 'Interpretation: Principles and Practice', in C. M. Hall and S. McArthur (eds) *Heritage Management in Australia and New Zealand: The Human Dimension*, South Melbourne, Victoria, Oxford University Press, pp. 88–106.

McBoyle, G. (1994) 'Industry's Contribution to Scottish Tourism: The Example of Malt Whisky', in A. V. Seaton (ed) *Tourism: State of the Art*, West Sussex: John Wiley and Sons Ltd, pp. 517–28.

MacCannell, D. (1976) *The Tourist: A New Theory of the Leisure Class*. New York: Schoken Books.

Macionis, N. (1996) 'Wine Tourism in Australia', in G. Kearsley (ed) *Tourism Down Under II: Towards a More Sustainable Tourism, Conference Proceedings*, Dunedin, New Zealand: Centre for Tourism, University of Otago, pp. 264–86.

Mallam, M. (1989) 'Can Heritage Charities be Profitable?', in D. Uzzell (ed) *Heritage Interpretation: The Natural and Built Environment*, London: Belhaven, pp. 44–50.

Morton, A. (2009) 'Divided Views Over "Clean Coal" Pilot Project', *The Age*, 9 July 2009: 5.

Oglethorpe, M.K. (1987) 'Tourism and Industrial Scotland', *Tourism Management*, 8(4): 268–71.

Pearce, P.L. (1990) 'Farm Tourism in New Zealand: A Social Situation Analysis', *Annals of Tourism Research*, 17: 337–52.

Pearce, P.L., Morrison, A.M. and Rutledge, J.L. (1998) *Tourism: Bridge Across Continents*. Roseville, NSW: McGraw-Hill.

Poria, Y. Reichel, A. and Brandt, Y. (2009) 'People With Disabilities Visit Art Museums: An Exploratory Study of Obstacles and Difficulties', *Journal of Heritage Tourism*, 4(2): 117–29.

PowerWorks (2009). Available online at: http://www.powerworks.com.au/ (accessed 31 July 2009).

Pretes, M. (2002) 'Touring Mines and Mining Tourists', *Annals of Tourism Research*, 29(2): 439–56.

Ruiz-Ballesteros, E. and Hernandez-Ramirez, M. (2007) 'Identity and Community – Reflections on the Development of Mining Heritage Tourism in Southern Spain', *Tourism Management*, 28: 677–87.

Shackleton, J.M., Spry, P.G., and Bateman, R. (2003) 'Telluride Mineralogy of the Golden Mile Deposit, Kalgoorlie, Western Australia', *The Canadian Mineralogist*, 41: 1503–24.

Shaw, G. and Coles, T. (2004) 'Disability, Holiday Making and the Tourism Industry in the UK: A Preliminary Survey', *Tourism Management*, 25(3): 397–403.

Stevens, T. (1988) 'Work Watching: The Growth of Industry Tourism', *Leisure Management*, 8(12) 40–43.

Super Pit (2009) Available online at: http://www.superpit.com.au (accessed 31 July 2009).

Swarbrooke, J. (1995) *The Development and Management of Visitor Attractions*. Oxford: Butterworth Heinemann.

Timothy, D.J. (2007) *The Heritage Tourist Experience, Critical Essays, Volume 2*. Hampshire: Ashgate Publishing Limited.

Towie, M. (2008) 'Spread of Mine Sparks Fears', *The Sunday Times*, 3 March.

Tourism Excellence (2009) Available online at: http://www.tourismexcellence.com.au/pdf/module5/IanSouthall.pdf (accessed 31 July 2009).

Vargas-Sánchez, A., de los Ángeles Plaza-Mejía, M. and Porras-Bueno, N. (2009) 'Understanding Residents' Attitudes Toward the Development of Industrial Tourism in a Former Mining Community', *Journal of Travel Research*, 47(3): 373–87.

Victorian Tourism Awards (2009) Available online at: http://victoriantourismawards.com.au/option.com_entrantsandresults/task.entrantdetails/business_id.92/ (accessed 31 July 2009).

Walker, C. (1997) 'A Bush Show Adds to Victoria's Attractions', *Sunday Age*, 26 January: 6.

Western Australia (2009) Available online at: http://www.westernaustralia.com/en/Destinations/Australias_Golden_Outback/Kalgoorlie//Pages/Kalgoorlie.aspx (accessed 31 July 2009).

Wheatley, A. (1989) 'Mine Site Tours – A Worthwhile Public Role', *Mining Review*, December: 25–26.

8 Developing a heritage tourism attraction in a working salt mine

The Kansas underground salt museum

Deborah Che

> In a world inundated with pre-fabricated, pre-programmed events, people crave authenticity. What better place to learn about salt and its importance than in the heart of an underground salt deposit? A journey into the exotic world of a working salt mine is possible in very few places on Earth. The Kansas Underground Salt Museum has a broad appeal for all who seek an indelible experience.
>
> (The Reno County Historical Society, Inc. 2006: 14)

Introduction

Mines can be venues for experiential, educational heritage tourism. A growing number of tourists are seeking new experiences where they can learn firsthand about a region's lifestyle, natural attributes, local history, and culture. The Travel Industry Association of America (TIA) considers 81 per cent of U.S. adults who had travelled in the past year, or 118 million people, to be historical/cultural travellers. Eighteen per cent of travellers in the past year reported that an educational trip was the main purpose of their travel while 30.2 million adults had taken such an educational trip in the past three years (Fermata Inc., 2006). In resource-dependent areas, mines formerly or currently in operation can provide this large subset of tourists with opportunities to learn about the unique culture, heritage, and settlement associated with extractive industries.

Salt mines which have been sites of resource extraction and production and, on occasion places for the aristocracy, celebrities and government officials to visit, are now tourist attractions for the wider public. Two of the famous sites include the Dürrnberg Mine near Salzburg, where 220,000 people who visit each year can ride down wooden slides that were built to transport miners, and the Wieliczka Mine, a World Heritage site near Cracow, where visitors can view both artifacts illustrating mining techniques spanning the thirteenth to the twentieth centuries, as well as its famed chapel with statues, altars, and chandeliers that miners carved out of rock salt. Other salt mines in Europe such as Salies de Bearn developed baths and health spas during the nineteenth century for therapies utilizing their brine springs to alleviate and cure rheumatism, arthritis, and circulatory ailments (Kurlansky, 2003).

Addressing economic restructuring through mining tourism

More recently, mining tourism has been viewed as a way to address economic restructuring in rural, European peripheral areas. Given shifts from Fordism to post-Fordism and the closing of mines in regions long dependent on mineral extraction, such sites have increasingly been viewed in terms of their cultural, heritage, aesthetic, and recreational values for post-industrial service activities such as tourism (Saurí-Pujol and Llurdés i Coit 1995). In Industrial South Wales, closed coal mines and the industrial landscape produced have become interpreted and marketed for heritage tourists (Wanhill, 2000). Likewise in Cardona, Spain, a major centre for potassic salt production has, since the shuttering of its mine, initiated the Salt Mountain project to capitalise on recreational uses of the rock salt outcropping and valley (Saurí-Pujol and Llurdés i Coit 1995). As in deindustrialized regions, such heritage tourism development has been largely supply led (Robinson, 1999) as part of government programs to address job losses in the mining sector.

As with productivist uses, the sustainability and viability of mining tourism attractions should be considered. Involving local communities and current and former coal miners and their descendants in the mine museums and historic sites can foster social sustainability as well as enrich and increase the authenticity of interpretative materials (Cole, 2004). Living histories from miners and their families can impart a 'real experience' to visitors (Jansen-Verbeke, 1999), key in experiential and heritage tourism. Local endorsement and/or ownership of the heritage sites is also key for social sustainability, although in the UK, fierce competition for government funding and tourists could lead to entertainment displacing education and historic authenticity (Cole, 2004). In such cases, achieving both social and economic sustainability may be more difficult.

Economic sustainability and viability are important for continued operations of mining heritage attractions. If the public sector finances the considerable capital costs needed to develop a museum that appeals to a broad segment of the public, mining tourism development can be feasible. Such development can also be viable as the total cost schedule will not include debt that needs to be serviced by project owners. A lower number of visitor days will be required. Yet, because such large-scale mining installations have high fixed repair, restoration, and upkeep costs, a large number of visits are still needed to cover operating costs (Wanhill, 2000). However, mining heritage tourism sites can generate surplus income to cover operating costs, particularly where the region or city has a cluster of attractions. Having a wider range of tourism attractions in a community can attract a greater number of visitors (Cole, 2004; Jansen-Verbeke, 1999). For the broader community, mining tourism attractions can also stimulate economic regeneration when there is a diverse economic base. However in many rural peripheral areas lacking services for tourists, tourism may have a limited multiplier effect. With little history of entrepreneurship, training and financial assistance are necessary as part of broader economic development plans for rural extractive regions to increase benefits from tourism (Cole, 2004). In this way mining tourism attractions, which heavily rely on volunteer, part-time and seasonal employment and cannot replace all mining jobs, can still generate greater economic benefits.

This chapter will now focus on the development of the only underground museum in a working salt mine in the Western Hemisphere, the Kansas Underground Salt Museum (KUSM). Located 650 feet below the prairie in Hutchinson, Kansas, this museum recognizes the heritage and current industry associated with salt mining in 'Salt City,' which grew by utilizing one of the world's major salt deposits formed over 200 million years ago when an ancient salt sea covered the area (Figure 8.1). This chapter will first examine the history of mining salt in Hutchinson and the development of mining heritage tourism in Salt City and at the KUSM which provides the unique opportunity to tour mined out areas and view their reuse for education/interpretation and for storing valuable business documents and Hollywood film negatives and costumes. This chapter then examines the challenges and opportunities for

Map 8.1 Major salt deposits and dry salt production sites in North America (The Salt Institute 2009).

the long-term growth, development, and viability of this heritage tourism site as it necessarily incorporates safety, environmental, heritage, and commercial concerns.

The development of Hutchinson, Kansas as Salt City

The evolution of Hutchinson, Kansas as Salt City has been a product of both prehistoric and historic forces. What is now Kansas lay under a shallow, ancient salt sea 230 million years ago. After the water evaporated, the salt left behind was compressed into solid rock salt. The Kansas Geological Survey has estimated that five trillion tons of salt are under the state's area (Elliott, 1995), with the richest deposits ranging from 300 to 400 feet in thickness lying 600 feet beneath Hutchinson (Figure 8.1). The salt industry in Hutchinson began with the discovery of rock salt on 27 September, 1887. Its exploitation was made possible by the city's steady supply of water from the Arkansas River and Cow Creek and its being on the Atchison, Topeka & Santa Fe; the Rock Island; and the Missouri-Pacific rail lines (Smith, 2003) which reduced shipping costs. By the 1890s, Hutchinson had 16 salt plants that produced two million barrels of salt annually, giving the city fame as 'the Salt City of the West' (City of Hutchinson, Kansas, 1893). Beef cattle production on Kansas and Nebraska prairie grasses also fuelled salt production as the mineral was critical for the production of dried, jerked, packed, and pickled meats and for processing hides. Kansas was the dominant supplier of salt to nearby meat packing cities like St. Louis, Omaha, St. Paul, Sioux City, Kansas City, Wichita, Denver, Oklahoma City, Ft. Worth, and Dallas (Zerger, 1945). Mining, processing, and shipping salt became Hutchinson's chief industry, which by the 1930s was consolidated into three operating companies, the Barton Salt Company, the Morton Salt Company and the locally-owned Carey Salt Company which in total employed approximately 600 men (Federal Writers' Project of the Works Projects Administration, 1984). New transportation developments funded by the Federal Highway Acts of 1916 and 1921 and the post-war 1956 Interstate Highway Act fuelled new demand for salt for de-icing roads (Smith, 2003).

Salt production led to the underground storage business in Hutchinson. As the dense rock salt was mined, long rooms were left between 20–50 feet thick columns of salt. The constant temperature of 68 degrees Fahrenheit, low relative humidity at the mining face, and lack of plant, animal, or insect life that could damage stored items provided ideal, secure storage conditions away from natural hazards, fire, and war/conflict zones (Ulrich, 2008). Underground Vaults and Storage (UVS), which leased mined out space from the Carey Salt Mine, opened in 1960 and is the repository for the formula for making Wrigley's Chewing Gum (Elliott, 1995) and Hollywood costumes, vintage and present-day films and television shows such as 'The Wizard of Oz', 'Gone with the Wind' and 'Friends' as well as data tapes and business and corporate papers (Kilbourn, 2009).

Hutchinson's salt also drew tourists. At the turn of the twentieth century, hotels and spas in Hutchinson, like their European counterparts, utilized brine water in their therapies. The F.A. Manda Hotel and Salt Water Natatorium and the Coliseum European Hotel and the connected Hutchinson Sanitarium offered free mineral water for guests; salt swimming pools; and Turkish, electric, salt-steam and other baths for treating rheumatism, kidney trouble, nervousness, etc. (Hutchinson, the Salt City in the Heart of the Great Kansas Wheat Belt, 1906, 1910). Hutchinson also attracted tourists who wanted to visit its salt mines. From 1927–1961, the Carey Salt Mine took school and teacher groups, chambers of commerce members, dignitaries, celebrities and Fulbright scholars as well as tourists underground (Ulrich, 2008). By the late 1950s, Carey had to restrict the number of visitors to 40 a day in the winter and 100 in the summer each Tuesday, Wednesday and Thursday as the tours utilized the same elevator to hoist salt up out of the mine (Heavy run of mine visitors, 1959). While the tourist committee of the Hutchinson Chamber of Commerce sought to develop a separate 600 foot 'tourist shaft' at the Carey Salt Mine ('Tourist' shaft proposed for mine, 1958), it was never built. In 1961, the Carey Salt Company cancelled its mine tours as guiding visitors around the mine interfered with efficient mine operations (Lawson, 1961). The Hutchinson Chamber of Commerce sent visitors to the other salt companies until the mid-1970s when Mine Safety & Health Administration regulations put an end to such official mine tours (Smith, 2000).

Transportation and economic changes pushed tourism development, including that centred on mining heritage, in Hutchinson. The city had grown wealthy from its being a key railroad hub for south central and western Kansas and a 'shipping and trading centre for the wheat belt.' But the city's 1963 Master Plan highlighted the dangers of being resource-dependent noting, 'Areas content to market their raw natural resources for processing elsewhere, or fail to adopt improved methods of production, fall out of step with economic progress and decline.' Hutchinson's economic and population growth flattened out as the city and Reno County, which did not push for connecting with the inter-states, was bypassed by them and the industries that needed a four-lane highway to distribute their product (Baker and Smith, 2004). While mining and agriculture were still important components in the city's economy, mechanized production in the salt industry required a relatively small number of employees. For instance, the Hutchinson Salt Company mine (the former Carey Salt Mine) needed only 25–30 year-round employees to produce 500,000 tons of rock salt for de-icing (Green, 2008). Post-industrial industries such as tourism grew in relative importance in the US and Great Plains economy. In this context, the Kansas Cosmosphere, a Smithsonian-affiliated space museum, and KUSM were seen as part of economic development and diversification efforts.

Developing a salt mining heritage tourism attraction

The KUSM was envisioned to be a unique, experiential, educational heritage tourism destination in Salt City. While some currently operating mines are open for visitors to view their operations and also as outlets for corporate public relations (Rudd and Davis, 1998), mining heritage museums located in working mines are unique. The KUSM would be a one of a kind attraction in the US as there were only 15 salt mines, none of which were open to tourists or the general public (The Reno County Historical Society, Inc. 2006). The Hutchinson Salt Mine, where the museum would be developed in areas away from mining operations, had artifacts left underground such as old salt cars, drilling equipment, and a train engine and tracks, as well as salt company employees who could help tell the story of salt 650 feet underground (Smith, 2000). This unique museum would not only celebrate the past, but the present too as the city had more than 67 miles of mined caverns and also had the Cargill (formerly Barton) and Morton evaporation plants which produced salt for pharmaceutical, baking, agricultural, and chemical uses. An underground salt museum would be a second attraction needed to complement the Kansas Cosmosphere and Space Center and annual events such as the Kansas State Fair that brought visitors to Salt City (The Reno County Historical Society, Inc. 2006).

The KUSM developed as a partnership between the private and public sectors. The first barrier to development, a small, eight-person elevator shaft, was solved when the parent of the future KUSM, the Reno County Historical Society (RCHS), the UVS, and the Hutchinson Salt Company agreed to share expenses in constructing a new mine shaft (The Reno County Historical Society, Inc. 2006). Developing the museum involved a three-way partnership (Ferrell, 2009) between 1) Hutchinson Salt Company, which owned the former Carey Salt Mine; 2) UVS, which had an exclusive lease of the Hutchinson Salt Company's mined out 980 acres; and 3) KUSM, which would lease space from UVS for five per cent of the museum's monthly gross income after the first six months of operation, rising to 10 per cent and then 15 per cent by 2009 (Clarkin, 2007b). The RCHS had risen over 60 per cent of the $7.8 million capital campaign goal which covered digging the mineshaft, installing the elevator, building the building, pouring the underground floors, and constructing subterranean walls (Underground Update, 2004). KUSM became a city project which RCHS operated under contract with the city of Hutchinson when Hutchinson issued a $4.8 million sales tax and revenues (STAR) bond to finance the remaining costs for electricity, power, lighting, water, sewer, telephone and trash, fire detection systems and security plans (Johnson, 2007). This combination of funds the RCHC raised and the city's STAR bond made the project feasible. Debt did not have to be serviced by the KUSM, thus lowering the visitor numbers needed for viability.

The KUSM, which opened on 1 May, 2007, offered a unique experiential, educational, heritage tourism attraction. Customer comments expressed amazement at how big the museum and mine were, as well as the experience of being 650 feet underground, walking where salt was mined and where it was a very pleasant 68 degrees Fahrenheit – quite a contrast to the 100+ degrees Fahrenheit topside on some hot Kansas summer days. According to Gayle Ferrell (2009), KUSM Director of Operations, the museum's best hook was when visitors stepped out of the hoist and saw how immense the museum was. They immediately saw the 600 foot long, 50 feet wide Great Room, with its 8–11' ceilings (Figure 8.1). Then beyond the open room when riding on the tram, visitors could view the vastness of the mined area, 300 foot narrows, and the checkerboard effect of support pillars as well as the galleries with exhibits. According to Myron Marcotte, mining supervisor at Hutchinson Salt Company, the lure of the museum was its taking visitors someplace different, starting from the time they

Figure 8.1 650 feet below the prairie in the Great Room of the Kansas Underground Salt Mine (Source: Deborah Che).

stepped on the elevator which took them 650 feet below Hutchinson. One had a sense of 'isolation from everything else,' particularly when the tour guides turned the lights off at one stop on the tram ride through the mined out caverns. The mine also had a thrilling 'aspect of danger' according to Marcotte even though actual mining took place 3,000 feet away from the museum and salt mines do not have methane or gases that could explode or suffocate. Additionally salt does not burn, although items taken down to the mine such as paper can burn. Consequently, smoking and open fires either for science experiments or steam tables were not allowed underground. Each visitor underground was required to wear a hard hat and carry a self-rescuer breathing device in case of fire which never has had to be used in the mine's entire history (Clarkin, 2008a). In the caverns, visitors could also pick and take a piece of rock salt home (Figure 8.2). According to Ferrell (2009), 99.9 per cent of visitors were satisfied. The fraction of one per cent of visitors who were disappointed in the KUSM desired more of an amusement park, commercialized experience with lights, active displays, a café, food to go, etc. – quite different from what the educational, heritage tourism attraction offered.

In addition to touring the caverns, the museum featured unique exhibits on the mining process; the documents, costumes, and original movie reels and old television shows that UVS stored (Figure 8.3); and on the oldest living creatures on earth, resuscitated bacteria extracted from fluid inclusions in a 250-million-year-old salt crystal which came from a New Mexico mine extracting from the same salt formation that is beneath Hutchinson. The scientists who conducted this research collected samples from the Hutchinson mine estimated to be 275 million years old, so if bacteria are found within they will likely predate the organisms previously found (Kansas Underground Salt Museum, 2008). KUSM also has hosted special events such as murder mysteries and Boy Scout campouts. Additionally the KUSM rented out space in the evenings for private parties, business meetings, and informational/sales presentations given

Figure 8.2 Visitors taking a bit of Kansas salt home (Source: Deborah Che).

Figure 8.3 Hollywood on display at the Kansas Underground Salt Museum (Source: Deborah Che).

by financial investment companies etc. In addition to providing the experience of being in a working salt mine, cell phone or text messaging distractions were eliminated as visitors were essentially on the 50th floor of a high rise in reverse (Ferrell, 2009).

The KUSM attracted a wide variety of general admission visitors. Families from Wichita as well as other metro areas such as Oklahoma City and Topeka dominated during the busy summer season. During the cooler times of the year, school groups studying geology, adult bus groups from regional metro areas, and retirees and empty nesters driving cross country or north–south visited. Importantly, the museum had access for disabled and limited mobility visitors. The KUSM provided wheelchairs and scooters to get to the tram and galleries which were helpful for the disabled as well as just tired visitors. KUSM's biggest challenge has been to attract local adults who believe that since they visited the Carey Salt Mine decades ago they have been to the museum, even though they have not seen exhibits such as the UVS display with Sony and Warner movie artifacts, and may downplay an attraction in their own backyard (Ferrell, 2009). Increasing visitation and income, which can finance new exhibits and marketing efforts, continues to be a goal of KUSM.

Opportunities, challenges and the viability of the Kansas Underground Salt Museum

While a unique attraction, the KUSM has faced challenges regarding its economic sustainability. Attendance for the first six months of operation in 2007 was 62,000, which fell short of the 100,000 originally projected. The museum consequently cut its staff as well as its 2008 marketing budget from $200,000, including $75,000 for television advertising, to $132,000, or 18 per cent of revenues even though it recognized that marketing was key to the museum's long-term success (Green, 2007). The city provided KUSM with $67,525 to cover its debt and operating expenses until the busy May–August 2008 season when losses could be recouped. To break even under the reduced budget, the museum had to pull in at least 54,000 visitors in 2008, not counting people attending special evening events under the reduced budget (Clarkin, 2008c). It surpassed the break-even point with 58,000 visitors (Ferrell, 2009). The KUSM thus covered its operating costs and was economically viable. For the next five years, it will be a new recipient of a recently renewed quarter-cent sales tax, a small component of the total 7.050 per cent sales tax rate in the city of Hutchinson, that was projected to generate $2 million and which would help KUSM develop exhibits and programming. Forty seven per cent of the quarter-cent tax receipts will go to street projects, 15 per cent to offset increases in property taxes, 33 per cent to the Kansas Cosmosphere, and five per cent, or $102,000, to the KUSM (Hall, 2008). In its appeal for renewal of the quarter-cent sales tax, KUSM highlighted its role in collecting sales tax as more than 80 per cent of the tickets it sold during its first year of operation were purchased by visitors from outside Reno County. Overall, 20 per cent of sales tax collected in Hutchinson came from individuals living outside the city (KUSM ticket prices to drop, 2008).

The KUSM's share of those sales tax revenues would help it increase its visibility, visitation, and income and grow by adding new exhibits, expanding its educational programs, and increasing its marketing efforts in order to guarantee its long-term viability (Schmitt, 2008). While the KUSM would like to be a Smithsonian affiliate like the Cosmosphere, but environmental and physical constraints make this difficult. For items exhibited underground, unless every last bit of salt dust is removed, corrosion is a problem when they are brought topside with the higher humidity. Additionally the hoist could only accommodate certain sized pieces, and unlike equipment that was cut apart to be reassembled underground, lenders would not permit artifacts to be altered in this manner (Ferrell, 2009). However, the KUSM can develop exhibits through other partnerships. A developing sister city tie with Bex, Switzerland and its salt museum may lead to shared exhibits (Kilbourn, 2008). Additionally through its partnership with UVS and Hollywood studios, the KUSM hopes to rotate memorabilia on display from Sony and Warner Brothers as well as to present another studio's memorabilia. In the future, the museum will convert oral histories of Carey Salt miners from 8 mm to DVD for a miners' stories exhibit (Ferrell, 2009). In addition to drawing return and local visitors, this future exhibit would incorporate the local community and thus speaks to issues of socially sustainability.

Additional events, attractions, and services can attract new, return, and local visitors. While the KUSM uses mining equipment and artifacts to tell the story of salt in Hutchinson, with its experiences of going 650 feet down in the elevator and the dark tram ride, it would like to be considered as an adventure or attraction rather than strictly as a museum. As a result of information gathered from its partnership with the Bex Museum, the KUSM aims to develop its museum concession area to host conferences and events (Kilbourn, 2008). Like the Cosmosphere with its space museum, planetarium, Dr Goddard's lab, and IMAX theater developed over the course of 20 years, the KUSM plans to have multiple attractions. In its longer-term planning to attract new visitors and grow, it would like to develop an interactive classroom experience underground with live cams, an underground train ride experience, as well as a topside dinner theatre for the general public as well as individuals who have restrictions going underground (i.e. families with children under four, smokers who do not want to go underground) (Ferrell, 2009).

While the KUSM is not off major interstate highways nor near major tourist cities, like the Dürrnberg and Wieliczka mines are (Salzburg and Cracow), the KUSM benefits from a wider range of tourism attractions than some mining attractions located in European rural, peripheral areas. Hutchinson, a regional hub with 40,787 residents, has a large number of quality community facilities funded by its philanthropists. The Davis Foundation gave the Kansas Cosmosphere and KUSM $25,000 for a joint marketing project, as it believed the two attractions could be more successful working together, rather than independently (Clarkin, 2007a). This grant paid for joint billboards around the state reading 'From 650 Feet Underground to the Moon' that highlight the two attractions and induce people to stop in Hutchinson. The Cosmosphere and KUSM also developed a joint ticket which saved visitors four dollars on the regular adult tickets to those attractions if purchased separately. This ticket has been extremely popular as it also allowed visitors to go to both of the '8 Wonders of Kansas' on one visit to Hutchinson or return within a year to see the other attraction (Ferrell, 2009).

Hutchinson's having multiple attractions also allowed the KUSM to specifically accommodate mine safety regulations on the number and age of tourists and limit the loss of visitorship. Given the availability of safety equipment, KUSM has a 200 person limit of visitors underground at any one time. Mine safety regulations also do not permit children under four years old. KUSM staff can accommodate walk-in visitors when the next available tour underground is not for a few hours by suggesting they go first to attractions such as the Cosmosphere (the space museum), Hutchinson Zoo, and Dillon Nature Center which have no limitation on the number of visitors. The KUSM can also assist families with a child under four years old by suggesting the family split in two groups going on separate tours a few hours apart. One parent can accompany older children on a KUSM tour while the other can take the two or three year old child who cannot go underground to one of the other attractions that does not have age restrictions, then switch places (Ferrell, 2009). Having multiple attractions can maximize visitors to KUSM by providing alternatives when tours are not available and by encouraging more and longer visits to Hutchinson and the KUSM.

For future social and economic sustainability and viability, KUSM, a unique experiential heritage attraction, must continue to attract a wide range of visitors. In their study of heritage tourists to the Wailing Wall in Jerusalem and Masada in southern Israel, Poria *et al.* (2004) characterized heritage tourists in terms of their having different links with a site, their perceptions of a site, and their reasons for visiting a site as 1) those seeking a 'heritage experience' who have a link to the heritage presented or have 'a desire to be involved in the heritage experience'; 2) those interested in 'learning history' who want to learn about the history and physical characteristics of a site; and 3) those seeking a 'recreational experience' who want entertainment or a day out. In KUSM's case, the 'heritage experience' visitors are those with connections to Hutchinson, the salt industry, and/or mining. For local visitors in particular, KUSM's inclusion of former miners as volunteers in interpreting and guiding them through the caverns augments the attraction's authenticity, connection to Hutchinson, and social sustainability. To overcome some of the challenges in getting more local visitors, KUSM charges Reno County residents the child's rate, which is a deep discount. By giving this discount, KUSM builds the local clientele by encouraging them to bring, not send, out-of-town relatives and friends (Ferrell, 2009). KUSM's 'learning history' visitors can be those who want to learn about salt mining in Hutchinson as well as the industries it spawned (i.e. underground storage). KUSM's 'recreational experience' visitors are there for the unique experience of being 650 feet below the prairie and travelling through mined-out caverns, at times in the dark.

Word of mouth from a friend or family member has been most effective in bringing visitors to KUSM, as many of the 'learning history' and 'recreational experience' tourists are not from Hutchinson but rather come from KUSM's regional markets (i.e. Wichita, Oklahoma City); other states outside the region, including most notably California, Indiana, Illinois, Texas, Oklahoma, and Florida; and foreign countries such as Australia and Denmark (Ferrell, 2009). However, additional marketing efforts are needed to ensure more people know about KUSM. As a non-profit organization, KUSM has to be as lean as possible with expenses to ensure it survives hard times and is around in the future, but it also needs to make sure it is better known. KUSM has utilized billboards, television ads, and the web. Additionally, it was featured in Kansas tourism commercials which showcased nine communities that were aired in neighbouring states (Clarkin, 2008b). Yet as *The Hutchinson News* notes, 'The Kansas Underground Salt Museum likely remains unknown to most of Kansas, not to mention out-of-staters, who do not know how spectacular an experience it is to go 650 feet underground. It is a first-rate tourist attraction with exhibits and the experiences it offers that simply needs to get known.' When Hutchinson hosted its annual high-volume special events, the Kansas State Fair and the National Junior College Athletic Association Tournament, the KUSM had record-breaking attendance (Critical Point, 2008). Those not in Hutchinson however may not know about the KUSM. In the critical early stages of KUSM's development and operation, Hutchinson's initial STAR bond investment and sales tax support helped make KUSM viable. Jansen-Verbeke (1999) notes that a top-down support policy of incentives created by public authorities that are willing to coordinate and support tourism initiatives in a region is needed for appreciation of industrial heritage sites. While a fundamental difference exists between a cultural asset and a cultural tourism attraction (McKercher, 2001), KUSM is a cultural heritage asset that also strongly appeals as a tourism product as the only underground museum in a working salt mine in the Western hemisphere. Like the Cosmosphere, Hutchinson's other major attraction which has been nurtured and supported by the city and private philanthropy, KUSM's marketing and development efforts also need to be supported to attract visitors to Salt City and its new, unique heritage site 650 feet below the prairie.

References

Baker, M.L. and Smith, J. (2004) *The Hutchinson Spirit.* Hutchinson, KS: Reno County Historical Society, Inc.

City of Hutchinson, Kansas [Souvenir of Hutchinson, Kansas] (1893) Hutchinson, KS: Ewing Stumm.

Clarkin, M. (2007a) 'Cosmosphere, Salt Museum to Share Grant', *The Hutchinson News*. Available online at: http//hsn.live.mediaspanonline.com/Newsbriefs/grant2007_12_18T22_19_07 (accessed 22 December 2007).

Clarkin, M. (2007b) 'County Nixes Museum's Fund Request; Officials Say Contribution Would Strain Taxpayers', *The Hutchinson News*. Available online at: http://hsn.live.mediaspanonline.com/Localregional/county2007_10_17T20_57_31 (accessed 25 October 2007).

Clarkin, M. (2008a) 'Attraction Aims to be Wonder Among Wonders', *The Hutchinson News*. Available online at: http://www.hutchnews.com/Todaystop/saltmuseumwonders (accessed 11 April 2008).

Clarkin, M. (2008b) 'A Boost Down Under: Recently, the Facility has been Exceeding Revenue Forecasts by $1,000 to $5,000 each Week', *The Hutchinson News*. Available online at: http://www.hutchnews.com/Todaystop/salmus (accessed 17 July 2008).

Clarkin, M. (2008c) 'Museum Taps a Vein of Tourists; Visits thus Far in 2008 Top Projections, Spark Hope Salt Museum will Break Even', *The Hutchinson News*. Available online at: http://www.hutchnews.com/Todaystop/salt2008-02-23T21-35-35# (accessed 1 March 2008).

Cole, D. (2004) 'Exploring the Sustainability of Mining Heritage Tourism', *Journal of Sustainable Tourism*, 20(6): 480–94.

'Critical Point: As Salt Museum Makes Strides, its Exposure Needs to Increase' (2008) *The Hutchinson News*. Available online at: http://www.hutchnews.com/Editorials/cri (accessed 27 April 2008).

Elliott, C. (1995) 'Salt Mines and Their Influence on Kansas', *Legacy: The Journal of the Reno County Historical Society*, 7(3): 20, 27, 28.

Federal Writers' Project of the Works Projects Administration (1984) *The WPA Guide to 1930s Kansas* (compiled and written by the Federal Writers' Project of the Works Projects Administration for the state of Kansas with a new introduction by James R. Shortridge). Lawrence, KS: University Press of Kansas.

Fermata Inc. (2006) *A Recreation Plan for the State Parks and State Forests in the Pennsylvania Wilds – Executive Summary*. Available online at: http://www.dcnr.state.pa.us/info/pawilds/recplan.aspx (accessed May 2008).

Ferrell, G. (2009) Interview, 9 July.

Green, J. (2007) 'Salt Museum Falling into Financial Hole; Officials Plan to Appeal to City for Financial Rescue', *The Hutchinson News*. Available online at: http://hsn.live.mediaspanonline.com/Todaystop/museum2007_09_18T22_09_39 (accessed 19 September 2007).

Green, J. (2008) 'Demand Skyrockets as Cities Stockpile after '07 Ice storm', *The Hutchinson News*. Available online at: http://www.hutchnews.com/Todaystop/salty (accessed 11 October 2008).

Hall, E. (2008) 'Sales Tax's Fate Looms for Officials; As Election Nears, City, Museum Leaders Planning Informational Campaign', *The Hutchinson News*. Available online at: http://www.hutchnews.com/Elections/taxes2008-08-10T20-55-59 (accessed 11 August 2008).

'Heavy Run of Mine Visitors' (1959) *The Hutchinson News:* 6 August.

Hutchinson, The Salt City: In the Heart of the Great Kansas Wheat Belt. (1906) Hutchinson, KS: Commercial Club.

Hutchinson, The Salt City: In the Heart of the Great Kansas Wheat Belt. (1910) Hutchinson, KS: Hutchinson Printing Co.

Jansen-Verbeke, M. (1999) 'Industrial Heritage: A Nexus for Sustainable Tourism Development', *Tourism Geographies*, 1(1): 70–85.

Johnson, N. (2007) '650 Feet Down', *Legacy: The Journal of the Reno County Historical Society*, 19(2): 4, 5, 8.

Kansas Underground Salt Museum (2008). *Exhibits*. Available online at: http://www.underground museum.org/exhibits/fluid_inclusion.shtml (accessed 3 September 2009).

Kilbourn, C. (2008) 'Salt Museum Looks to Gain a Sibling; Officials Hope to Share Exhibits, Information with Swiss Sister City, Institution', *The Hutchinson News*. Available online at: http://www.hutchnews.com/Todaystop/swiss (accessed 25 August 2008).

Kilbourn, C. (2009) 'An Idea with Depth: Firm Celebrates Soldier's Notion 50 years ago to Turn Salt Mine into Storage Fields', *The Hutchinson News*. Available online at: http://www.hutchnews.com/Localregional/ideadepth (accessed 11 February 2009).

Kurlansky, M. (2003) *Salt: A World History*. New York: Penguin Books.

'KUSM ticket prices to drop' (2008) *Legacy: The Journal of the Reno County Historical Society*, 20(3): 19.

May, E.L. (1961) 'Salt Mine Tours Halted', *The Hutchinson News:* 25 May.

McKercher, B. (2001) 'Attitudes to a Non-viable Community-owned Heritage Tourist Attraction', *Journal of Sustainable Tourism*, 9(1): 29–43.

Poria, Y., Butler, R. and Airey, D. (2004) 'Links between Tourists, Heritage, and Reasons for Visiting Heritage Sites', *Journal of Travel Research*, 43: 19–28.

The Reno County Historical Society, Inc. (2006) *Underground Bound! Where Excitement Rises as you Descent into the Only Museum of its Kind in the Western Hemisphere! Kansas Underground Salt Museum.* Hutchinson, KS: Reno County Historical Society, Inc.

Robinson, M. (1999) 'Tourism Development in Deindustrializing Centres of the UK: Change, Culture and Conflict', in M. Robinson and P. Boniface (eds) *Tourism and Cultural Conflicts*, New York: CABI Publishing, 129–59.

Rudd, M.A. and Davis, J.A. (1998) 'Industrial Heritage Tourism at the Bingham Canyon Copper Mine', *Journal of Travel Research*, 36(3): 85–89.

The Salt Institute (2009) *Major Salt Deposits and Dry Salt Production Sites in North America.* Map.

Saurí-Pujol, D. and Llurdés i Coit J.C. (1995) 'Embellishing Nature: The Case of the Salt Mountain Project of Cardona, Catalonia, Spain,' *Geoforum*, 26(1): 35–48.

Schmitt, L. (2008) 'Vote for Sales Tax . . . and Help KUSM', *Legacy: The Journal of the Reno County Historical Society*, 20(3): 13.

Smith, J. (2000) 'Daring Mighty Things: The Underground Museum Project', *Legacy: The Journal of the Reno County Historical Society*, 12(2): 24, 27, 28.

Smith, J. (2003) 'Precious Cargo: The Story of Salt Transportation', *Legacy: The Journal of the Reno County Historical Society*, 15(1): 5–12.

'Tourist' shaft proposed for mine' (1958) *The Hutchinson News:* 11 April.

Ulrich, B.C. (2008) *The Carey Salt Mine.* Chicago: Arcadia Publishing.

'Underground Update' (2004) *Legacy: The Journal of the Reno County Historical Society*, 16(2): 20.

Wanhill, S. (2000) 'Mines – A Tourist Attraction: Coal Mining in Industrial South Wales', *Journal of Travel Research*, 39(1): 60–69.

Zerger, C. R. (1945) 'Historical Notes on Kansas Salt', Unpublished M.A. thesis, University of Kansas.

9 Visitor interpretation of the environmental impacts of the gold rushes at the Castlemaine Diggings National Heritage Park, Australia

Warwick Frost

Introduction

Mining heritage is often preserved in national parks, managed and interpreted by national park agencies and enjoyed by tourists as part of the national park experience. However, as national parks are generally intended to protect natural values, the inclusion of cultural heritage (such as mining landscapes) is incidental and often problematic. Indeed there may even be a tendency for land managers from a natural heritage background to ignore or downplay cultural heritage in national parks (Frost and Hall, 2009; Griffiths, 1996). An important exception is the Castlemaine Diggings National Heritage Park, which protects one of the main sites of the Australian Gold Rushes of the 1850s. Established in 2002, this national park is 7,500 hectares (18,000 acres) in size and is located 120 kilometres north-west of Melbourne. It was the first national park in Australia to be specifically designated as 'heritage'. Whereas all national parks up to this time had been declared for their natural values, this was established primarily for its historical or cultural landscape. As promoted in its visitor guide, The Castlemaine Diggings National Heritage Park is a rare place where you can still see authentic traces of the great Victorian gold rushes of the 1850s, sites and relics that harbour secrets and tell stories about how life really was on the diggings (Parks Victoria, 2004).

While the Castlemaine Diggings National Heritage Park contains a large number of historic sites, it is also nearly entirely covered by Box-Ironbark eucalypt forest. On the surface, the national park presents the appearance of a virtually untouched natural landscape, yet the forest is itself an artefact of the Gold Rushes in two ways. First, Box-Ironbark forests grow on auriferous ground. As such, in the nineteenth century the government reserved large areas of it from freehold sale. Without the Gold Rushes, it would probably have been sold and cleared for grazing. Second, the structure of the forest has been highly modified by mining. In particular, nearly all the trees have been cut down a number of times and what is visible today is coppice regrowth (Figure 9.1). It is likely that before the Gold Rushes the forest was more open, with larger trees more widely spaced apart (Environment Conservation Council, 2000).

For many years, the Box-Ironbark forests were the only major vegetation type not protected in national parks. This was due to the potentially large and valuable amounts of gold they still contained and to their use for timber production. However, following a long campaign by the Victorian National Parks Association, in 2002 the Victorian Government, through its agency Parks Victoria, created a network of six Box-Ironbark national parks covering 76,500 hectares (180,000 acres). Of the six, only Castlemaine Diggings was established for its historic values, though the others also contain extensive historic mining sites.

In establishing Castlemaine Diggings, Parks Victoria consciously placed a cultural landscape within a system of national parks which had evolved primarily with natural landscapes

Figure 9.1 Coppice regrowth of eucalypt tree cut during Gold Rush, Castlemaine Diggings National Heritage Park (photo Stephen Frost).

in mind. Whereas previously cultural landscapes might be a minor part of a national park, at Castlemaine Diggings, they became for the first time – the central feature. There is also an issue in that this was the first national park in a well developed tourism destination. Branded as the Goldfields, this destination had previously been exclusively marketed in terms of cultural heritage tourism (Frost, 2003).

As Parks Victoria develops its management plan for Castlemaine Diggings, it has the opportunity to develop effective interpretation for visitors on the social and environmental history of the Gold Rushes. Most of the research on visitor interpretation of Gold Rush heritage mining sites has focussed on social history, particularly in Australia (Clark and Cahir, 2003; Evans, 1991; Frost, 2001, 2003, 2005, 2006, 2007; Garton Smith, 1999; Mayne, 2003; Ham and Weiler, 2004), but also in California (De Lyser, 1999). Indeed, this social focus is also apparent at heritage mining sites in general (Cameron and Gatewood, 2000; Edwards and Llurdés i Coit, 1996; Hewison, 1987; Wanhill, 2000). In contrast, issues relating to the environment are rarely examined, with usually no more than some incidental comments. A good example of this occurred in Edwards and Llurdés i Coit (1996). Their study commenced with the argument that the environmental scars of coal mining had given Wales a negative destination image, but this was barely discussed again in the article.

This chapter aims to examine issues relating to the development of visitor interpretation of the environmental impact of the Gold Rushes at the Castlemaine Diggings. The unique status of this national park demands that effective interpretation needs to help visitors understand the inter-relationships between its cultural and natural landscapes. However, this uniqueness causes difficulties, which are particularly magnified by the lack of similar prior developments elsewhere. The national park managers cannot simply use previously developed interpretation, their challenge is to develop new interpretation for a new type of national park.

This chapter is divided into two sections. The first focuses on the difficulties of developing interpretation in an area where there has been little historical research. In interpreting the social history of the Gold Rushes, the difficulty would be in choosing between the wide ranges of historical source material available. In particular, *visitor* interpreters would likely need to make decisions between often conflicting *historical* interpretations and approaches. In contrast, for interpreting environmental history the difficulty may be the paucity of historical research. This problem is compounded by the dual nature of the national park, which has led to differing perspectives on the balance of emphases between the natural and cultural landscapes. Taking these difficulties into account, the second section examines a range of options as to how this interpretation might be structured.

Heritage or history?

It is curious that Castlemaine Diggings was designated as a *Heritage* National Park. The alternative could have been to call it an *Historic* National Park. After all, a key section was previously designated as the Castlemaine Historical Reserve; a number of goldfields in Victoria are still called historic parks (Beechworth) or reserves (El Dorado) and in California a number of goldfields are protected as State Historic Parks (including Coloma, Columbia, Malakoff Diggings and Bodie).

What's in a name is more than an idle question. The distinction between history and heritage troubles researchers in tourism and heritage interpretation. Hewison argued bluntly that 'heritage is not history'. Rather, heritage for Hewison was a 'distortion of the past' which promoted 'fantasies of a world that never was' (1987: 10). Lowenthal noted that history was often seen as factual and 'real' and therefore unchanging. In contrast, he characterised heritage as biased and 'bad history', at its worst 'a partisan perversion' (1998: 102–3). Timothy and Boyd described history as 'the recording of the past as accurately as possible', whereas heritage was 'often the re-creation of the selective past' (2003: 4,237).

However, historians have a different point of view as to their role. Rather than 'recording the past as accurately as possible', their primary interest is in *interpreting* the past, developing theories to explain and analyse historical processes and causes. As Davison, in rejecting the distinctions made by Hewison and Lowenthal, argued:

> Even before the [history] discipline was exposed to the influence of postmodernism and poststructuralism, historians had largely abandoned the pretence of objectivity. Any history, they would cheerfully admit, was written from a point of view and, while they might eschew deliberate fabrication and distortion, the past they portrayed reflected as much of themselves as their subjects.
>
> (2000: 120)

These issues are further complicated by consideration of the visitors. Moscardo (1996) argued that heritage attractions needed to encourage 'mindfulness' amongst visitors. Mindful visitors, she argued, were 'active, interested, questioning and capable of reassessing the way they viewed the world' (Moscardo, 1996: 382). To achieve this, effective interpretation needed to be 'multisensory . . . personally relevant, vivid or affectively charged . . . unexpected or surprising; [and] questions are used to create conflict or ambiguity' (Moscardo, 1996: 384).

Rather than interpretation being constructed or produced by national park agencies for consumption by tourists, it is often a 'co-construction' between visitors and interpreters:

> The resulting narratives are contested by tourists and become subject to negotiation. During the performance of the story, tourists are not passive readers of the text. Rather, they are actively engaged by using their prior background, negotiating, filling gaps, and imagining. Hence, service providers do not simply teach history and tourists do not only learn about the past.
>
> (Chronis, 2005: 400)

The nature of the visitor experience and understanding is affected by the physical heritage on display. This may range from ruined 'ghost towns' managed under policies of 'arrested decay', for examples Bodie and Hill End (De Lyser, 1999; Mayne, 2003), to fully reconstructed living-history museums such as Sovereign Hill (Evans, 1991; Frost 2005). However, in many cases all that remains may be a pile of rocks and a hole in the ground, providing little for the imagination of the visitor (Figures 9.2 & 9.3).

These differing views of heritage and history raise questions for those developing visitor interpretations. Do they limit their interpretation to just objective facts? How they deal with contrasting historians' interpretation? Should they choose just one, or seek to show multiple and contested perspectives? How do they encourage visitors to be mindful and co-construct the interpretation? The volume and variety of research into the social history of the Gold

Figure 9.2 Bodie State Historic Park, California. A policy of arrested decay keeps the building standing and allows tourists to imagine what the town was once like (photo Stephen Frost).

Figure 9.3 The remains of an 1850s puddling machine are barely discernible through the regrowth at Castlemaine Diggings (photo Stephen Frost).

Rushes greatly complicates visitor interpretation, requiring multiple approaches and perspectives (Frost, 2005, 2007). However, for the environmental impact, the paucity of historical research creates a different problem.

The environmental impact of the Gold Rushes: a historiography

In 1976, J.M. Powell wrote *Environmental Management in Australia, 1788–1914*, the first broad-based environmental history of Australia. As a pioneer in the field he took the opportunity to predict future directions for research. In particular he argued:

> We urgently require a thorough evaluation of the profits and losses to the Australian environment which were registered in the major years of pastoral expansion [1830s–1840s] and in the succeeding confusion of mining activity during the 'fifties [1850s] . . . Hasten the day when biogeographers, geomorphologists and historical geographers combine to interpret and communicate the ecological impact of both the 'Squatting Age' and the 'Golden Age'.
>
> (Powell, 1976: 32)

Since he wrote those words, there has been a great deal of research into the environmental changes which resulted from pastoralism (Barr and Cary, 1992). In contrast, there has been far less research into the impact of the Gold Rushes. Powell devoted just four pages to the

Gold Rushes (1976: 37–41). Later general environmental histories of Australia allocated only a similar amount and usually covered similar material (Barr and Cary, 1992: 53–55; Bolton, 1992: 69; Lines, 1991; Pyne, 1991: 195–98). Some otherwise excellent environmental histories even ignored the Gold Rushes; for example, Bonyhady (2000) examined nineteenth century environmental issues through various paintings and artists, but neglected the Gold Rushes despite an extensive range of contemporary paintings of the goldfields. In the quarter of a century after Powell wrote, the most detailed exploration of the environmental impacts of Gold was to be in Weston Bate's (1978) study of Ballarat. Though particularly valuable for its focus on what the environment was like immediately prior to the discovery of gold, this work was confined to one particular locality.

The celebration of the 150th Anniversary of the Victorian Gold Rushes in 2001 finally stimulated greater research. It resulted in specialised articles by the historian Don Garden (2001) and archaeologist Barry McGowan (2001) on the environmental impacts. However, both were limited by trying to cover a large geographical range and timespan within the confines of articles in celebratory volumes. In addition, the more general histories published around the 150th Anniversary were particularly adept at weaving environmental issues into their narratives (Annear, 1999; Hocking, 2000).

Nonetheless, despite these advances, the literature on the environmental effects of the Gold Rushes remains limited. The recent works are primarily descriptive case studies from which some generalisations are drawn. The problem, in essence, is that the sheer scale of the Gold Rushes works against more comprehensive analysis. The Rushes occurred in all Australian states, in a large number of places and involved hundreds of thousands of people. They spanned over fifty years, during which new Rushes were continually occurring. The environments they affected included mountains and plains, rainforests and eucalypt woodlands, arid grasslands and alpine heath. Gold mining techniques varied and can be divided into a number of types based on where the gold was located and how it was extracted. The main types include fossicking, surface (or shallow) alluvial, deep alluvial, deep reef, sluicing and dredging. Some goldfields were dominated by one type of mining, at others a range could occur in close proximity and at others a number could be tried at the same site over the years. Finally, the evidence of environmental change is enormous, ranging from numerous detailed contemporary books, letters, reminiscences, sketches, paintings and photographs, to still existing landscapes crammed with diggings, ruins, pits, shafts, water channels and mullock heaps.

What is needed is some way to make sense of this overwhelming variety and complexity. Unfortunately, as historians have not yet achieved this, developing visitor interpretation at Castlemaine Diggings is problematic. If the environmental impact is to be featured in this interpretation, then both basic historical research and the development of a more theoretical explanatory framework will have to occur at the same time as the visitor interpretation is created.

Visitor interpretation at Castlemaine Diggings

As its management plan is still under development, little visitor interpretation has been specifically created for Castlemaine Diggings. At this stage this consists of a basic visitor guide comprising two A4 pages of descriptive text and a double page map (Parks Victoria, 2004). A visitor centre is planned; this will not be within the national park, but in the 1850s market building in the town of Castlemaine. However, there is interpretation which dates from before the establishment of the national park. This includes some interpretive panels, a 95-page guidebook to the goldfields at Castlemaine and nearby Maldon (Mount Alexander

Diggings Committee, 1999), brochures for existing walking trails (Mount Alexander Shire Walks and Trails Advisory Board, 2000) and general guides to the Box-Ironbark forests (Calder and Calder, 2002).

As this interpretation is updated, there is an opportunity to systematically develop interpretation focussing on the environmental impacts of the Gold Rushes. Given the lack of definitive historical research, how might that visitor interpretation be structured? Three possible options are considered below.

Option 1: incidental interpretation

The current situation is that visitor interpretation is based on trails, resulting in disjointed topics which arise due to the order of features along these trails. There has been no attempt at thematic interpretation like that utilised at nearby Sovereign Hill (Ham and Weiler, 2004). References to environmental impacts are scattered throughout the interpretation, though their placement is incidental rather than systematic or planned. Interpretation of environmental impacts only occurs because there is some feature which needs explanation. Often such references are sandwiched between technical or social explanations, with little linkage. For example, the Eureka Reef Walk consists of 20 interpretative stops; but only one specifically deals with tree cutting and coppicing modifying the forest and two mention silt from ore processing plants (Mount Alexander Diggings Committee, 1999: 25–36). In contrast, interpretation for the Garfield Wheel and the trail from it to Expedition Pass, does not discuss any environmental impacts, even though there is abundant physical evidence of mines and mullock heaps (Figure 9.4); (Mount Alexander Diggings Committee, 1999: 37–38; Mount Alexander Shire Walks and Trails Advisory Board, 2000). It is possible that future interpretation will duplicate such patterns, relegating environmental impacts to a minor place.

Option 2: themed by mining type or time periods

A conventional approach might be to order interpretation by time periods or mining types (and this could equally apply to both social and environmental perspectives). Some sites lend themselves to such an approach. Forest Creek, which currently has no interpretation, could be presented in terms of 1850s alluvial mining, especially as it is the location of a number of famous 1850s photographs (Fauchery and Daintree, 1983; Hocking, 2000: 211). One possibility would be to reproduce these photographs on interpretive boards at or near their original locations. Nearby, the Garfield Wheel operated from 1887 to 1914 and so represents a later period of more industrialised mining.

However, these are exceptions. The majority of sites encompass a number of time periods. At the Forest Creek Gold Mine, the Eureka Reef and the Wattle Gully Mine, mining extended from the 1850s until after World War Two – furthermore the Eureka Reef contains Aboriginal rock wells from before European settlement (Mount Alexander Diggings Committee, 1999: 23–36 & 41–42). At these sites, the choice would be between focussing on one time period and trying to represent environmental impacts from a number of time periods. A similar problem arises in trying to link environmental impacts to types of mining. At some sites, this is possible as they represent one type of mining. However, many of the sites contain multiple types of mining. Common combinations are alluvial surface, underground and sluicing.

Nonetheless, it is recognised that this may be the most likely approach. These are complex sites and existing interpretation focuses on explaining the different types of mining and the

Figure 9.4 Uninterpreted site on trail to Expedition Pass. Remains are of a Cornish horizontal chimney (photo Stephen Frost).

time periods in which they operated. This framework could be extended to cover environmental impacts. The main drawback of this could be a tendency to pitch the interpretation at a predominantly technical level. This could lead, for example, to explanations that this type of mining at a certain time led to this type of mullock heap or silt.

Option 3: themed by type of impact

Another possibility is to consider a framework of activities by which the Gold Diggers had an impact on their environment (Table 9.1). The activities considered here are listed in the rough order as undertaken by miners. They cover a wider range of effects than just the act of mining for gold and extend over a far wider geographical area than just mine sites.

Focusing on activities may be a way of encouraging 'mindfulness' amongst visitors. A litany of mining techniques and machinery might fascinate a few, but be too technical for most visitors. In shifting the emphasis to the broad range of activities undertaken by miners, there is greater scope for making more personal and relevant connections. The great advantage of this option is that it focuses on people, exploring how they behaved, what they affected and how they thought. The spotlight is on people rather than technology. Such an approach is already widely utilised in interpreting the social history of the Gold Rushes. Sovereign Hill, for example, works well because it represents an entire town and community, not just the miners or the machinery. In addition, this people-centered approach to environmental impacts can be used at any mining site throughout the world.

Table 9.1 Miners' activities and their impacts on the environment

Activity	*Impact*	*Visible evidence*
Travelling	Travelling to and from Melbourne, created roads and in the 1860s a railway.	Roads and tracks. Towns which were established as stopping places.
Digging	Extracting ore from ground. Either surface or underground.	Holes, open-cuts, shafts etc.
Processing	Puddling, sluicing and crushing. Huge demand for water.	Machinery and buildings. Processed ore in mullock heaps and silt dams. Water races.
Timber-cutting	Diggers burnt trees for warmth and cooking. Lit fires to aid prospecting. Later, more industrial mining created enormous demand for firewood.	Extensive vegetation modification. Nearly all trees coppiced. Large trees all gone. Vegetation denser. Conversely, forests protected to provide firewood.
Eating	Need for food stimulated farming. Wildlife shooting.	Farming-mining nexus. Good example is Chinese market gardening.
Waste	Human waste, including sewerage and rubbish.	Bottle dumps. Cemeteries (high mortality rate).
Building	Huts, houses and mine buildings towns.	Much abandoned and in ruins. Existing towns.
Contemplating	Did miners think about their impact?	Archival material: newspapers, paintings, letters, books etc.

Conclusion

The landscape at Castlemaine Diggings is neither natural nor cultural. Rather, it is a combination of both. It contains many historic buildings, sites and ruins, but they are situated within a natural environment of extensive Box-Ironbark forest. In turn, this forest is also an historic artifact, conserved and highly modified by its gold mining history. This double nature of the landscape is recognised in its designation as Australia's first National Heritage Park.

This duality creates operational challenges. Whereas other historic tourist attractions, such as Port Arthur and Sovereign Hill, are managed by specialised organisations, Castlemaine Diggings is the responsibility of Parks Victoria. A major land manager in Victoria, Parks Victoria is primarily concerned with natural environments. It does have properties with secondary cultural landscapes, but Castlemaine Diggings is its first national park with the main emphasis on history.

The complex inter-relationship between environment and history at Castlemaine Diggings requires effective interpretation for visitors. However, encouraging an understanding of the environmental impacts of the Gold Rushes should not be just an addition to more conventional approaches emphasising social and technological history. Instead, the link between gold and the natural environment is central to the national park and in turn, the tourist experience. To fully understand and appreciate the national park, visitors need to be told a new story of the environmental impact of the Gold Rushes.

References

Annear, R. (1999) *Nothing but Gold: the Diggers of 1852*. Melbourne: Text.

Barr, N. and Cary, J. (1992) *Greening a Brown Land: The Australian Search for Sustainable Land Use*. Melbourne: MacMillan.

Bate, W. (1978) *Lucky City: The First Generation at Ballarat 1851–1901*. Melbourne: Melbourne University Press.

Bolton, G. (1992) *Spoils and Spoilers: A History of Australians Shaping Their Environment*. Sydney: Allen and Unwin, 1st Publication, 1981.

Bonyhady, T. (2000) *The Colonial Earth*. Melbourne: Miegunyah.

Calder, M. and Calder, J. (2002) *Victoria's Box-Ironbark Country – A Field Guide*. Melbourne: Victorian National Parks Association, 2 edn, 1st Publication, 1994.

Cameron, C.M. and Gatewood, J.B. (2000) 'Excursions into the Un-Remembered Past: What People Want from Visits to Historical Sites', *The Public Historian*, 22(3): 107–27.

Chronis, A. (2005) 'Constructing Heritage at the Gettysburg Storyscape', *Annals of Tourism Research*, 32(2): 386–406.

Clark, I.D. and Cahir, D.A. (2003) 'Aboriginal People, Gold and Tourism: The Benefits of Inclusiveness for Goldfields Tourism in Regional Victoria', *Tourism, Culture & Communication*, 4, 123–36.

Davison, G. (2000) *The Use and Abuse of Australian History*. Sydney: Allen & Unwin.

De Lyser, D. (1999) 'Authenticity on the Ground: Engaging the Past in a California Ghost Town', *Annals of the Association of American Geographers*, 89(4): 602–32.

Edwards, J.A. and Llurdés i Coit, J.C. (1996) Mines and Quarries: Industrial Heritage Tourism. *Annals of Tourism Research*, 23(2): 341–63.

Environment Conservation Council (2000) *Box-Ironbark Forests and Woodland Investigation: Draft Report for Public Comment*. Melbourne: Environment Conservation Council.

Evans, M. (1991) Historical Interpretation at Sovereign Hill. *Australian Historical Studies*, 24(96): 142–52.

Fauchery, A. and Daintree, R. (1983) *Sun Pictures of Victoria: the Fauchery-Daintree Collection 1858*. Melbourne: Currey O'Neil Ross.

Frost, W. (2001) Golden Anniversaries: Festival Tourism and the 150th Anniversary of the Gold Rushes in California and Victoria. *Pacific Tourism Review*, 5(3/4): 149–58.

Frost, W. (2003) 'A Pile of Rocks and a Hole in the Ground: Heritage Tourism and the Interpretation of the Gold Rushes at the Mount Alexander Diggings', in R. Black and B. Weiler (eds) *Interpreting the Land Down Under: Australian Heritage Interpretation and Tour Guiding*, pp. 204–18. Golden USA: Fulcrum.

Frost, W. (2005) 'Making an Edgier Interpretation of the Gold Rushes: Contrasting Perspectives from Australia and New Zealand', *International Journal of Heritage Studies*, 11(3): 235–50.

Frost, W. (2006) 'Heritage Tourism at Eureka Stockade in an International Context', in A. Mayne (ed) *Eureka 1854–2004: Reappraising an Australian Legend*, pp. 77–89. Fremantle: Fremantle Arts Centre Press.

Frost, W. (2007) 'Refighting the Eureka Stockade: Managing a Dissonant Battlefield', in C. Ryan (ed) *Battlefield Tourism: History, Place and Interpretation*, pp. 187–94. Oxford: Elsevier.

Frost, W. and Hall, C.M. (2009) 'National Parks, National Identity and Tourism', in W. Frost and C.M. Hall (eds) *Tourism and National Parks: International Perspectives on Development, Histories and Change*, pp. 63–77. London and New York: Routledge.

Garden, D. (2001) 'Catalyst or Cataclysm? Gold Mining and the Environment', *Victorian Historical Journal*, 72[1/2 (special edition *Victoria 150 years of Gold*)]: 28–44.

Garton Smith, J. (1999) 'Learning from Popular Culture: Interpretation, Visitors and Critique', *International Journal of Heritage Studies*, 5(¾): 135–48.

Griffiths, T. (1996) *Hunters and Collectors: The Antiquarian Imagination in Australia*. Cambridge: Cambridge University Press.

Ham, S.H. and Weiler, B. (2004) 'Diffusion and Adoption of Thematic Interpretation at an Interpretative Historic Site', *Annals of Leisure Research*, 7(1): 1–18.

Hewison, R. (1987) *The Heritage Industry: Britain in a Climate of Decline*. London: Methuen.

Hocking, G. (2000) *To the Diggings! A Celebration of the 150th Anniversary of the Discovery of Gold in Australia 1851–2001*. Melbourne: Lothian.

Lines, W.J. (1991) *Taming the Great South Land: A History of the Conquest of Nature in Australia*. Sydney: Allen and Unwin.

Lowenthal, D. (1998) *The Heritage Crusade and the Spoils of History*. Cambridge: Cambridge University Press.

Mayne, A. (2003) *Hill End: An Historic Australian Goldfields Landscape*. Melbourne: Melbourne University Press.

McGowan, B. (2001) 'Mullock Heaps and Tailing Mounds: Environmental Effects of Alluvial Goldmining', in I. McCalman *et al.* (eds) *Gold: Forgotten Histories and Lost Objects of Australia*, pp. 85–100. Cambridge: Cambridge University Press.

Moscardo, G. (1996) 'Mindful Visitors: Heritage and Tourism', *Annals of Tourism Research*, 23(2): 376–97.

Mount Alexander Diggings Committee (1999) *Discovering the Mount Alexander Diggings*. Castlemaine: Mount Alexander Diggings Committee.

Mount Alexander Shire Walks and Trails Advisory Board (2000) *A Walk from the Garfield Wheel to Expedition Pass*. Brochure.

Parks Victoria (2004) *Castlemaine Diggings National Heritage Park: Visitor Guide*. Brochure.

Powell, J.M. (1976) *Environmental Management in Australia, 1788–1914. Guardians, Improvers and Profit: An Introductory Survey*. Melbourne: Oxford University Press.

Pyne, S.J. (1991) *Burning Bush: A Fire History of Australia*. Sydney: Allen and Unwin.

Timothy, D.J.and Boyd, S.W. (2003) *Heritage Tourism*. Harlow: Prentice Hall.

Wanhill, S. (2000) Mines – A Tourist Attraction: Coal Mining in Industrial South Wales. *Journal of Travel Research*, 39(1): 60–69.

10 Sustaining the benefits of heritage mining for site, city and region

Exploring the success of Sovereign Hill outdoor museum

Mary Hollick

Introduction

Sovereign Hill outdoor heritage museum has received numerous national and state tourism awards for its innovative presentation of the gold rush period and as a best practice attraction based on mining. It has grown to occupy, quite intensively, a site of 64 acres. It attracts an estimated 560,000 visitors annually. International tourists account for approximately 28 per cent of visitors and intra state arrivals represent 60 per cent of all visits (Sovereign Hill, 2007). Its host city of Ballarat, as the place of Australia's only civil uprising (the Eureka rebellion), has an enviable reputation as one of Australia's most richly endowed heritage cities. Sovereign Hill tells the story of Ballarat's mid-nineteenth century gold rush boom.

In over 35 years of operation Sovereign Hill has developed hundreds of new product innovations – refurbishments and extensions, major new constructions and interpretative enhancements to sustain its operation and visitation. The following case explores the masterful management of its most successful product rejuvenation, the *son et lumière*, *Blood on the Southern Cross* (BOSC) based on the events leading up to the Eureka rebellion. Its development and management reflects the approach which has been taken to manage the organisation as a whole and therefore provides a detailed insight into management, usually lacking in a brief case analysis. BOSC has been instrumental to the sustainability of the organisation by lifting visitation, generating substantial income and contributing to wider regional benefits generated by Sovereign Hill.

Success factors in visitor attractions

Considerable research has recognised that the factors that characterise successful attractions, including those based on mining themes, are complex, interrelated and finely balanced. Swarbrooke (2001) suggests four elements are critical to success: the organisation and its resources (i.e. prior experience, human resources, finances); the product (i.e. based on a unique idea, accessible location in a strong catchment area, variety of onsite attractions, visitor facilities, value for money); the market (i.e. targeting growth markets, has positive word of mouth recommendation) and the management of the attraction (i.e. experienced, professional managers, investment in professional marketing, quick response to changes in the business environment, monitor performance regularly and plan for the future).

Scholars have acknowledged the complex of factors in heritage attraction success specifically (Swarbrooke 2002; Benckendorff and Pearce 2003; Bramley 2001, 2003; Bramley and Curren 1997; Fyall *et al.*, 2003). Bramley (1997, 2001) has observed that notions of success and failure are complex; that 'success', especially for public sector enterprises such as museums, can be qualified by social good objectives, i.e. if the organisation's objectives are satisfied. He includes the need for long-term vision; determination in business planning; the value of developing complementary facilities, such as tourist accommodation, to increase yield in addition to a number of factors pertinent to cultural attractions in particular; the importance of historical accuracy in heritage product development; limiting the product focus to the 'period of significance' and the need for well trained interpretative staff.

From this research, a number of critical success factors for sustainable attraction success emerge:

Ownership structure

Swarbrooke (2001) has considered the nature of ownership (i.e. public, private, not for profit) and its influence on the organisation's priorities and values. Sovereign Hill is structured as a Company Limited by Guarantee. This structure combines private sector operating capacity, enabling rapid response to changing external conditions, with a 'limit by guarantee' which ensures profits are ploughed back into the venture rather than dispersed to shareholders, for example, and so provides a sustainable funding base for regeneration. The Limit by Guarantee structure also enables the organisation to access philanthropic and government grant monies in addition to incentives funding normally unavailable to the private sector.

Location

This refers to both proximity to a major source market and relevance to growth markets.

Community orientation

The local people should perceive that the attraction adds value to the local community.

Determination and professionalism in management

This includes the capacity for long-term vision, professional skills and taking a strategic approach to all management functions.

Developing complementary facilities such as tourist accommodation to increase yield

The necessity for clustering of facilities serves important functions in terms of increasing visitor length of stay as well as providing vital additional income streams for organisations.

Theme capable of reinvention and extension

To maintain appeal to the changing needs of the market and as a source of refreshment. Additional factors, appropriate to heritage attractions, including mining attractions.

Historical accuracy in heritage product development

The setting is important in adding value to the visitor experience and provides an important context for learning (Hollick, 2008).

Limiting the product focus to the 'period of significance'

Coherence of thematic presentation is needed for contemporary audiences, unfamiliar with the time period and or processes represented.

Well-trained interpretative staff

Informative staff support visitor learning and help guide their experience of sites.

Most of these factors have been fundamental to Sovereign Hill's long-term success both as an attraction and as a contributor to wider regional development. The following case extends our understanding of these aspects by exploring the strategic management of BOSC, to reveal the degree of complexity and level of strategic skill required to successfully represent mining heritage to a changing audience over time.

Tourism in the case study area

Ballarat is located in the Goldfields tourism product region of Victoria. It has a population of 86,000 and is located one hour west of Melbourne, the state's capital and major source market. Ballarat receives approximately 2.3 million visitors annually comprising 68 per cent domestic day-excursionists, 31 per cent domestic overnighters and 1 per cent international (Hollick and Lynch, 2007). It promotes itself as an iconic heritage and culture destination. Its gold-seeking founders left the city with a rich heritage of high Victorian architecture, sculptures, gardens and wide boulevards that inspired the nineteenth century English wanderlust, Anthony Trollop, to refer to the city as 'the Florence of the South'.

Tourism directly employs over 2,444 people (6 per cent compared to 5.4 per cent for the state) of all persons employed (TTF, 2005). The Ballarat attractions product reflects a pattern across much of Australia and beyond: a dominant attraction, surrounded by many low profile ones. The local Council supports tourism via a Tourism and Economic Development Levy, based on property valuations for all businesses within the city. The tourism component of the funds is used for destination marketing and promotion in addition to administrative and tourism activities including salaries, secretariat for the tourist board and LTA and several accredited Visitor Information Centre's (VIC).

Sovereign Hill was developed on a vast, old quartz-mining site on the eastern side of the city on Crown land. Disused for many decades, the mining relic was considered unsightly and was about to be the subject of an Apex service club clean up. On high ground overlooking the city, the site ran parallel to the original Main Road which brought the gold seekers from the south coast and whose story it would tell. From the mid 1960s a highly skilled, community-based

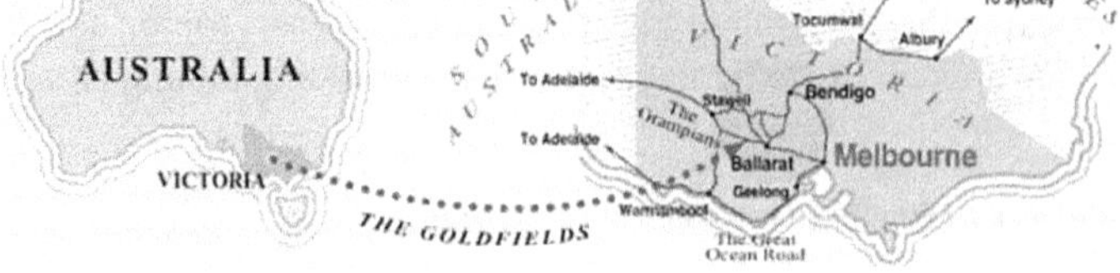

Map 10.1 Study area location, Ballarat in relation to Victoria, Australia.

Figure 10.1 Original site of Sovereign Hill (Source: Sovereign Hill Education Unit, Teacher resources).

committee, concerned that the city's rich history would be lost, formed an association. They harnessed community and government support and defined the period of representation, key themes and source of funds. In a unique agreement a peppercorn lease from local government to the association specified that the ownership of the site was ultimately vested with the people of Ballarat and has remained so.

At the time of Sovereign Hill's 1970 opening, the state of Victoria was in the midst of a severe drought and economic downturn set to continue to the end of the decade. The local economy was struggling to adjust to the restructure of Australian manufacturing occurring at this time. Ballarat had lost most of its manufacturing industry; 6,000 jobs were lost in the 1970s (Elias, 1993) and with them much of its capital base. Aside from visiting friends and relatives (VFR) tourist activity was generated by two annual community-based events which created shallow spikes in visitation but no sustained impact.

The development of Sovereign Hill had an immediately positive and substantial impact on the destination. As external shocks became the norm in tourism management, Sovereign Hill's largely recession-proof performance has been sustained over economic downturns that occurred during 1982–83 and 1990–92, a lengthy, national pilot's strike in 1989; the Asian currency crisis of 1998; the loss of a major domestic carrier, Ansett Airlines; the terrorist attacks of 11 September 2001; the threat of pandemic viruses and terror; in addition to drought and further recession post 2003.

As noted previously, Sovereign Hill is registered under the Victorian Companies Act as a non-profit, community based organisation. Strategic responsibility rests with an elected, skills based Board of 21, who serve voluntarily. A Chief Executive Officer and his management team carry out strategic planning and daily management of Sovereign Hill. Over 220 equivalent full-time staff are supported by over 275 volunteers (Johnson, 2007).

The outdoor museum component comprises a representative Goldmining Township and its *son et lumière*, BOSC. Sovereign Hill also operates its formal Gold Museum which draws over 170,000 visitors p.a. (Johnston, 2007). It has a pastoral property, Narmbool, approximately 15 kilometres south of the main site in addition to accommodation capacity on site for over 200 guests. Of Sovereign Hill's 560,000 visitors, over 470,000 are day excursionists and BOSC generates 85,000 night visitors (pers.com. 2007).

Sovereign Hill is a major economic driver for destination tourism in Ballarat and regional Victoria. Its annual turnover is $20m with an economic multiplier for the local economy

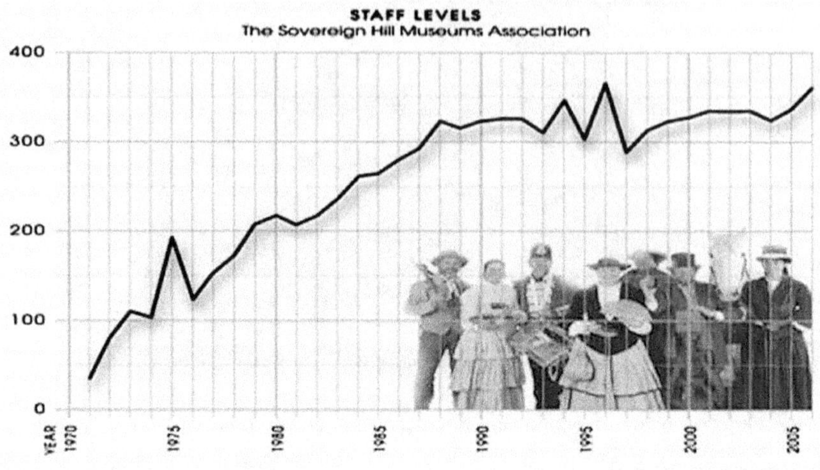

Figure 10.2 Employment 1971–2005 (Source: Johnson, J. 2007).

conservatively estimated to be \$50m p.a. As Figure 10.2 reveals, Sovereign Hill has generated strong employment growth for its region despite constant volatility affecting tourism at all levels of the economy through its history. Sovereign Hill generates a wages bill of \$10m p.a. into Ballarat's economy and a further \$10m in purchases for consumables, supplies and services (Johnson, 2007). Attendances at BOSC alone generate 76,000 bed nights with a further 100,000 bed nights estimated to be generated through visitation to Sovereign Hill alone (Johnson, 2007).

Product management

Strategic marketing is especially critical in tourism business performance, as it is the management function that provides the key linkages between customer expectations and needs and the organisation, which in turn influences all aspects of the product development and management. Rapidity of changes in consumer market tastes, trends and expectations has also reinforced the importance of strategic marketing.

One of the major functions of marketing within organisations is the management of the product life cycle (PLC). It is important for organisations to recognise when they need to 'relaunch' their business with new products and get the timing right, as a major development involves significant investment over and above the normal cost of marketing, especially for attractions where fixed costs are high.

Products can pass through several stages during their lifetime: a new product is introduced; it enjoys a growth phase as sales volumes increase, it reaches maturity and declines or is rejuvenated. At each stage Kotler suggests (Kotler, cited in Swarbrooke, 2001) tourism attraction products and their markets have different characteristics relating to sales volume, costs per customer and the nature of customers and competitors that require different strategic marketing responses at each stage of the life cycle. Offsetting decline requires adjustments in strategy and tactics (Smallwood, 1973) in relation to these and is therefore a useful means of understanding the nature of product management within organisations.

Product development of BOSC

For Sovereign Hill management there were clear commercial and historical reasons for developing a night time experience. Visitor numbers had stagnated between 1988 and 1993 and were trending down (BHPA, 2000). The cost of operating and maintaining the attraction continued to rise. The organisation had not covered its operational costs from gate takings since 1982 (BHPA, 1981–1982).

Qualitative research during 1991 (Yann *et al.*, 1991) had shown that a sound and light show could be another catalyst for bringing people back.

Sovereign Hill represented a huge, under-utilised capital asset. Improvements to the land by 1992 were capitalised at $25 million (BHPA, 1991–1992) and the site was only used intensively between 9.30 am and 5 pm. The underlying consideration, however, was to add to the interpretation of the period being represented in novel ways and on a continuing basis. The Eureka event was right in the middle of the historical period represented by Sovereign Hill, but it was being interpreted poorly as an annual daytime pantomime within Sovereign Hill and via a rundown diorama at the original site in Ballarat East. The Eureka rebellion had come to be seen as a key event in the creation of Australian nationhood and identity and therefore had strong emotional appeal.

The Eureka story had the ideal product marketing attributes to capture the public imagination: excitement, drama, passion, and violence. It was considered to be Ballarat's most contentious event. The Eureka rebellion began as a revolt of gold miners against the police and military forces over the licensing system, which operated on the goldfields. Resentment against the high license fee fed a political movement for universal suffrage and voting by ballot. To protect themselves against reprisals the miners built a stockade and under a flag

Figure 10.3 Blood on the Southern Cross (BOSC) (Source: Tourism Victoria).

featuring the Southern Cross and proclaimed a 'Republic of Victoria'. The attack from troopers and military did come on 3 December 1854 and it was all over in 15 minutes. It resulted, nonetheless, in 30 deaths and drew more power from its symbolism as the 'Eureka Rebellion' and subsequent use of the flag (which was designed during the confrontation) by many protest groups representing everything from land rights to union boycotts.

As a product, BOSC typified major developments at Sovereign Hill in general: a lengthy research process prior to construction, a need to justify itself historically and commercially and lengthy fit out and installation. After opening, the management of new products was further characterised by ongoing, extensive market research and market testing, clear organisational goals to drive decision-making and a partnership with government to fund further infrastructure development. It was understood from the outset that the product would need to be constantly improved, adding value through product extensions and packaging. Local support was vital in addition to responsiveness to the markets as they diversified. These objectives provided the direction for the product management and underpinned its success.

The 90 minute production opened in January 1993 utilising the whole Sovereign Hill site as a set and moved visitors over three precincts by vehicle transporters. Visitors were initially settled into Sovereign Hill's Orientation Centre for a visual narrative that provided contextual information. Visitors were then led by a staff member in concert with sound and light to the gold Diggings where the drama and intrigue of the events leading up to the battle were then conveyed. A key distinction from international *son et lumière* at this time was the way technology was used to trigger the power of imagination. The filth and squalor of mining conditions underground were brought to life by cinematic quality surround-sound technology. Bellowed conversations between wet and weary men across the Diggings and the galloping hooves of horses carrying the traps (police) in the hunt for gold licences triggered the 'movie in the mind' of visitors. No actors were needed. Technological interpretation was attractive because it provided historical quality control and a strong ongoing revenue stream while overcoming crushing labour costs. Initially there were four staff employed to run the shows of 110 visitors per show and as capacity was increased to 200 visitors per show only six staff were employed (Zulic, 2003).

From the Diggings area visitors are then transported to the southern end of the site to experience 'The Battle' and the events surrounding it. Visitors decamp into a 'roadside hotel' where even the most cynical 12-year-old is awed by the orchestral score as the final stage of the show opens onto a panorama of night time in 1854, Ballarat depicted as a thousand twinkling lights where the final dramatic events unfold. At the show's end visitors are transported back to the township to learn of the impacts of the rebellion. The use of one live actor at this point offered a novel twist to the show's presentation. In serving the needs of two major audiences, adults and children, the product had to combine a powerful and coherent narrative for adults who were interested in the events and the 'wow' factor for children raised on an entertainment diet of big budget Disney productions. Appropriate opportunities therefore had to be found in the narration for runaway coaches and burning hotels for children while conveying a complex storyline involving unfamiliar government officials and diggers for the adults.

BOSC product life cycle results

From the outset BOSC proved to be an instant success. For its first three months of operation from January 1993, 20,000 visitors came to see the show. Total visitation to Sovereign Hill increased to 512,443 representing an increase of 13.1 per cent on the previous year's arrivals of 453,221. The results of forward bookings from the US the following year were up 4 per cent

Table 10.1 BOSC packages sold through RACV

1991–92	$20,097
1992–93	$35,920
1993–94	$116,747
1994–95	$160,321
1995–96	$231,711

Source: Sovereign Hill marketing department archive.

and internationals then represented one in four visitors. Overall income grew by 2.6 per cent while the wider Australian economy was still in recession (BHPA, 1992–1993).

The increase in visitor numbers was directly attributed to BOSC. As Table 10.1 demonstrates, the yield from packages with accommodation was a key factor in the success of the product. Sovereign Hill received 69,000 bookings in the first full year, 18 per cent above projections (Sovereign Hill, 1994). During 1993–1994 a total of 53,000 packages were sold through all outlets representing $1.1million.

By 1995, BOSC was acknowledged as a key profit generator for Sovereign Hill and without it the organisation would have been in jeopardy. A growing proportion of the show's patrons elected to take packages incorporating overnight accommodation and daytime entry. In 1995, sold packages alone totalled $1.8 million in revenue. The growth of BOSC package sales sold by the Royal Automobile Society of Victoria was indicative of its contribution.

Show capacity was limited to 90,000 per annum and therefore the PLC for BOSC was projected to fall by 1996 and another product innovation would be urgent. Without it the product would be in decline by 1997–99. By the end of 1996, however, the forecast decline had not eventuated – 275,844 people had seen the show and it had registered its fourth consecutive year of full capacity attendance. BOSC attendances grew by 11 per cent while the wider positive impact saw Victorian resident visitation to Sovereign Hill up by 21 per cent.

The only factor slowing stronger growth was an inability to increase capacity in line with demand. Numbers per show were approximately 110 in 1993 and by 1997 capacity had been increased to an average of 150 per show. The major marketing objective for 1997 was to widen audience access through the installation of infrared transmission of the narrative in Mandarin, Cantonese, Japanese, Korean and German. In that year, over 90,000 people saw the show with a 39 per cent increase in visitors from mainland China. Visitation from other Asian countries remained at 1996 levels despite a major Asia currency crisis. By 1999, audiences of close to 100,000 a year were being drawn to BOSC and profits for the whole attraction increased. Continued energetic management and creative reinvestment extended its cycle and enabled it to continue to defy projected stagnation (Table 10.3, Appendix 1).

In its eighth year of operation, BOSC still defied decline. Table 10.2 shows that record numbers achieved three years previously were exceeded in 2000 with 92,847 visitors for the show (Sovereign Hill, 2000).

As Figure 10.4 illustrates, there were indications of maturity and possible stagnation in 2000 but numbers took off again in 2001 boosted in part by another royal tour, which featured Sovereign Hill and global 'free ink' publicity. Figure 10.4 demonstrates the extent to which BOSC revived Sovereign Hill daytime visitation, in addition to lifting overall performance. It is very evident that BOSC immediately arrests the market decline in numbers from its launch in 1993 and that overall visitation from that point mirrors the performance of BOSC.

BOSC's initial success resulted from a whole of organisation response involving operations, human resources and marketing. Positive word of mouth was dependent on the consistency

Table 10.2 BOSC attendance 1993–2000

Year 30th June	*Attendances*	*No. of Shows (approx.)*
1993 (half year)	38,429	610
1994	68,951	600
1995	77,526	650
1996	90,938	620
1997	91,850	630
1998	88,707	600
1999	88,171	925
2000	92,847	615
2001	85,000	600
2002	91,000	610

Source: Sovereign Hill internal statistics, 2004.

and reliability of its operations management: technical excellence with minimal downtime, an efficient computer telephone booking system, simple pre-payment requirements, an all weather operation, and experienced booking staff selling the show enthusiastically and including a dinner, accommodation and day entry package options (Sovereign Hill, 1996). Initial success was also attributed to the novelty factor and aggressive marketing which included strong publicity in eastern state press, TV holiday program exposure and trade shows. These attributes remain pertinent over a decade later.

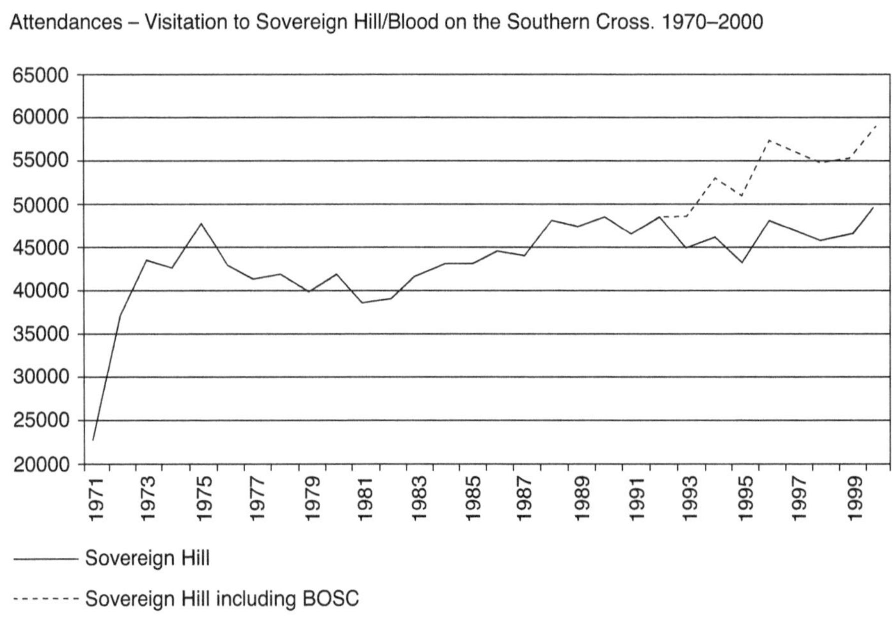

Figure 10.4 Visitation to BOSC and Sovereign Hill 1970–2000.

Source: Sovereign Hill, *2001 Victorian Tourism Awards* Submission p. 10.

The success of the product was never taken for granted and a trigger for a major reinvestment in the product in 2003 was put down to

> *... nothing but our own sense that it was time to do it. The need for renewal wasn't particularly driving it – the visitor numbers the year prior to the redevelopment were the best we'd had.*
>
> (Sullivan, 2001)

The corporate marketing plan for BOSC for 2000–2001, however, indicated that for the marketing staff at least, product revival was a key issue

> *The show is in its ninth year of production which implies that reinvention/reinvigoration is now more of an issue than ever before. The technical upgrade of the ... equipment in 1997 and 1999 has put the show on a stronger footing in maintenance terms but ... the script requires renewal.*
>
> (Sovereign Hill, 2000–2007)

The need to introduce a new range of special effects for children and further personalise the narrative for adults were set as priorities. This saw a cycle of research and market testing recommence from 2000–2003. The result was another $2 million upgrade of the existing storyline and show launch in mid 2003. The new show featured a state-of-the-art technical upgrade (Sovereign Hill, 2003) and a huge purpose-built aqua screen that allowed larger-than-life images to appear and disappear. A new, longer script strengthened the human appeal of the story, offering greater insight into the goldfields life and the minds of the central characters so a more powerful engagement with the conflicting perspectives and issues was possible.

A partnership with Government in the provision of infrastructure funding was again integral to the redevelopment. The $2 million transformation was substantially funded by a $1.2 million grant from the Victorian State Government Regional Infrastructure Development Funding Programme. BOSC achieved synergy between day-time visitation, accommodation and night-time utilisation.

Product management and the product life cycle

The following section analyses the product management of BOSC in relation to key marketing characteristics – product, customers, competitors and product marketing in order to explain its sustainability.

Product characteristics: sales and costs

Under normal product marketing conditions sales are low during the introductory phase of a new product, rising rapidly through the growth stage to peak and then decline. This was not the case with BOSC for several reasons. It came to the market with the quality product reassurance of its parent attraction, which had an established brand reputation of over 20 years and not as a stand-alone product. Sovereign Hill enjoyed 92 per cent awareness by Victorians and 72 per cent of them had experienced the product (Sovereign Hill, 1992). The usual challenge of building brand awareness was therefore not an issue. The new product could build on this and the marketing message, '*come to see something new*', was easier to sell to the domestic market than starting over as a standalone or new product. The extent of market research during the concept and product development phases also meant that the product had been pre-tested with core markets to some degree.

Costs borne in pre-opening negotiations with the wholesale trade to pre-sell the product also meant that positive word of mouth could generate sales in a quicker time period than in waiting to create these links after opening. From the outset and through the presentation of the product as part of a package, yield continued to increase ahead of numerical sales.

Private sector attractions generally need to achieve high visitor numbers early in their new product life cycles because they often have to pay back their capital costs over a relatively short period (Swarbrooke, 2001). In order to get the show up and running by 1 January 1993, \$2.5 million was invested. The scale of the production and investment represented a considerable risk. Of the total start up costs for BOSC, \$700,000 came from a state government grant, \$400,000 from sponsorships and the balance of \$1.4 million from Sovereign Hill reserves (Sovereign Hill, 1996). When looking at the issue of the internal rate of return, it becomes clear why Sovereign Hill senior management took pains to secure government support and sponsorship. Government support for projects was the equivalent of a donation and allowed the organisation to extend its capitalisation considerably. Because external funds from government and donors were directed to infrastructure while Sovereign Hill funded the non-recurrent capital investment, the organisation was able to maximise the return on its capital investment.

Donations also made the extent of Sovereign Hill's contribution a bit misleading. The securing of external funds in effect reduced the pressure of getting a full return from the \$2.5 million investment, as would be the case for a private sector operator where the funds had been sourced from investors or borrowed. Extending the capitalisation through external funding such as donations, sponsorships and bequests enabled the organisation to generate a cash profit on its capital more easily because they did not have to repay interest or loan funds. This model was applied to all major projects. The organisation also avoided the impediment of not being able to proceed with further regeneration until the loan was cleared. In that sense Sovereign Hill freed itself of a financial 'ball and chain' and enabled it to regenerate at a rate its private sector competitors in the local attractions sector could not match. Costs per customer were also minimised because Sovereign Hill also took control of the distribution system and acted as wholesaler by taking bookings directly.

Customers

Through the introductory, growth and mature life stages of BOSC there was naturally strong support from the domestic markets of Melbourne and regional Victoria and a concerted effort remained to sustain these markets.

From the outset, however, groundwork was invested in the Asian market. BOSC was not isolated from the rest of Sovereign Hill and part of the appeal for Asian wholesalers was the provision of other products and services offered by Sovereign Hill by day. Value adding through free Gold Museum entry and provision of gold samples for Chinese New Year were among a range of enticements offered. Through a concerted effort in Asia and a concentration on mainland China there was an increase of 48 per cent in the Chinese market. This was the largest international market for Sovereign Hill by 2000 with 50 per cent of Chinese visiting Victoria coming to Sovereign Hill.

The main target markets, aside from Asia, were Victoria (family and VFR), New Zealand, southern NSW, seniors and schools because of their reliability, growth potential and/or response to marketing initiatives. Tens of thousands of school children stay for up to a week for various school programs and have remained a vital anchor market.

Competitors

Competition is a constant reality and has required continued monitoring and strategic responses. Sovereign Hill was able to price its BOSC entry at a premium during the introductory stage when it was novel and there were few competitors but as competition increased it had to hold its price and provide added value to the visit. Changes in the competitive environment as the life cycle progressed however came not so much from the growth of *son et lumière* productions elsewhere but from trends in the wider leisure market and developments in technology based home entertainment in particular.

BOSC was competing against the growth of night-time leisure offerings in the State's capital, Melbourne for its major market. The added competition from night-time AFL football matches was directly challenged by a Sovereign Hill campaign in 1997 that included TV, print and promotional media inviting visitors to, *Come and see the hottest show under lights at Sovereign Hill.* While the wider economy remained flat throughout the 1990s, the synergies between BOSC and its own accommodation were saving the organisation from stagnation and certainly decline.

BOSC product marketing: objectives and strategies

BOSC marketing objectives supported those of Sovereign Hill in general and focused on yield: maintaining day visitors to Sovereign Hill lift BOSC awareness and visitation and raise the sales of accommodation and packages (Sovereign Hill, 2003).

Because of the high brand awareness in the Victorian market, objectives focused on maximising market share and profit from the outset. At first, management thought of BOSC as simply an extension of the product line – as 'another thing to do at Sovereign Hill' at night so the emphasis was not on marketing it as a stand-alone product. The marketing effort was therefore concentrated in Melbourne, regional Victoria and to a lesser degree the international trade. The initial strong response from the market then came as a shock: '*We weren't quite sure where the market was going to come from or how big it was. When it opened it was a case of, 'Hell this is a lot bigger than we thought*' (Johnson, 2003).

After 1995 and positive response from the market, staff developed a more aggressive strategy with major tour bus companies presenting BOSC as a stand-alone product. The years 1999–2000 saw another big jump in visitation in a direct response to the strength of marketing and macro environmental conditions. Despite volatile environmental conditions for most of the 1990s the year 2000 was a high point for Sovereign Hill generally. The then Deputy CEO, Jeremy Johnson (2003), reflecting years of management experience explained

> *Our market projections are always conservative to account for the unexpected. When I wrote the forecasts for that year we didn't know what was coming.*

Sustainable management

How then did this product manage to defy decline over its decade plus life cycle to date and support the performance of the attraction as a whole? The answer lies partly in the 'whole of organisation' approach to product management and investment. Table 10.3 (Appendix 1) demonstrates that Sovereign Hill introduced 28 new attractions and/or refurbishments to the day-time experience. BOSC was acknowledged as a key contributor to Sovereign Hill's cash flow so there was vested interest in shoring up its reliability.

The operational and technical reinvestment that contributed to increased reliability was fundamental to positive word of mouth. Table 10.3 reveals that the product was never left to the whim of market forces. There were major reviews of BOSC in 1994, 1996, 1997, 2000 and 2003 that resulted in significant product improvements. Market research was a constant source of investment, often triangulated and used to inform all decision-making.

In light of the organisation's dependence on the continued success of BOSC, Sovereign Hill has been careful to spread the management risk also. Most of the senior management team have taken responsibility for managing it over relatively lengthy periods. BOSC is now utilised as a training opportunity for future potential managers for Sovereign Hill (Sullivan, 2001).

The interesting question is how Sovereign Hill management knew when investment was required and what form it should take. To understand this it is necessary to appreciate the formal and informal reporting mechanisms that operated within Sovereign Hill overall and their role in bringing about a shared understanding of the need for change and readjustment. Market satisfaction operated continually as a key driver of performance and there was a range of formal and informal mechanisms in place for data gathering. The focus on visitor satisfaction moderated a management fixation with costs and economic return beyond anything else and acted as a stimulus for new investment.

A range of institutions and consultants undertake research studies constantly and reflect an awareness of the need to keep in touch with changes in the market, customer needs and expectations. Sovereign Hill either drove the studies or had an overseeing role to ensure precise and applicable feedback. A formal daily monitoring, reporting and feedback system was instituted in 1995, when Jeremy Johnson (the current CEO) was brought into the organisation to improve the operational reliability of BOSC. In conformance with evidence-based systems management, a detailed running sheet was introduced and reported on what worked well, what did not, visitor behaviours and the audience mix. This was important feedback that came in daily so that every element was monitored to assist with improvements and adjustments that were needed. The monitoring system was valued by staff because it was valued by management, '*because we know that their [the staff] intervention is a part of the visitor experience*' (Sullivan, 2001).

The culture of the organisation reflected an integrated approach to product management woven together via a series of ongoing 'strategic conversations'. These comprised ongoing discussions about the performance of established or new developments or future possible projects in the context of a five-year strategic plan, a five-year budget and capital budget.

Enculturation of local destination operators also occurred as they were brought in to supply accommodation capacity as demand grew. Sovereign Hill set the requirements for facilities and service, price, margins and reservation arrangements, lifting the quality of local product in the process. The introduction of BOSC was accompanied by a positive impact on occupancy rates in Ballarat with motels and accommodation houses showing a lift in 1995 of 13 per cent (Dermott, 1995). Later research has confirmed Sovereign Hill's assessment that of all BOSC visitors who stay overnight, two thirds stayed in Ballarat not Sovereign Hill.

Conclusion

The development of 'Blood on the Southern Cross' demonstrated how Sovereign Hill 'extended the product line' by re-mining familiar imagery and storyline to professionalise

Table 10.3 Investment in BOSC product and results 1991–2002

Year	*Marketing*	*Operations (technical)*	*HR*	*Results*
1991	Two market research studies & product planning; meet with international, interstate and local trade	Development of site plans; trial concept with late night shopping and lamplight tours		n.a.
1992	Focus group perceptions and concept testing research study; research on profile of visitor market in local accommodation houses and caravan parks; research on schools and tour group interest; establish booking system. Develop marketing plan. Forward bookings from US created	Work on synergies with other profit centres – e.g. catering and accommodation (Severe lightening strikes)	Appointment of two professional staff development officers, technical services manager and BOSC manager	n.a.
1993 Launch BOSC January	Capital city press National news; host 200 ATC trade show delegates	Extend accommodation facility (more severe lightening strikes)	Appointment of bookings manager and technical assistant	38,429 (half year)
1994	Capital city press National TV coverage in four travel programs, two new brochures, saleable souvenirs produced, three outdoor signs installed; trade show attendance ATE, ITOA; trios to US & S E Asia	Extension to accommodation (eight rooms); 84 bed barrack accommodation for schools, recreation facilities for schools/ families	Appointment of four Mandarin and Cantonese speaking guides for Asian groups; booking operators trained in telemarketing; computer programmes rewritten	68,951
1995	National/Interstate: TV travel show, commercial for night and day experience; joint posters, press kit for Melbourne & interstate; new brochure. Local: billboards, banners, flags, accommodation house & VIC displays & famils. Aggressive marketing to local conferences; TV print campaign	Preventative maintenance for vehicles, equipment technical, operations; replacement of projection-based presentation for disk based storage	Training: preventative maintenance for vehicles, equipment technical, operations; enhancements in tele-computer bookings system & new marketing options; computer systems training	77, 526 set growth rate target of 10% for Vic market and 5% for NSW; achieved 10% overall

Table 10.3 (continued)

Year	*Marketing*	*Operations (technical)*	*HR*	*Results*
1996	TV commercial for night and day experience; aggressive marketing to local conferences; TV print campaign; wide promotion in Asia; evaluation surveys	Preventative maintenance for vehicles, equipment technical, operations; extension to accommodation (12units, $1,154,000m); technical & content enhancements to show	Appointment of full time technical manager	90,938 Set growth rate target of 10% for Vic market and 5% for NSW; achieved 10% overall
1997	Extensive Victorian press & RACV packages promotion; Discover Ballarat package to SA and Melb press for SH and BOSC; joint attractions with VTOS for Vic attractions; promotions to seniors; visited all NMRA outlets and RACV packages promotion; aggressive promotion into New Zealand; mainland China promotion with TV & ATC; hosting Asian tour operators and VJP; overnight packages for Singapore Evaluation surveys	Install infra-red transmission for delivery in five languages; refurbishing gift shop; heating seats in outdoor auditorium; all electrical and computer cabling system replaced with fibre optics to safeguard against lightening strikes	Appointment of two professional actors; computer software training; tele-computer training for bookings staff; electronics course for technical assistant; site electrical licences for two technical staff; safety training for BOS staff	91,850 NSW increase 5% BOSC visitor numbers plateau at 90,000 Asia currency crisis

(continued on next page)

Table 10.3 (continued)

Year	*Marketing*	*Operations (technical)*	*HR*	*Results*
1998	Victorian print media; journalists newsletter, direct telecasts; RACV packages promotion; NMRA advertising to NSW; direct day tour and operator promotion into New Zealand; special packages and facilities for Asian market; mainland China promotion with TV & ATC; hosting Asian tour operators and VJP; extended packages for Singapore; bi-annual newsletter to industry e.g. inbound tour operators, coach companies etc. Evaluation surveys	Enhance special effects; soundtrack transferred from CD to digital to improve reliability; increase show capacity; online computing with five partner motels for BOSC; extend and update computer systems; improved safe operational and emergency procedures up to four shows per night	Employment of technician to be on-site during all shows; tele-computer training for bookings staff; safety and emergency training for BOSC staff	88,707 Target: 10% increase in Asian market, 30% increase in NZ; Mainland China numbers up 39% NZ up 17% (132% increase during the three tenors and 86% during Bledisloe Cup events in Melbourne) SH mkts: Vic 56%; OS 24% (Asia 44% of that) Competition from Grand Prix and Casino
1999	As above; TV 'Footy' advertisement Melbourne and regional Victoria; South Australian TV campaign for BOSC	Installation of new technology; enhanced lighting and special effects on diggings; sets cleaned and repainted; gas heating installed in outdoor auditorium; screen expanded; major upgrade of computer technology; revised script		88,171

Table 10.3 (continued)

Year	*Marketing*	*Operations (technical)*	*HR*	*Results*
2000	As above; focus group research in Melbourne and Ballarat; Clemenger's press and TV campaign to Melbourne market; update of Ballarat Welcome pass; RACV press and packages; three day group tour for seniors; 2/3rds of BOSC patrons stay in Ballarat overnight	Preventative maintenance program to ensure reliability	Tele-computer training for bookings and reception staff; safety and emergency training for BOSC staff; Chinese and Japanese cultural training	92,847 Royal visit; good Easter weather; BOSC and accommodation principle reasons for SH success Vic 45% (80% of Vic's have visited); increase by 4.5% =11,000 visitors; internationals up by 15.4% or 16,000; 50% increase in Chinese arrivals (50% of Chinese to Vic include SH in their visit)
2001	As above; families for interstate and international travel agents and tour operators; new international student market emerging; research evaluation surveys; packages to Melb market with Ballarat Lodge	Preventative maintenance program to ensure reliability; accreditation process with better business; refurbish ensuite facilities in SH Lodge (S11attacks)	Tele-computer training for bookings and reception staff; safety and emergency training for BOSC staff	85,000 Victoria 45% of visitation; interstate market static from 1999 $300,000 in RACV packages alone

(continued on next page)

Table 10.3 (continued)

Year	*Marketing*	*Operations (technical)*	*HR*	*Results*
2002	As above; two market evaluation research studies of likelihood of return visits; > $1 million marketing budget; continued extensive domestic and international program focus on interstate and international markets; Melb TV, press, radio; Regional Vic TV; new TV commercial; Easter promo BOSC across regional Vic, Melbourne press; TV domestic lifestyle and holiday programmes; up degraded website; focus on inclusion in Asian wholesaler itineraries	$2 million show upgrade included new script, special effects and state-of-the-art lighting; extensive computer system upgrade; work commenced on new foreign language service for BOSC	Employment of second full-time electrician; mobile phone back up	91,000 27% Victorian; Regional Vic 15% of day market and 25% of night market; 73% internationals; 5% increase in international; 70,000 Asian, 50,000 Chinese (mainly Asian); BOSC generates over 1400 bed nights/week in Ballarat region
2003	Sovereign Hill's advertising thrust moves away from the traditional 'here we are – come and see us' style, to a more events-focused promotion. Aims to attract higher domestic & local visitation, as well as to shorten the return visit cycle	A $3 million upgrade over 18month period. Show remained open to avoid detriment to local economy		Numbers hold at 81,000 in face of hostile environmental conditions internationally and domestically

Source: BOSC reinvestment data sourced from Annual reports 1991–2002, Tourism Award Submissions 1992,'94,'96,'97,'98, 2001,'03 and interviews with executive managers

the city's key event. BOSC allowed Sovereign Hill to lift the region's brand. In national, unprompted awareness studies since 2001 Ballarat had the highest awareness of destinations in Victoria and among Victorian respondents Sovereign Hill is the most frequently endorsed attribute (Hollick and Lynch, 2007).

BOSC was Sovereign Hill's most significant product development of the 1990s, demonstrating that challenging themes associated with mining and social history can be presented with wide audience appeal reversing the product life cycle trend. It took the museum into theatre, history further into entertainment and retail marketing into wholesaling. It could claim to have done more to educate the public about the Eureka event than anything before it. The company structure under which the attraction operated was vital in its sustainability. Extending their capitalisation through external funding not normally available to the private sector enabled the organisation to generate a cash profit on its capital more easily because they did not have to repay interest or loan funds. This was critical in freeing funding for reinvention, as issue that challenges all attractions.

A greater understanding of what constitutes 'sustainable product management' is revealed in the BOSC case. Tourism product management is not the sole preserve of marketing. The success of BOSC in defying decline was closely related to factors beyond the product attributes of strong narrative, clever and novel, high technology and integrity of theme. Success has also hinged on rigorous business management: succession planning; quality monitoring and assurance systems; investment in research and staff training; all of which kept the product reliable and relevant so as to ensure positive word of mouth, the most powerful force in marketing success.

Volatile environmental conditions continue to challenge both the attraction and destination. The disassociation of an increasingly urbanised society from its past perhaps poses one of the greatest challenges for heritage attractions such as Sovereign Hill, having a direct bearing on the way it continues to reinterpret its mining themes. The values and practices which have underpinned the attraction's sustainability continue; success is now embedded but not taken lightly or for granted.

References

Benckendorff, P. and Pearce, P. (2003) 'Australian Tourist Attractions: The Links between Organisational Characteristics and Planning', *Journal of Travel Research*, 42: 24–35.

BHPA Annual Reports 1981–1999 Ballarat Historical Park Association.

Bramley, R. (2001) 'So You Want to Build a Hall of Fame?', in C. Pforr and B. Janeczko (eds), *CAUTHE 2001: Capitalising on Research: Proceedings of the Eleventh Australian Tourism and Hospitality Research Conference: Hosted by the University of Canberra, 7–10 Feb. 2001, Canberra, ACT* Belconnen, ACT : Tourism Program, University of Canberra, 17–27.

Bramley, R. (2003) 'Why Government-initiated and Funded Tourism Enterprises Fail: A Critical Evaluation of The Roles of Government in the Development of Visitor Attractions in Queensland, Australia' (PhD dissertation) School of Tourism and Hospitality Management, Lismore, Southern Cross University.

Bramley, R. and Curren T. (1997) 'Cultural Theme Attractions – Heritage Interpretation or Just an Edifice Complex?', *Tourism Research: Building a Better Industry – Australian Tourism and Hospitality Research Conference*, Sydney, Bureau of Tourism Research 6–28.

Dermott, B. (1995) Brian Dermott & Associates, *Sovereign Hill and Ballarat, Marketing Strategy, Final Report* Melbourne: BDA.

Elias, D. (1993) 'Projections of Employment for Regional Victoria and Returns from the New Tourism Initiatives', *The Age* [CD-ROM], 23 August 1993.

Fyall, A., Garrod, B. and Leask, A. (eds) (2003) *Managing Visitor Attractions: New Directions*, Oxford: Butterworth-Heinemann.

Hollick, M. and Lynch, D. (2007) *Tourism Impact Model, Report for Ballarat City Council*, Ballarat University of Ballarat.

Hollick, M. (2008) 'Browsers and Seekers: Understanding the Nature of Free Choice Learning in Outdoor Museums', *ICOM/CECA 07*, Vienna, (20–24 August 2007).

Johnson, J. (2003) Executive Director, Sovereign Hill, *Interview* 30/9/2003.

Johnston, J. (2007) *Submission by Sovereign Hill: Parliamentary Inquiry into Regional & Rural Tourism*, Submission no.65 Ballarat: Sovereign Hill.

Mottram, M. (2004) 'Nobody Saurus', *The Age*, 4th January, 2003. Available online at: http://www.theage.com.au/articles/2003/01/03/1041566221700.html?oneclick=true (accessed 10 July 2004).

Smallwood, J. (1973) 'The Product Life Cycle: A Key to Strategic Marketing Planning', *MSU Business Topics* 21, Sovereign Hill Education Unit, Teacher resources. Available online at: http://www.sovereignhill.com.au/?id=historicphotogallery (accessed March 2009).

Sovereign Hill (1992) *Tourism Awards Submission 1992*, Ballarat: Sovereign Hill.

Sovereign Hill (1994) *Victorian Tourism Awards 1994*, Submission, Ballarat: Sovereign Hill.

Sovereign Hill (1996) *Blood on the Southern Cross 1996 Victorian Tourism Awards*, Ballarat: Sovereign Hill.

Sovereign Hill (2000) *Sovereign Hill Corporate Plan 2000–2001*, Ballarat: Sovereign Hill.

Sovereign Hill (2000–2007) *Annual Reports*, Sovereign Hill Museums Association Ballarat: Sovereign Hill.

Sovereign Hill (2003) *Tourism Award Submission 2003*, Ballarat: Sovereign Hill.

Sullivan, T. (2001) Deputy CEO and Director, Museum Services, *Interview*, 28 August.

Swarbrooke, J. (2001) *The Development and Management of Visitor Attractions* (2nd edn) Oxford: Butterworth-Heinemann.

Tourism Transport Forum (TTF) (2005) Tourism Employment, Campaign Region of Goldfields.

Yann, Hoare, Wheeler (1991) Marketing Research Consultants, *Blood on the Southern Cross Research Study*, Study J7 198, Sovereign Hill Marketing Department Files.

Zulic, J. (2003) Head of Design Department, Sovereign Hill, Interview, 28 July.

PART IV

Traditional mining attraction destinations

11 Flogging a dead horse or creating cultural capacity?

The development and impact of mines as alternative tourist destinations: a case study of South Wales, UK

Andrew L. Jones and Kathryn Flynn

Introduction

The main focus of this chapter is the development of mines as tourist attractions, and particularly the social impact and relationships between mine-based attractions and their host communities. The chapter will present findings from qualitative research undertaken by the authors in 2007 and updated in 2009.

Secondary research involved an overview of the decline of UK heavy industry with a particular focus on redundant Welsh mining industries and their reasons for decline, with emphasis on the direct impacts on respective communities. Primary research examines the tourism redevelopment of a range of former redundant mines and associated properties in South Wales. These former mines have now largely been redeveloped as tourist attractions and the chapter includes research findings based upon Cefn Coed Colliery, Dolau Cothi Gold Mine, Big Pit National Mining Museum, Rhondda Heritage Park, Llewernog Silver and Lead Mine and the National Waterfront Museum in Swansea. For each case example, a generic overview of the attraction and facilities is outlined giving an initial review of their recent past and redevelopment. The main focus of the chapter, however, is the discussion and evaluation of the linkages that now exist between each mining attraction, and their relationships with their local host communities, particularly the positive and negative redevelopment issues and experiences.

The chapter concludes with a summary of evidence on the primary results. These particularly highlight and raise discussion on issues such as the positive and negative experiences of, for example, social impacts and redevelopment. The results also correlate such experiences of impact against such criterion as the type of attraction, their status, size and location in order to determine common and differentiated issues. The chapter reviews key conclusions from the case studies by assessing the fundamental challenges and outcomes faced by the redevelopment of each mining attraction in South Wales and in turn draws broader more general conclusions on the growth of mine-based attractions and the current advantages and challenges that proposed future mining attractions might face.

Industrial heritage/mine tourism concepts and developments

From the early 1970s the concept of heritage has, as a direct result of its multi-faceted definition of 'all things inherited', been a growing focus for tourism. According to Brisbane (1996: 4) the broad classification of tourism attractions that have been classified as heritage include a diverse range of subjects namely: *'landscapes, music, theatre, clothing, food, language, architecture....and more contemporary associations with past industry'*. This is underpinned by Williams' (1998) typology of heritage attractions, namely artistic, socio-cultural, built, transport, history, landscape, science, military, and industrial attractions (Williams, 1998). In recent years, many tourism development projects have emerged within such classifications.

The modern concept of industrial heritage-based tourism originated largely in France. There the concept materialised in the late 1960s to early 1970s in the form of the 'eco museum', which focused on industrial working practices and community life (Verbekke, 1999). Following the French success in heritage tourism, throughout the 1970s, other countries followed. This was often particularly attributed to generous European policy focus and funding available at that time (Ball, 2000).

By the 1970s, the concept of industrial heritage was significantly adopted by government policies for regeneration across a broad range of western industrial nations. This too was an early concept adopted by Welsh tourism agencies notably the Welsh Development Agency and the Wales Tourist Board. In this respect the consequence of Wales' strong associations with heavy industry provided the 'back drop' for industrial based tourism which soon featured strongly as a means of socio-economic regeneration. The aim of such strategies was to increase the local profile and economy, which had over several preceding decades, experienced socio-economic problems following the decline of heavy industry and especially that associated with coal mining. For Welsh Government the socio-economic potential of industrial heritage tourism was strongly embraced through the encouragement of public sector based industrial heritage tourism attractions. By the late 1960s and early 1970s Welsh industrial heritage tourism was considered as a broader remedy for widespread unemployment and regeneration (Rees 1991 cited in Macdonald and Thomas, 1997).

One of the earliest policies included the 1967 White Paper 'Wales the Way Ahead' published by The Welsh Office in Cardiff. Later significant strategies included in 1983 'Realising the Tourism Potential of the South Wales Valleys', a commissioned report to the then Wales Tourist Board (WTB) focusing on Welsh Valley's heritage, through a working relationship between the host and visitor (Travis, 1983). In 1988 'Programmes for the Valleys' was launched, a partnership initiative between national and local government agencies. Unlike the rest of the United Kingdom where the focus on heritage and culture was broader based, by the 1980s Welsh strategies for regeneration were primarily rooted in industrial heritage tourism which tended to highlight the country's heavy industry and mining past (Wanhill, 2000).

More recent examples of Welsh tourism initiatives featuring heritage/mine tourism and primarily focused on socio-economic regeneration and economic development of communities have proliferated. They have included:

- 'Tourism 2000: A Perspective for Wales' (WTB, 2000)
- 'The First Cultural Strategy for Wales' – (Welsh Assembly Government, 2000)
- 'A Cultural Tourism Strategy for Wales' (WTB, 2003)
- 'Herian: Industrial Heritage Tourism Development Initiative' – (WTB, 2003)
- 'National Tourism Strategy 2006–2013' – (Welsh Assembly Government, 2006)
- 'Review of Cultural Tourism in Wales' – (Welsh Assembly Government, 2008)

It is fair to say therefore that during the 1970s, tourism, particularly industrial/mine heritage and cultural-based tourism was at an early stage of development; however, since the mid 1980s and 1990s, its growth has been substantial. The extent of this growth can now be illustrated by the increase of industrial/mine based tourist attractions across Wales over the last two decades. These have included, for example, Rhondda Heritage Park, Big Pit National Mine and Mining Museum, Dolau Cothi Gold Mine, Cefn Coed Colliery Museum, South Wales Miners Museum, Llewernog Silver and Lead Mine, Llechwedd Slate Mine and Caverns and other attractions associated with mining such as Kidwelly Tin Plate Museum, Swansea Vale Railway, Blaenavon Steel Works and Swansea's National Waterfront and Maritime Museum – see Figure 11.1.

Welsh heritage/mine tourism – growth and comparative trends

Tourism has been fundamental in the drive to realise the regeneration of cultural assets in Wales. In 2005, tourism in Wales totalled 10 million domestic staying visits, and one million overseas staying visitors, which amounted to 7 per cent of the gross domestic product. Of particular importance, such tourism figures sustained 100,000 jobs in the Welsh economy, 9 per cent of the overall work force (WTB, 2005). In this context heritage/mine based tourism, has played a significant role in contributing to this growth, as heritage/mine tourism attractions have accounted for a large percentage of the overall tourism mix (57 per cent). These trends are evidenced from the mid part of this decade. For example, the WTB visitor statistics from 2005 categorise over 50 per cent of Welsh tourism attractions as being cultural or heritage/mine based. These of course include a broad range of historic sites such as mines, museums, galleries, steam railways, and workplaces (WTB, 2005). (Figure 11.2)

The 2004, WTB figures for overseas visitors show the spend from overseas tourists as £311 million, a 13 per cent increase from 2003 (WTB, 2005). Additional research outlined in the WTB's Cultural Tourism Strategy for Wales (2003a) identified the most popular activity for

Figure 11.1 Visitors to Big Pit – National Museum of Wales, Blaenavon, South Wales experience a visit to a deep coal mine.

Figure 11.2 National Museum of Wales – National Waterfront Museum, Swansea – opened 2006.

overseas visitors was visits to heritage sites, (85 per cent of the market) and the third most popular activity was visits to arts and heritage exhibits (48 per cent of the market). Significantly, figures also show that 37 per cent of all respondents visited heritage sites as their main tourism purpose. More recent research to support these trends includes the WTB's surveys 'Visits to Heritage Attractions: 2003, 2004, 2005', where the highest proportion of visitors to historic properties, museums/galleries and workplaces is derived from the overseas tourist market. From this it can be deduced that overseas visits can make a real positive contribution to Welsh heritage/mine tourism industries and in turn provide a multiplier effect for their respective host communities.

Specific research into the motivations of heritage and cultural visitors is ongoing by Visit Wales however, to date, the most current and comprehensive available research is covered by Visit Wales 2008. In this respect a survey commissioned by Visit Wales (2008) showed that visits to attractions stood at almost 10,700,000, up less than one per cent on the volume of visits in 2007 which stood at nearly 10,600,000 thus reflecting the general downturn in the Welsh, UK and global economies during this time.

The report nevertheless shows that visits to free attractions increased by just over 4 per cent and visits to free attractions now account for almost 49 per cent of all visits. In terms of the category of attraction the report showed that themed attractions enjoyed the largest increase in visitor volumes (this includes historic themed attractions such as Big Pit Mining Museum and Rhondda Heritage Park) up nearly 28 per cent during that period. This compares to historic properties which actually experienced a decline in visitor numbers, down nearly 8 per cent during the same time period (Beaufort Research, 2009). These figures are illustrated in Table 11.1.

Despite the present economic downturn future predictions outlined since 2002, and more recently in 2008, by Visit Wales and the Welsh Assembly, demonstrate positive expectations

Table 11.1 Comparison of yearly visits 2008/2007 – visitor attractions Wales

Sample	*Visits 2008*	*Visits 2007*	*% Change*
Historic properties	2.97 million	3.2 million	–7.6
Museums and art galleries	3.77 million	3.5 million	+8.2
Industrial/craft attractions	0.75 million	0.7 million	+2.9
Gardens/natural attractions	1.97 million	2.1 million	–6.1
Wildlife	0.19 million	0.2 million	–6.0
Themed attraction/leisure park	0.23 million	0.18 million	+27.8
Railway /tramway	0.7 million	0.6 million	+18.7

Adapted from Beaufort Research, 2009.

predicting that tourism will continue to grow at an average rate of 6 per cent per annum over the next 10 years with heritage/mine based and more general cultural attractions strongly contributing to this growth and in turn making a significant contribution to the Welsh economy. (Visit Britain, 2008; Welsh Assembly Government, 2002).

The research investigation

The aim of the research was to evaluate the community impacts that industrial heritage/mine tourism attractions have had on their respective communities and the potential operational factors that may influence the levels of impact and benefit. Specifically, the study aimed to meet three primary objectives.

- To evaluate the social impacts and benefits of industrial heritage/mine tourism attractions.
- To investigate the issues and challenges associated with the operations of industrial heritage/mine tourism attractions.
- To evaluate future prospects for industrial heritage/mine tourism attractions and to recommend strategies to facilitate positive social synergy.

As part of the early stages of the research, secondary data was collated and this provided a contextual underpinning to the study. Such data included a review of societal trends and patterns from an historical context through to more contemporary heritage issues. Research data was gathered from a range of sources, namely journal articles, government based reports and websites. In addition to secondary data which provided the benchmark for comparative analysis, primary data collection was also undertaken. In order to minimise the remoteness between the researcher and the interviewee it was considered more appropriate to select a phenomenological research paradigm (qualitative approach) as the issues and impacts investigated would reveal more in-depth perspectives, attitudes and perceptions. A further reason for the selection of the phenomenological paradigm, supported by Denscombe (2003), was the sensitivity of the research subject area in probing complex economic and social issues. In order to fully investigate the research objectives set, a case study location of the South Wales Valleys was chosen. Specifically the selection of attractions included: i) West Wales locations including Carmarthenshire and Ceredigion; ii) East Wales locations including Torfaen, Rhondda, Merthyr; and iii) Mid-central Wales locations including Neath, Port Talbot and Swansea.

In conjunction with a phenomenological research method, multiple cases were selected to ensure validity and reliability, namely twelve industrial/mine heritage attractions, seven community groups and six heritage/mine tourism associations. A semi-structured interview approach was adopted. This approach, supported by Veal (2006) allowed interviewees flexibility in relation to expressing their personal views and feelings from open ended questions.

The target sample included: i) the attractions themselves (selection of twelve), focusing on contrasting industrial/mine practices including coal, tin plate, and silver and encompassing varying attractions namely: public, voluntary and private; ii) the key organisations and associations responsible for the strategic development of industrial heritage/mine tourism (selection of six); and, finally, iii) local community representatives in the surrounding localities (selection of seven). The above provided a triangulated sample for formulation of data based from three diverse perspectives. To ensure reliability and validity the same set of questions was asked of each interviewee. This ensured that aims and objectives of the research were broadly met. The sample included Table 11.2.

In relation to the feedback gained from the communities in which the attractions were based, data was gained from the 'Communities First Partnership'. In this context, Communities First Partnerships were devised by the Welsh Assembly Government as part of a flagship programme to make improvements for people in communities across Wales. Representatives from such organisations helped provide a 'voice' from the local community.

Upon completion of the interview data a non-quantifying method of analysis was selected primarily using the visual display technique of matrices and content analysis. As Robson (2002) suggests, such a method makes it possible to group together non-statistical related themes which can clearly establish the similarities and differences between the data sets.

In any research there are limitations and this study is no exception. It is appreciated that if wider objectives were set (for example, geographical and attraction based) and lengthened time scales and availability of funds were permitted, then more rigorous and more broadly applicable results could have been generated. Such observations do not illustrate weaknesses, but present some considerations for future research. However these points aside, it

Table 11.2 Survey target sample

West Wales	*South Wales*
Dolau Cothi Gold Mine – National Trust – Voluntary Kidwelly Tin Plate Museum – Trust / Local Authority – Public / Voluntary Llewernog Silver and Lead Mine – Private	South Wales Miners' Museum – Trust – Voluntary Cefn Coed Colliery Museum – Local Authority – Public National Waterfront Museum – NMGW – Public Aberdulais Falls – National Trust- Voluntary Swansea Vale Railway – Private – Private
East Wales	*Organisations*
Rhondda Heritage Park – Local Authority – Public Blaenavon Iron Works – CADW / Trust – Public / Voluntary Big Pit Mining Museum – NMGW – Public Brecon Mountain Railway – Private – Private	WTB Tourism Partnership – South West Wales (South West Wales Tourism Partnership) WTB Tourism Partnership – East Wales (Capital Regional Tourism) CADW, Welsh Assembly Government, Herian, NMGW Community Locations in Industrial South Wales: Swansea, Rhondda, Neath and Port Talbot, Carmarthen, Merthyr, Blaenavon, Ceredigion

is, nevertheless, felt that the results generated illustrate both a valid and reliable insight into the impacts of industrial heritage/mine tourism attractions on their respective communities.

Engendering social and cultural capacity

In terms of social impact on communities the study specifically focused on questions that were linked to employment prospects, community self-esteem, impact of special/staged events, educational gain and improved environmental amenity. In short, industrial heritage/mine attractions had varying levels of impact on their respective communities in this respect.

The results highlighted that both positive and negative social impacts were largely influenced by general factors that were determined by, for example, the scale of the attraction, its location and its type. As well as these, more critical operational factors associated for example with i) the sophistication of organisational structure; ii) financial support and funding; and iii) variety of product market focus particularly the tensions associated between conservation or commercialisation were key determinants in differentiation.

Clearly, evidence from the research shows that there were more positive outcomes from the two larger national museum attractions, Big Pit and the National Waterfront Museum, especially for the greatest positive community impact. The positive impact proved largely attributable to the significant levels of government support (both in terms of funding and strategic management) available to both these attractions.

One key area that both the national museum attractions were able to have most significant impact on was employment. This was essentially linked to both their large operational scales, which required a relatively large workforce, and their major strategic policy remits for broader regional regeneration. For example, Big Pit concentrated efforts on employing ex-miners principally to address community regeneration, following the high levels of unemployment from many local mine closures in the 1980s, but also to ensure product quality and authenticity. The National Waterfront Museum also significantly influenced the local employment market, employing 200 directly at the museum, and 1,930 indirectly from expected tourism 'spin offs' linked to new inward investment. New employment opportunities at both attractions also contributed positively to community self-esteem which had and continues to have strong community associations. Further examples of positive impact, particularly associated with esteem were generally linked to local events. These were often associated with local school education networks and strongly supported through government education policy on curriculum (National Curriculum – Curriculum Cymraig).

Another significant positive impact was that concerning improved environmental amenity. For example Big Pit, located in Blaenavon, South Wales, was strongly associated with the Blaenavon Partnership Initiative. This partnership has promoted many local regeneration projects centred on environmental enhancement, urban renewal and improving local amenity. This in turn has significantly benefited the local community by creating a platform for inward investment and inward migration. Similarly, the National Waterfront Museum has significantly influenced the urban aesthetic and visual amenity of Swansea's South Dock and waterfront area. From its impressive glass fronted architecturally designed structure and landscaped surrounds, it has contributed to inward investment, and has encouraged newly developed infrastructure (Figure 11.2). The positive impacts of the national museum sites have resulted in a largely receptive and positive community view, which also has assisted in the strengthening of overall community identity and self respect.

With respect to the remaining, somewhat smaller, industrial heritage/mine attractions, whether public, private or voluntary, the scale of positive community impact proved limited.

Instead, community views and opinions largely reflected apathy and cynicism and, in general, such attractions failed to assist greatly in strengthening community self-esteem and identity. For these attractions, limited impact was largely attributed to their small-scale and relatively limited levels of government support.

With regard to employment, it is fair to say that the remaining attractions did have an impact although overall employment figures proved small, when compared to the national museum sites. In this respect differences were also significant especially when the numbers of voluntary/paid staff, part-time/full-time staff, and seasonal/fixed-term contracted staff were considered. In this respect paid full-time employment opportunities were limited. This was particularly noted for the National Trust voluntary sector properties namely; Aberdulais Falls and Dolau Cothi Gold Mine and the private sector properties including Swansea Vale Railway and the voluntary independent properties such as Kidwelly Industrial Tinplate Museum where reliance was primarily placed on part time seasonal and voluntary staffing. In this respect voluntary work was perceived to have important social impacts by encouraging community synergies. However, from an economic perspective such attractions failed to have any significant regeneration or employment effect. This is perhaps a particular disappointment and reflects a broader policy failure as original remits for many such attractions were to primarily focus efforts on regeneration.

On another disappointing note employment opportunities, even including the larger national museum sites, remained largely insignificant especially when former levels of mining industry employment are considered. Current levels of employment have certainly not replaced the number of former mining jobs lost over the last three decades. This is also compounded by additional employment structural changes especially with respect to often less generous contractual terms, remuneration, and work career opportunities. These issues can be further complicated by the 'fickle' nature of the tourism industry which is particularly subject to uncertainty and seasonality.

Education impacts for the remaining industrial heritage/mine attractions also proved limited. However, impacts did vary between sites. This was particularly the case for fee based attractions, as they experienced direct competition from the larger free entry public sector attractions. This issue was further influenced by the fact that, for many attractions, little or no funding was available to appoint full-time on-site educational support staff. To compete, evidence suggests that many of the smaller attractions had made attempts to enhance their market products through diversification. This for example included promoting small-scale events and introducing basic interactive marketing methods. However, broader issues were raised relating to maintaining authenticity contrasted with the need to remain commercially viable.

In summary it is evident that the larger attractions, particularly those with strong operational and financial support, have achieved greatest success in terms of positive social impacts and product market development. In contrast, many of the smaller attractions have become a little pessimistic and cynical in their views towards engendering wider community benefit as their efforts have had to be largely concerned with business sustainability and viability.

Ensuring effective operational structures

In terms of operational challenges, the study specifically focused on questions that were linked to organisational structures and product markets and development. In this respect issues and tensions associated with the increasing need to be commercially viable and to maintain historical relevance were explored.

The results found that there was a significant imbalance in the support mechanisms available to the small- to medium-sized public, private and voluntary sector attractions. In contrast central support, planning and management for the larger government-based attractions was far more extensive, particularly funding. This was evidenced from extensive grant allocations having been made since 2000 to both national museum attractions: Big Pit and the National Waterfront Museum. In this context, sharp contrasts can be made with the levels of public funding and Welsh Government support available to the smaller public, private, and voluntary sector attractions. This was particularly relevant for the private attractions which were largely exempt from public sector funding support.

As a result of restricted organisational support, many attractions, particularly the smaller public, private and voluntary sites, were obliged to pursue alternative means of income generation. Principally, this was via diversification in the form of commercially based marketing activities and product development which tended to include merchandise, food and beverage, events, and interactive exhibits. Results showed that diversification and product enhancement can potentially act as a positive for the attraction, especially increased visitor numbers. However, results also showed that diversification can also potentially lead to over commercialisation of an attraction, negatively influencing product authenticity and in turn be detrimental to broader community objectives based on building cultural capacity.

All attractions appreciated such consequences and, where possible, strived to consider a balance between conservation and commercialisation. However, for many, diversification was considered essential for viability and sustainability. Notably, results showed that many voluntary and private attractions had considered alternative means for market sustainability as opposed to more commercial options. One means in particular included networking and collaborating on marketing. However, many of the attractions, particularly in the voluntary or private sectors, claimed that there was reluctance by the public sector to pursue joint opportunities. Moreover, it was found that irrespective of the existing national cultural tourism strategies for heritage/cultural tourism promoted by Welsh Assembly Government, certain local authorities were more receptive and supportive of promoting alternative tourism developments associated with perceived 'more innovative' and 'economically viable' tourism markets such as markets associated with adventure and outdoor pursuits.

In summary the results from the survey generally show that there remain various challenges for the sector including, issues associated with i) community engagement, ii) addressing management competencies; iii) ensuring effective strategic and organisational support; and iv) developing innovative markets that remain educationally and culturally relevant. For the majority of attractions these are of particular concern, due to their original remits of being associated with economic, environmental and community regeneration. In this context the results suggest that building such competencies to ensure the sustainability of such attractions will depend on the success of tackling broader measures. These can largely be attributed to i) encouraging strategies that address community support and tackle apathy; ii) further embrace the 'market place'; iii) address sector differences; iv) create new market opportunities; and v) further target educational markets.

Flogging a dead horse? Or ensuring cultural capacity and opportunity?

Clearly, from the results there was a bias towards the two larger national museum sites. Such observations are strongly corroborated by the predictions of, for example, Morris and Wilkinson (1993) and Dicks (2000), and contested by authors such as Bianchini (1990),

Colenutt (1991), and Rogers (1991 cited in Smith, 2003). In this respect positive impact proved largely attributable to the significant levels of government support (funding, strategic and organisational) available to both the national museum sites. The results particularly corroborate the views of Richards (1996) and Worpole (1991 cited in Smith, 2003), and Smith (2003), who suggest that culture and heritage has become strongly advocated by government for regional benefit.

One significant area that both national museum sites were able to influence was employment. This was essentially associated with their operational scale, which required a relatively large workforce, and in broader terms linkages to national policy remits for economic and cultural regeneration. Improved employment figures at both sites also positively influenced community self-esteem and in turn engendered strong community associations with both the attractions. Further examples of positive esteem were also linked to community based events. Clearly the larger attractions have had a fairly significant impact on the overall perceptions of the areas within which they are located. In this respect, improvements to local infrastructure and local amenity are of particular note. In turn this has led to more positive and receptive attitudes toward investment and property price stability which in turn has significantly benefited the local communities. These observations go some way to support the views of authors such as Bianchini (1993) and McGuigan (1996 cited in Smith, 2003) and Hall (2004) who suggest that positive impacts from such attractions can result in cultural capacity building by encouraging receptive and positive community views and assisting in strengthening overall community identity and local self-esteem.

With the smaller attractions, whether public, private or voluntary, the scale of positive community impact in relation to social indicators including employment, local self-esteem, events, education and environmental amenity was less noticeable. Instead, community views proved at best less positive and at worst largely negative and cynical and thus failed to assist in greatly strengthening community self-esteem or building cultural capacity. This largely contradicts the contentions of authors such as Marks (1996 cited in Fevre and Thompson, 1999), Dicks (2000) and Hall (2004) who would argue the opposite contention. In the case of this study the smaller attractions had limited impact which was largely attributed to their small scale and their limited access and support from public resources. The polarisation between larger publicly supported institutions and the smaller often private or voluntary institutions was thus a marked contrast in this respect.

The creation of employment opportunities across the range of attractions surveyed was also a fairly mixed blessing. Apart from the larger national attractions employment impact was restricted. Voluntary work did appear to have important social impacts, with respect to community morale and engagement, however, from an economic perspective, particularly in relation to community regeneration, it failed to have significant positive effect. This, perhaps, is of particular concern and somewhat of a disappointment since the original *raison d'etre* for such attractions was one primarily focused on economic, community and cultural regeneration.

Even when positive employment impacts are considered, particularly in the case of the national attractions, new employment opportunities generated remain fairly insignificant, when the former levels of mining and other associated industry based employment are considered. This too is compounded by the nature of new employment opportunities which have tended to be based on shorter contractual terms, lower remuneration and less career based opportunities.

It is clear that the smaller industrial/mine heritage attractions have had less social or cultural impact within their host communities. This, as the evidence points, was largely due to limited funding and support from the 'public purse'. In this respect such attractions had to concentrate their efforts on rationalising their often limited resources and prioritising commercial viability thus offsetting and detracting from broader objectives for social and community engagement and synergy. In this context efforts were thus largely seen as being concerned with business sustainability and viability. Despite such pressures however, there was some evidence of a desire for community engagement which is supported by theorists including Witt (1991), who suggests that positive community support and impacts are achievable, provided appropriate planning, management and support is in place. The results thus remain encouraging but nevertheless highlight the fact that the availability of resources, particularly public ones, are a major deterrent to achieving broader cultural and community based objectives.

Clearly the research has highlighted the issue that there is a significant imbalance in support elements available to the small- to medium-sized public, private and voluntary sector attractions. For the larger public sector attractions central support, planning and management is comparatively generous, particularly for public funding. Thus some correlation does exist between strong public support and positive effects on levels of community support and cultural capacity building. Such concepts support the views of authors such as Mansfeld (1992), Edwards and Llurdés i Coit (1996), and Ball (2000) amongst others, who have advocated that such sharp contrasts between levels of public funding and support available to smaller public, private and voluntary sector attractions can be a major determinant of broad success or failure within their respective communities.

The results conclude that there are a variety of challenges for the development and sustainability of South Wales' industrial/mine heritage tourism attractions. These include challenges associated with:

1 engaging and supporting local communities
2 combating apathy both within the management structures of each attraction and more broadly based apathy associated with the communities at large
3 challenging existing local and regional support
4 ensuring effective professional management
5 balancing commercial viability with cultural relevance
6 ensuring that broader objectives related to building cultural regeneration are realised
7 combating the growing polarisation of opportunity between public, private and voluntary sector attractions and
8 ensuring that the development of all such attractions 'sit' within a broader strategic framework for cultural capacity building across Wales, particularly within the context of the existing nationwide Cultural Tourism Strategy (Welsh Assembly Government, 2008).

Lessons from the research would suggest that such directions are required if the impacts and benefits from industrial/mine attractions are to improve community engagement, build cultural capacity, sustain community self-esteem and fully embrace cultural and economic regeneration.

In summary all industrial/mine heritage attractions are invaluable in creating a cultural setting and image for Wales. This is particularly the case for those attractions which are rare and unique such as Blaenavon Ironworks and Big Pit, now a World Heritage Site, Cefn Coed Colliery, in its prime the world's deepest mine, Dolau Cothi Gold Mine a unique example of a Roman period excavation, and Llewernog Silver and Lead Mine being a unique asset to Wales. These attractions form an important contribution to the cultural fabric of Wales' past. However research clearly shows that the key message for tourism and economic policy makers is one that requires a more strategic inclusive and government led approach if such attractions are to be sustained and evolve into cultural assets for the regions and communities in which they are located. Within this context cultural resources and assets are still deeply rooted in the former mining communities that still exist in many parts of Wales. However, mine based tourism is not a panacea for economic sustainability nor is it, as many would suggest, 'flogging a dead horse'. When integrated with properly balanced approaches to regeneration such tourism initiatives can show great benefits and sustain pride in once proud communities that over the last three decades have, to some extent, lost their way and purpose.

Acknowledgements

Our thanks go to the communities that helped compile data for this research.

References

Ball, R. (2000) Time for Celebration – Industrial Heritage Tourism as a Force for the Regeneration of Coalfield Communities: Feasibility – Desirability Analysis, Staffordshire University, UK.

Beaufort Research (2009) *Visits to Tourist Attractions 2008 Report for Visit Wales*, Beaufort Research Ltd, UK.

Brisbane, M. (1996) *A Future for our Past?: An Introduction to Heritage Studies*, English Heritage, London.

Denscombe, M. (2003) *The Good Research Guide: For Small-Scale Social Research Projects*, Maidenhead: Open University Press.

Dicks, B. (2000) *Heritage, Place and Community*, Cardiff: University of Wales Press.

Edwards, J, A and Llurdés i Coit, J.C. (1996) *Mines and Quarries Industrial Heritage, Tourism Annals of Tourism Research*, Oxford: Pergamon Press, 23(2): 341–63.

Hall, D. (2004) *Tourism and Transition: Governance, Transformation and Development*, Cambridge: CABI Publications.

Macdonald, R. and Thomas, H. (1997) *Nationality and Planning in Scotland and Wales*, Cardiff: University of Wales Press.

Mansfeld, Y. (1992) 'Industrial Landscapes as Positive Settings for Tourism Development in Declining Industrial Cities – the Case of Haifa Israel', *Geo Journal*, 28 (4): 457–63.

Marks, D. (1996) 'Slate Heritage – The History of a Welsh Industry or a Trip Down the Light Fantastic', in Fevre and Thompson *A Nation, Identity and Social Theory: Perspectives from Wales 1999*, Cardiff: University of Wales Press.

Morris, J. and Wilkinson, B. (1993) *Poverty and Prosperity in Wales: An Analysis of Socio-Economic Divisions*, HTV Current Affairs Department, 'Wales this Week' Unpublished report.

Robson, C. (2002) *Real World Research: A Resource for Social Scientists and Practitioners/Researcher*, Oxford: Blackwell.

Smith, M. (2003) *Issues in Cultural Tourism Studies*, New York: Routledge.

Travis, T. (1983) *Realising the Tourism Potential of the South Wales Valleys, A Report to the WTB by the Leisure and Tourism Department of the Centre of Urban Regional Studies*, Birmingham: Birmingham University Press.

Veal, A.J. (2006) *Research Methods for Leisure and Tourism, A Practical Guide*, London: Prentice Hall.

Verbekke, J. (1999) 'Industrial Heritage a Nexus for Sustainable Tourism Development', *Tourism Geographies*, Department of Social and Economic Geography, Catholic University of Leuvan, Belgium, 1(1): 70–85.

Visit Britain (2008) *The UK Tourist, 2008 Visitor Statistics*, London: TNS.

Wales Tourist Board (WBT) (2000) *Tourism 2000: A Perspective for Wales*, Wales Tourist Board, Wales.

Wales Tourist Board (2003) *A Cultural Tourism Strategy for Wales*, Wales Tourist Board, Wales.

Wales Tourist Board (2005) *Visits to Tourist Attractions 2004*, Wales Tourist Board, Wales.
Wanhill, S. (2000) 'Mines – A Tourist Attraction, Coal Mining in Industrial South Wales', *Journal of Travel Research*, England: Sage Publications, 39(1): 60–69.
Welsh Assembly Government (2000) *Cultural Strategy for Wales*, Welsh Assembly, Wales.
Welsh Assembly Government (2002) *The National Economic Development Strategy of the Welsh Assembly Government – A Winning Wales*, Welsh Assembly, Wales.
Welsh Assembly Government (2006) *National Tourism Strategy 2006–2013*, Welsh Assembly, Wales.
Welsh Assembly Government (2008) *Review of Cultural Tourism in Wales*, Welsh Assembly, Wales.
Williams, S. (1998) *Tourism Geography*, London: Routledge.
Witt, S. (1991) *The Impact of Tourism on Wales*, Swansea University Press: Wales, UK.

12 Mining heritage and tourism in the former coal mining communities of Cape Breton Island, Canada

Kim Lemky and Lee Jolliffe

Introduction

Cape Breton Island, once world renowned for its coal deposits and as home to the largest self contained steel mill in the world, is now more well known as a scenic destination area for tourism, exemplified by the Cabot Trail. The island is also recognized for its vibrant culture in music, especially the Cape Breton fiddle playing, as demonstrated by the annual Celtic Colours International Festival (Brown and Geddes, 2007). The last underground coal mine and steel mill closed in 2001, leaving behind a vast industrial infrastructure and a very weak economy. Heritage vestiges of the mining industry are still visible in the architecture of the communities such as company houses, the monuments, museums and a tourism route linking this heritage infrastructure. The community heritage of coal mining lives on here through culture, literature, story-telling, and music (Mining Culture Symposium, 2005) as in other mining communities (Pretes, 2002). However, this mining heritage is not the main focus of tourism on the island.

This chapter examines tourism on Cape Breton Island providing an overview of the coal mining district, reviewing institutional arrangements for managing and marketing tourism, and the challenges to exploiting this mining theme in the tourism industry. Information for the chapter was obtained from relevant literature, and from questioning tourism operators in a form of action oriented participant observation (Cole, 2005). The conclusions reflect the role of the lead author as a tourism coordinator in the region. A case study approach, noted as useful in exploratory research (Beeton, 2005) is taken in investigating the mining related visitor attractions of the region.

Cape Breton, a 10,311 square kilometre island in eastern Canada was originally settled by the Mi'kmaq native people thousands of years ago. In the 1600s both English and French garrisons were established on Cape Breton Island (Lamb, 1975). The French were the first European group to settle extensively in the area in the early 1700s. With the siege of the Fortress of Louisbourg in 1758, they lost their stronghold to the English. By the 1800s thousands of Scottish settlers expelled during the Highland Clearances arrived developing a strong culture and changing the landscape (Brown, 1979). By the mid 1800s coal was being mined and people immigrated from all over the world to work here. The mining industry, along with steel production, had reached its peak by the mid 1940s (Tennyson, 2000; Millward, 1985). In 1955, a causeway was built to connect the island to mainland Nova Scotia. This facilitated travel for residents and tourists alike.

Easy access to energy fuelled the steel industry for decades. Coal mining shaped the economy, the transportation system, the development of the urban landscapes of Cape Breton Island and the culture. The island economy, since the closure of the coal mines and steel mill, comprises some light industries, call centres (centres equipped to handle high volumes of customer care or marketing telephone calls), a pulp and paper mill, some gypsum extraction, lobster and crab fishing, agriculture and tourism. All of these industries are precarious to some degree and rely on a strong economy to survive. Tourism is present in many communities throughout Cape Breton Island and often, especially around the Cabot Trail, it is the main source of income in the summer for local businesses. For Cape Breton Island as a whole, tourism accounts for approximately 10 per cent of the labour force (www.gov.ns.ca).

Many events influenced tourism development here. The arrival of Alexander Graham Bell, the inventor of the telephone, in 1885 in Baddeck, cemented this community as a tourism destination and a location of innovation as Bell continued his experiments here. In the early 1920s American fly fishermen and their families flocked to the Margaree River for the salmon fishing. In the 1930s Casey Baldwin, an MP for Victoria County, lobbied for the establishment of the Cabot Trail, a 300 kilometre scenic route, and for the formation of Cape Breton Highlands National Park. The first was completed in 1932, the latter founded in 1936. The primary impetus for these developments was to increase visitation and length of stay (Bethune, 2009). Nova Scotia developed a scenic travel way (or scenic highway) system throughout the province to guide visitors through the culture, history, and natural features of unique areas in the province. Cape Breton Island, because of its diversity is divided into five scenic travel ways (NSDTCH, 2008), the Cabot Trail being the most well known. As well, the Colliery Route developed in 2002 links the mining-related attractions in the North Eastern section of the island. These themed routes are not unique to Nova Scotia. As Silbergh *et al.* (1994: 123) note in their examination of theme trails in Scotland, trails have a long history and can be a significant tool for conservation, education and tourism marketing.

In the 1960s, two significant attractions were developed in the north-eastern coal mining region. In 1967, the Cape Breton Miner's Museum (a community museum) opened in Glace Bay and in 1969 Fortress Louisbourg, a National Historic Site of Canada; the largest historical reconstruction in North America (the original fort was built in 1713 by the French and destroyed by the English in 1758) was opened. Restored by the federal government as a make-work project for displaced coal miners (Key, 1973), it operates under Parks Canada, the federal agency responsible for designation, reconstruction, and management of significant sites. Together these two visitor attractions attract almost 100,000 visitors in total each year (NSTHC, 2008) to the mining district of Cape Breton Island.

The economic impact of Cape Breton Island's tourism industry was estimated at 208 million dollars (Canadian) in 2008 (Nova Scotia Department of Tourism, Culture and Heritage, 2009) and it employs over 5,500 people on Cape Breton Island. Visitation to the island is thought to be close to three quarters of a million. With traditional industries in decline, tourism will likely continue to be nurtured as an economic activity.

Mining and tourism contexts

As identified by the Geological Survey of Canada (Zaslow, 1995), there were two major coal fields in the region, the largest in the north eastern section and a smaller deposit on the west coast of Cape Breton Island near Inverness. Industrial development was centred around the most populated area of Cape Breton Island, around Sydney (e.g. Sydney Mines, New Waterford, Dominion, Glace Bay, to name a few) (Table 12.1).

During World Wars I and II, the Sydney area was one of the most important in the country, as a staging ground for the convoys traveling to Europe because of its strategic location on the St. Lawrence Seaway (Tennyson, 2000). It supplied the steel and coal required for the

Table 12.1 Key milestones in Cape Breton mining and tourism

1672	French explorer Nicolas Denys noted the coal deposit near Sydney Harbour
1720	1st Coal mine shaft sunk at Port Morien
1857	Glace Bay's first commercial coal mine opens
1877	Caledonia Mine is world's first to use an underground telephone system to dispatch trains in the mine
1891	First railroad came to Cape Breton Island
1899	Dominion Iron & Steel Co. opens in Sydney (1 July)
1902	Sydney Mines Opens Nova Scotia Steel & Coal Co.
1907	1st Dragger mine equipment installed at #2 Colliery, Glace Bay
1923	Striking Sydney coal workers and families attacked after church on July 1st, by 1100 troops, this day is henceforth known as Bloody Sunday
1925	Coalmine owned stores closed and never reopened, William Davis Day established as 11 June, in recognition of a coal miner killed by police during a protest
1932	Completion of the Cabot Trail
1967	Plans made to close Cape Breton coal mines due to lack of productivity and low prices of coal, DEVCO established
1967	Paving of the Cabot Trail
1967	Cape Breton Miner's Museum opened in Glace Bay
2001	Cape Breton Islands last underground coal mine is closed
2001	North Sydney Heritage Museum, Opened its Fossil Centre

Source: CBC News on line, in depth; Cape Breton Coal, 14 December 2005; destination Cape Breton Island Factsheet 2001.

war effort. In the 1940s to 1960s the industrial economic base was slowly faltering as mines closed. Subsequently mining communities sought to develop tourism products related to mining in Cape Breton.

In 1967, the Cape Breton Development Corporation (DEVCO) was created to manage the phasing out of the coal industry in the Sydney/Glace Bay region and to find alternative employment for 3,500 miners. While Wallace (1991) notes DEVCO achieved neither objective, Brown and Geddes (2007) observe that DEVCO did contribute to developing the basic infrastructure for tourism in the region leading to the development of a tourism cluster marketing culture, heritage and entertainment. Today over 40 museums varying from large National Historic Sites to provincial museums to a Mi'kmaq cultural centre (Wagmatcook Cultural Centre) and a Gaelic college (Gaelic College of Arts and Crafts) interpret diverse cultures on Cape Breton Island.

The main tourism attraction is now the Cabot Trail a scenic highway constructed in 1932 and with its subsequent paving in the 1960s Cape Breton Island was firmly established as a tourism destination. In the later part of the twentieth century, over 400,000 visitors annually travelled to Cape Breton Island to visit the Cabot Trail and Cape Breton Highlands National Park (Parks Canada, 2009). The park was named the second best national park in North America by National Geographic in 2005, for its preservation of nature and support and integration of local communities (http://www.nationalgeographic.com).

Organization of tourism

Institutions concerned with tourism on Cape Breton Island are federal, provincial, regional, municipal governments and agencies, and community-based volunteer groups. The federal Enterprise Cape Breton Corporation (ECBC) is a crown corporation created to promote and assist the financing and development of Cape Breton Island. ECBC's small geographical focus, local control and flexible mandate enable the Corporation to devise initiatives specific to local needs and priorities. (http://www.ecbc-secb.gc.ca). When policy permits, ECBC is one of the main organizations working to develop tourism here with their 2003 'Cape Breton Tourism Road Map Destination Development Plan'. This study recommended further developing the Colliery Route, a thematic route linking mining attractions, enhancing the links between attractions, and expanding the mining-related museum and the heritage district in Glace Bay. ECBC has fluctuated in its financial support of marketing for Cape Breton Island, making it difficult for Cape Breton Island to achieve a continuous and sustained presence as a destination in the world market place.

The Nova Scotia Department of Tourism, Culture and Heritage is the overarching body providing marketing support. It supported the development of specific destination areas within Cape Breton Island as well as leading tourism destination marketing for Nova Scotia. The primary website www.novascotia.com promotes Cape Breton Island, and the four travel ways that guide visitors in their trip planning. They also publish printed matter in the form of a driving guide, a Tourism Regions Map, and the Doers and Dreamers Guide which provides a listing of businesses and things to do. The Culture and Heritage division looks at preserving and communicating culture reflected by 'A Treasured Past a Precious Future: A Heritage Strategy for Nova Scotia – 2008–2013' (Nova Scotia Department of Tourism, Culture and Heritage, 2008a). The report recognizes the provincial government role to preserve, maintain, develop and protect Nova Scotia's heritage. Tangible heritage assets here include the artifacts, buildings and records and intangible include folklore, language and oral history including those related to mining heritage and history.

Destination Cape Breton Association (DCBA) was formerly a membership driven organization which accessed funds to develop a Cape Breton wide website and for marketing Cape Breton Island. It is currently restructuring to have a stronger role, and the main funding agency is Enterprise Cape Breton Corporation (ECBC). DCBA manages ten visitor information bureaus located strategically all over the island. The Cape Breton Island website promoting tourism is www.cbisland.com. Printed matter includes the Cape Breton Travel guide providing detail on the five tourism drives on Cape Breton Island including the Colliery Route, a Cape Breton Map and a Cape Breton Artisan Trail Map. This organization collaborates with the province to ensure that the island is appropriately marketed.

The Regional Development Agency is another level of administration with the mandate to facilitate economic development, community strategic planning and major core community renewal. It is the coordinating organization charged with leading economic development at the regional level under the authority of the Nova Scotia Regional Community Development Act (http://www.strait-highlands.ns.ca). The two regional development agencies on Cape Breton Island are the Straight Regional Development Authority, responsible for the counties of Inverness and Richmond, and The Cape Breton County Regional Development Authority responsible for the economic development surrounding Cape Breton's mining district. A façade/beautification program was implemented by the Cape Breton County RDA to stimulate the economy. This was implemented in collaboration with ECBC. Mining towns such as Glace Bay and New Waterford now have more attractive downtown cores; this has led to the establishment of more businesses, and community spirit in areas that have lost their prime economic means in the coal mines (pers.comm. Eileen Lannon Olford 9 September 2009).

A final level of government organization on Cape Breton Island is municipal. Cape Breton Island is divided into four municipal counties: Cape Breton, Victoria, Inverness, and Richmond. The municipalities are dedicated towards developing their local communities. The Cape Breton Regional Municipality covers an area of 2500 square kilometres and requires five regional coordinators for recreation and the local festivals. Victoria, Richmond and Inverness each have one Director of Tourism. These Tourism Directors work with both the RDAs and individual businesses to promote tourism, recreation and festivals.

Mining-related museums in Cape Breton Island are all community driven and run. They do have access to some provincial funding but they are managed by community volunteer boards of directors, taking their guidance from these boards. These community groups are also responsible for the individual marketing of their museums.

All of these organizations to some extent work together to market tourism. However, partnerships are limited by the mandates that each uphold and their function. For example, ECBC, the province and municipalities are the funding partners, while the destination development organizations, the regional development authorities and the community groups are responsible as both proponents and implementers. This results in a disproportionate reliance on not for profit groups and organizations without either organizational or financial stability to promote the island's tourism assets that include the rich mining heritage.

Visitation to mining attractions in Cape Breton

The visitor numbers to the mining-related museums compared to other attractions on the island are indicated in Table 12.2. Sites receiving the most number of visitors are those known throughout Canada and that benefit from national marketing strategies, e.g. the Alexander Graham Bell National Historic Site and the Fortress of Louisbourg National Historic Site. The mining-related attractions receive much less visitation than the federally managed ones. Nonetheless, the Cape Breton Miner's Museum is one of the most visited museums on the Island, despite it being off the main travel routes. The mining-related attractions are limited by geography and marketing and will undoubtedly have stable visitor numbers as long as the current marketing strategy is in place. Cape Breton Island and the province have used the Cabot Trail as an overall marketing point for decades, thereby channelling visitors to this area of the island.

The geography of the transportation routes in Cape Breton encourage visitors to travel up the TransCanada Highway on Route 105, or along the east coast on the Fleur-de-lis Trail or the west coast along the Ceilidh Trail (Map 12.1).

Cape Breton Island receives 32 per cent of Nova Scotian visitors (Table 12.3). Of these, 26 per cent pass through Baddeck and 20 per cent to North Sydney and the end of the TransCanada Highway, until crossing the ferry to Newfoundland. Approximately 20 per cent visit the Cabot Trail with 5–16 per cent of visitors stopping. In the Colliery Route area where the majority of the mining-related tourism products are located, the numbers of visitors travelling to the area are low. Only 1–3 per cent of visitors travel to the area, with only 2 per cent of visitors stopping. In contrast, all of the 9 per cent of visitors travelling to the town of Louisbourg also visit the Fortress of Louisbourg (Corporate Research Associates Inc., 2005).

Cape Breton Island has a wide breadth and depth to its mining-related tourism attractions and activities (Table 12.4), including built infrastructure in the form of museums, monuments, and scenic travel ways through former mining towns, to annual festivals celebrating the coal mining heritage (e.g. Bartown Days, North Sydney; Coal Dust Days New Waterford; Johnny Miles Festival, Sydney Mines). The coal mining industry is featured in many museums; related elements such as coal mining geology are highlighted in the Fossil Museum. The transportation system is interpreted at two railroad museums; this rail system was essential to coal mining industry development. Former miners to some extent are incorporated into

Table 12.2 Visitor numbers to mining related attractions in Cape Breton

Attraction	*Location*	*Visitor #s 2004–05*	*Visitor #s 2005–06*	*Visitor #s 2006–07*	*Visitor #s 2007–08*
Community					
Cape Breton Miner's Museum Largest museum, constructed on top of a mine shaft, guided tour of colliery with former miners, mining village	Glace Bay	22, 915	20,004	18,855	18,757
Glace Bay Heritage Museum Coal mining exhibit in a former town hall with rotating exhibits	Glace Bay	n/a	n/a	n/a	n/a
Dominion Heritage Museum Restored school house exhibits history of town and coal mining	Dominion	n/a	n/a	n/a	n/a
Sydney Mines Heritage Museum and Fossil Centre Focus on steel making & coal mining, fossils from coal field in a former railroad station	Sydney Mines	2,626	3,743	2,651	3,151
Inverness Miner's Museum Presents coal mining history of area in a former railroad station	Inverness	7,068	8,559	6,227	5,227
S&L Railroad Museum Restored railway station (1895), original freight shed with exhibits	Louisbourg	2,626	3,743	2,651	3,151
Colliery Route Thematic route linking attractions	Starts in Glace Bay	n/a	n/a	n/a	n/a
Provincial					
Highland Museum	Iona	21,538	21,995	19,818	19,359
Cosset House Museum	Sydney	5,067	5043	4157	4133
Museum of Industry	Stellerton, mainland Nova Scotia	47, 435	36,076	32,891	27,728
Federal					
Cape Breton Highlands National Park	Cabot Trail	311,000	340,000	134,000*	118,000
Alexander Graham Bell Museum	Baddeck	102,000	97,000	97,000	93,000
Fortress of Louisbourg	Louisbourg	83,000	77,000	74,000	71,000

Source: NSTCH 2009, NSTCH 2006–2009.

* Numbers were calculated by number of visitors instead of visitor days.

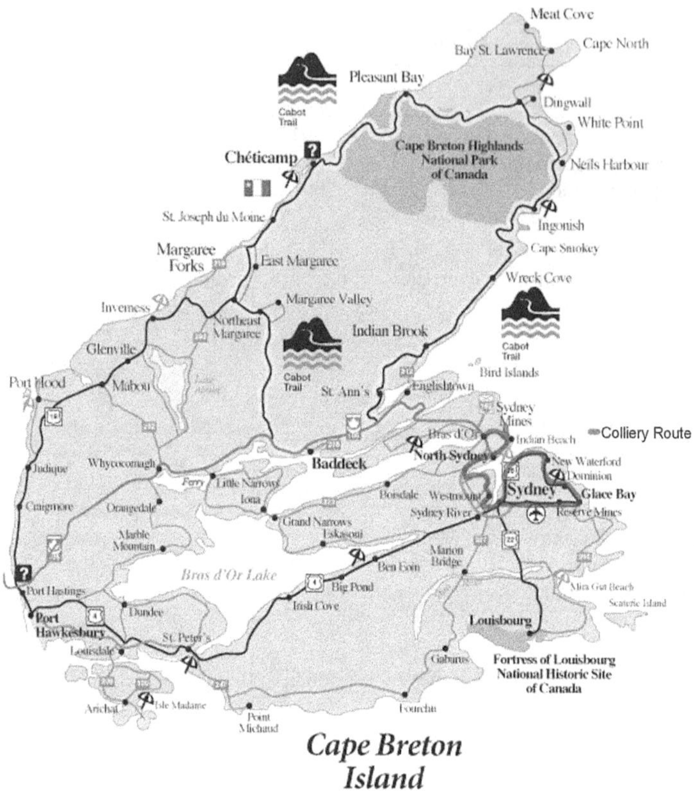

Map 12.1 Cape Breton Island with the Colliery Route located in the North Eastern Section.

Source: Nova Scotia Department of Tourism, Culture and Heritage (reprinted with permission).

the tourism industry as informed guides, providing an authenticity immediately obvious to visitors, e.g. at the Cape Breton Miner's Museum.

Case study: Cape Breton Miner's Museum

The Cape Breton Miner's Museum in Glace Bay opened in 1967 as a centennial capital project. These projects commemorated the centennial of Canada's establishment as a country in 1867. Key (1973: 194) refers to this project as 'one of the most successful monuments to confederation . . . . a centennial capital project which miraculously achieved a distinction of being architecturally well designed and technically functional'. Constructed as a stylized abstraction of colliery surface buildings, the building has a central memorial tower in the form of a head frame (Figure 12.1).

The museum is representative of both the socio-cultural and productive types of mining attractions identified by Edwards and Llurdés i Coit (1996: 349) providing the visitor with a perspective into the life of the coal miner and coal mining in both the nineteenth and twentieth centuries. The museum website describes this attraction as paying 'tribute to the region's long and rich history of coal mining' (www.minersmuseum.com). The Museum is also home to the famous choir of miners – The Men of the Deeps. Adjacent to the main museum building

Table 12.3 Where do Nova Scotia's visitors travel?

Community	*Percentage in the area (%)*	*Percentage who stopped (%)*	*Percentage who stayed at least one night (%)*	*Average number of nights stayed*
†‡Baddeck	26	16	9	2.2
†Cheticamp	20	13	6	1.8
*Glace Bay	3	2	1	4.7
†Ingonish	19	11	5	1.5
*Inverness	13	4	1	3.1
Iona	1	1	<1	3.1
Louisbourg	9	9	3	2.0
*Mabou	13	3	1	3.5
†Margaree Forks	21	4	1	2.9
*New Waterford	2	1	<1	5.5
‡North Sydney	20	6	4	2.8
‡Port Hastings	32	6	2	5.5
Port Hawkesbury	12	4	2	2.0
*Port Morien	1	<1	<1	6.6
St. Peters	10	4	1	5.5
Sydney	17	7	4	5.4

Source: Adapted from NSTCH 2005.

* Former mining community along the Colliery Route.
† Cabot Trail.
‡ Trans Canada Highway.

a miners village has been developed, with the architecture reflecting the typical workers' houses provided by mining companies (Figure 12.2). One building accommodates a restaurant serving typical local cuisine.

The main museum offers a section on geology, the mines and mining technology. Machines formerly used in mining are arranged outside on the grounds with interpretive signage reflecting the processing type of mining visitor attraction noted by Edwards and Llurdés i Coit (1996: 349). The museum was built on top of a coal seam and a purpose built coal mine was dug out by real miners. An underground mine tour, limited to 20 people at a time is conducted. It includes visits to various caverns mined for coal, ranging from a seam that is a comfortable height for standing to one barely a foot in height. Interpreters are former coal miners, a trend that Pretes (2002) noted as contributing to providing an authentic perspective on coal mining and the conditions that miners worked under. Visiting a former coal mine and touring underground is typical in the heritage mining tourism industry (Pretes, 2002).

Unlike other areas where the visitors to heritage mining attractions may stay in the locale, in Glace Bay the corresponding tourism infrastructure around the mining museum has not been developed. Currently Glace Bay receives only 3 per cent of Cape Breton Island drive through visitors (Nova Scotia Tourism Culture and Heritage, 2004). This infrastructure is limited to attractions. Seven accommodation providers on the Colliery Route provide a total of 19 rooms in Bed and Breakfasts (Nova Scotia Tourism Culture and Heritage, 2009). Visitors from outside the area come to view the mining attractions but would most likely overnight in either Louisbourg or

Table 12.4 Top 10 visitor activities on Cape Breton Island

Top 10 visitor activities	*% Activity*	*Actual numbers in 2004*
Leisure walking/hiking	80	574336
Visiting craft shops/studios	69	495364.8
Visiting a national/provincial historic site	67	481006.4
Nature observation	64	459468.8
Visiting museums	51	366139.2
Beach exploring/beach combing	42	301526.4
Visit a national or provincial natural park	42	301526.4
Live performing music or dance	28	201017.6
Visiting night clubs, lounges, or pubs	25	179480
Birding	25	179480

Source: Nova Scotia visitor exit survey.

Sydney where there is more diversity of accommodation with hundreds of rooms to choose from (pers. comm. Lou Leith Destination Cape Breton Association 17 September 2009).

Case study of railway related sites

Transport attractions are acknowledged as a type of industrial heritage attraction (Edwards and Llurdés i Coit, 1996). In Cape Breton this type of attraction has a focus on the railway history

Figure 12.1 Cape Breton Miner's Museum, Glace Bay. Source: Kim Lemky.

Figure 12.2 Cape Breton Miner's Museum Village. Source: Kim Lemky.

related to mining in the area as many small railways were built from the coal mines to the ocean ports for shipping. The first railroad was built in 1834 from Sydney Mines to the coal pier in North Sydney, a distance of 4.8 kilometres, the first one to be built with iron rails and a standard gauge. The railroad for freight and passenger traffic was completed in 1890 from Port Hastings (now the Canso Causeway) to Sydney by the Intercolonial Railway, initiated to join the Atlantic Provinces to the rest of Canada. The Inverness Miner's Museum, the North Sydney Heritage Museum and Fossil Centre, the Orangedale Railway Museum and the S & L Railway Museum in Louisbourg are all housed in former railway stations. The stations built in the early 1900s to service the coal industry and passengers were vital to the coal mining industry. All these lines closed and none except for Orangedale is currently linked to a railway line.

In Inverness County the former railroad track (Inverness and Richmond Railway) that brought coal to Port Tupper, from Inverness and Mabou, was constructed in 1901. In the 1970s it was decommissioned. In 2008 it was transformed into the TransCanada Trail, a multi-use trail. It extends from the Canso Causeway to Inverness along the western coast of Cape Breton Island. This demonstrates one way in which the heritage related to coal mining is being reinvented for recreation and tourism.

The Sydney & Louisburg Railway (S & L Railway) was the most significant railway for the coal and steel industries on Cape Breton Island. It had its inaugural run in 1895 and rails were removed in 1968. It had 185 kilometres of track. The railway was vital to the development of the coal mines and the steel plant as well as linking the local communities to Sydney. In World War II coal trains arrived at local shipping piers at the rate of one every 15 minutes, and more than 6,000 coals bearing ships left Sydney Harbour between 1940 and 1945. After the 1940s there was a slow demise of the coal industry due to cheaper oil and gas from the Middle East. This railway was phased out, but is commemorated in the S&L Railway Museum.

There is still one railroad in existence, the Cape Breton and Central Nova Scotia Railway operating between Truro and Sydney with spurs to Port Hawkesbury/Point Tupper, owned by Rail America. It carries freight but no passengers. That this infrastructure still exists might offer some opportunity in the future for recreating and interpreting the rich railroad history related to mining operations in the area.

Case study of the Colliery Route

The Colliery Route developed in 2002 by a local community group provides a self-guided tour of the sites of interest in the rich coal mining history of north-eastern Cape Breton (Map 12.1). The lead author travelled this route with a handout provided at the Cape Breton Miner's Museum. It was a picturesque drive along the coast and provided an interesting perspective of the architecture of the mining towns, the monuments to coal mining in the region, and linked a number of mining and other museums. The route was demarcated by distinctive trail blazer signs (Figure 12.3). It started in Glace Bay and followed Route 28 along the coast through the

Figure 12.3 Colliery Route Signage. Source: Kim Lemky.

towns of Dominion and then through New Waterford, New Victoria, Victoria Mines through to Whitney Pier and Sydney, Sydney River, Westmount, Edwardsville, North Sydney, and Sydney Mines. The trail ends in Little Bras D'Or where it meets the TransCanada Highway. The Colliery Route demonstrates how the mining heritage and sites of the region can be bundled together as a product for prospective heritage tourists (McKercher and du Cros, 2002).

The potential for further development of the route has received some attention in terms of regional tourism studies. As previously noted a 2003 ECBC report recommended further developing the route enhancing links between attractions. If acted upon, the further enhancement of the route could be critical to nurturing mining related tourism in the region. Currently the pamphlet used to promote the route is out of print; this should be reprinted and available for visitors.

Discussion

The diversity of the tourism product based on mining on Cape Breton Island varies from museums dedicated to mining, railway attractions, an educational travel way through mining communities, to festivals and monuments. In addition a golf course has been developed on reclaimed mine tailings. Edwards and Llurdés i Coit (1996) classify such attractions in the four types of productive, processing, transport and socio-cultural. In Cape Breton's mining region all types are present (Table 12.5) demonstrating the trend for post-industrial regions to use industrial remains for heritage and to market themselves as heritage destinations (Timothy and Boyd, 2003).

The successes of developing a tourism product based on mining in Cape Breton are varied. With over 40 museums and more than 60 heritage organizations only a small segment of heritage attractions are related to mining. One of the challenges that community museums are faced with is that they are managed by volunteer boards of directors, there is no secure funding, corporate memory or continuity. As with the mining attractions in Norway (Swarbrooke, 2002) many of those related to mining in Cape Breton are small and seasonal with a mainly domestic market, only receiving a few thousand visitors annually (Table 12.1). The exception to this visitation trend is the Cape Breton Miners Museum receiving about 20,000 visitors annually. With active mining no longer present in Cape Breton Island the strong visitor performance of this museum is a sign that mining heritage related tourism is evolving. As Timothy and Boyd (2003) indicate, natural resource industry attractions are particularly appealing because of their outdoor settings where the relationship between humans and nature is immediate and vital.

Table 12.5 Types of mining attractions in Cape Breton

Type	*Cape Breton Example*
Productive	Cape Breton Miner's Museum
Processing	Cape Breton Miner's Museum
Transport	S & L Railway Museum
Sociocultural	Glace Bay Heritage Museum Dominion Heritage Museum Sydney Mines Heritage Museum Cape Breton Miner's Museum Village

Source: After Edwards and Lurdés i Coit (1996) and Timothy and Boyd (2001).

Conclusion

In conclusion, the decline of the coal and related steel industries in Cape Breton led to the establishment of heritage attractions related to community mining histories, such as the Cape Breton Miner's Museum. Regional economic incentives offered by the federal government directly influenced the development of and/or improvement of other tourist attractions in the region (i.e. the Cabot Trail and Fortress Louisburg).

There are six main museums related to representing and interpreting the mining culture in Cape Breton Island, and except for the Inverness Railroad Museum these are all close together geographically and linked thematically in "the Colliery Route". However, mining-related attractions and sites make up only a small part of the heritage tourism product of the region, and the one product that bundles these together, the Colliery Route, while promoted locally is not promoted by the Province of Nova Scotia as a travel way (although it is mentioned in tourism literature, as a route). Individually, visitor numbers at the majority of community museums are small. The central museum in the route is the Cape Breton Miners Museum in Glace Bay, where the unique tour with a former coal miner provides the visitors with an authentic underground coal mine experience. Various levels of governments have worked together to support this industry, through funding and policies promoting heritage. A unique Cape Breton Regional Economic Development Authority program to gentrify the downtown cores for many of these former mining communities has led to some economic development.

To date, the marketing strategy for Cape Breton Island has focused primarily on key attractions such as the Cabot Trail; however, to encourage visitors to stay longer, a comprehensive strategy is needed to provide visitors with a greater understanding of the diversity of products, including those related to mining history, available for sampling on Cape Breton Island. To this end, the Destination Cape Breton Island Inc. is at the time of writing developing a new strategy in collaboration with Enterprise Cape Breton Corporation to revitalize the marketing of Cape Breton Island as a destination area.

The attraction of the Cape Breton Miner's Museum in collaboration with the ambassadors in the choir 'The Men of the Deeps' has been somewhat successful. Unfortunately, the tourism mining attractions as a whole are not at the centre of the marketing strategy for Cape Breton Island either in spirit or geographically. Nonetheless, as the current trend in tourism is towards heritage tourism and authentic experience, there may be an increased energy towards a stronger marketing of these attractions nurturing mining-related tourism here for the benefit of former mining communities.

References

Beeton, S. (2005) 'The Case Study in Tourism, a Multi-method Case Study Approach', in B.W. Ritchie, P. Burns, and C. Palmer (eds) *Tourism Research Methods: Integrating Theory with Practice*, Oxfordshire, Wallingford: CABI Publishing.

Bethune, J. (2009) *Images of our Past, Historic Baddeck*, Halifax, Nova Scotia: Nimbus Publishing.

Brown, K.G. and Geddes, R. (2007) 'Resorts, Culture and Music: The Cape Breton Tourism Cluster', *Tourism Economics*, 13: 129–41.

Brown, R. (1979) *A History of Cape Breton Island*, Bellville, Ontario: Mika Pub. Co.

Cole, S. (2005) 'Action Ethnography: Using Participant Observation', in B. W. Ritchie, P. Burns and C. Palmer (eds) *Tourism Research Methods: Integrating Theory with Practice*. Wallingford, Oxfordshire: CABI Publishing.

Corporate Research Associates Inc.(2005) *2004 Nova Scotia Visitor Survey Exit Study Final Report*, Nova Scotia: Nova Scotia Department of Tourism, Culture and Heritage.

Edwards, J.A. and Llurdés i Coit, J.C. (1996) 'Mines and Quarries: Industrial Heritage Tourism', *Annals of Tourism Research*, 23(2): 341–63.

Enterprise Cape Breton Corporation. Available online at: http://www.ecbc-secb.gc.ca (accessed 17 September 2009).

Key, A. (1974) *Beyond Four Walls: The Origins and Development of Canadian Museums*, Toronto: McClelland and Stewart Limited.

Lamb, J.B. (1975) *Hidden Heritage: Buried Romance at St. Ann's Cape Breton Island*, Wreck Cove, Nova Scotia: Breton Books.

McKercher, B. and du Cros, H. (2002) *Cultural Tourism: The Partnership Between Tourism and Cultural Heritage Management*. New York: Haworth.

Millward, H. (1985) 'Mine Locations and the Sequence of Coal Exploitation on the Sydney Coal Field', in K. Donovan (ed) *Cape Breton at 2000 Historical Essays in Honour of the Island's Bicentennial 1785–1985*. Nova Scotia, Sydney: Cape Breton University Press, 1720–980.

Mining Culture Symposium. Cape Breton University, Nova Scotia: Sydney, June 2005.

National Geographic. Available online at: http//www.nationalgeographic.com/traveler/features/2005 (accessed 17 September 2009).

Nova Scotia Department of Tourism, Culture and Heritage (2008) *A Treasured Past a Precious Future. A Heritage Strategy for Nova Scotia 2008–2013*. Unpublished Internal Government Document.

Nova Scotia Department of Tourism, Culture and Heritage (2009) Penny Harvey, Provincial Museum Stats for Cape Breton. E-mail (8 September 2009).

Parks Canada (2009) Visitor Experience Assessment Cape Breton Highlands National Park of Canada, 27, 29 May 2009 (Unpublished Document).

Pretes, M. (2002) 'Touring Mines and Mining Tourists', *Annals of Tourism Research*, 29(2): 439–456.

Silbergh, D., Fladmark, M., Henry, G. and Young, M. (1994) 'A Strategy for Theme Trails', in M. Faldmark (ed) *Cultural Tourism*, London: Donhead.

Swarbrooke, J. (2002) *The Development and Management of Visitor Attractions*. Oxford: Butterworth-Heinemann.

Tennyson, B.D. (2000) *Guardian of the Gulf: Sydney, Cape Breton, and the Atlantic Wars*. Toronto: University of Toronto Press.

Timothy, D.J. and Boyd, S.W. (2003) *Heritage Tourism*. Harlow, Essex: Pearson Education Limited.

Wallace, I. (1991) *A Geography of the Canadian Economy*. Don Mills, Ontario: Oxford University Press.

Zaslow, M. (1975) *Reading the Rocks: The Story of the Geological Survey of Canada 1842–1972*. Ottawa, Ontario: The Macmillan Company of Canada Limited in association with the Department of Energy, Mines and Resources and Information Canada.

13 Geotourism site development in Slovakia

Geza M. Timcak, Pavol Rybár and Jana Jablonská

Introduction

This chapter describes some of the most interesting geotourism sites in Slovakia. Apart from surface phenomena, caves and mining heritage sites are also mentioned. A short description of the geological setup of the country is followed by a brief outline of the history of geotourism in Slovakia. To indicate the development dynamics of geotourism, the aims and impacts of the relevant government policies are discussed addressing the question of the sustainability of geotourism.

Slovakia as a destination

Slovakia is situated in Central Eastern Europe (Figure 13.1). The country has a wide range of landscapes, from high and rocky to flat lowlands. As a state it is a relatively new – it was formed in 2002 after the partition of the former Czechoslovakia. Czechoslovakia in turn was formed in 1918 after the splitting up of Austro-Hungarian monarchy. It is a multi-ethnic state. From its 5.5 million inhabitants, nearly 5 million are Slovaks and the largest minority is Hungarian (about 0.5 million). Lesser minorities are the Ruthenians, Germans, Poles, Ukrainians, Czechs and Gipsies. Thus the country has a rich spectrum of cultures that used to live together peacefully.

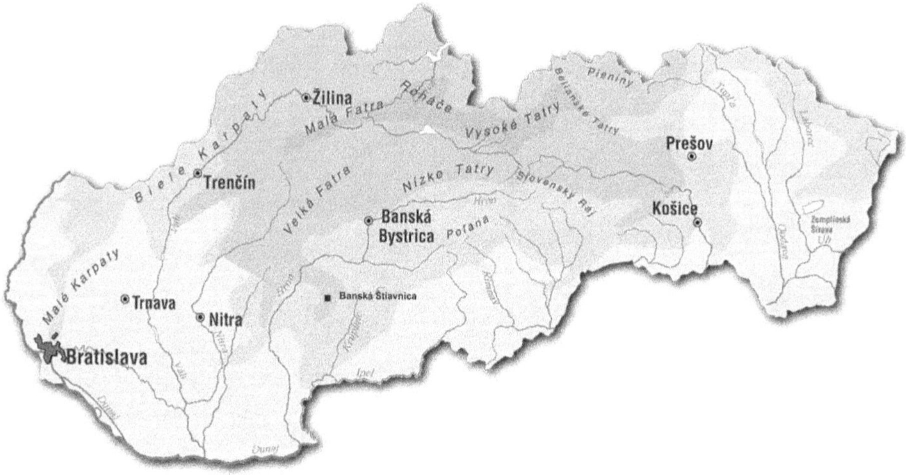

Map 13.1 Map of Slovakia.

Nevertheless, Slovakia is still not sufficiently well known as a tourist destination. In 2008, only 0.5 per cent of EU citizens have chosen Slovakia as their holiday destination. According to an EU survey from February 2009 (Ministry of Economy, 2009) up to 56 per cent of EU citizens organize their holidays individually using Internet resources. Thus, the national tourism portal of the Ministry of Economy (Slovak Travel Portal) may help Slovakia to become better known as a destination. The same survey indicates that 31 per cent of EU citizens go to various destinations because they prefer local natural and cultural specialties and flavours. This trend would give Slovakian geotourism a great chance to attract visitors.

Formal aspects of geotourism development in Slovakia

Tourism is governed by the tourism section at the Ministry of Economy. Tourism in turn forms a part of regional development, which is regulated by the Ministry of Building and Regional Development. The marketing of Slovakia as a tourist destination is done by the Slovak National Tourist Board (SACR) state agency. As a part of the efforts of the Ministry of Economy to allocate appropriate tourism development funds to the various regions of Slovakia, it developed a system of 'regionalisation', where the defined parts were ranked and allocated local, regional, national or international importance (www.economy.gov.sk). As it can be seen at the Slovakian Travel Portal (see references), the administrative and tourism related regions do not overlap, as the latter follow the historical division of regions. In the 'New strategy of tourism development of the Slovak Republic up to 2013' (Ministry of Economy, 2007) mining heritage tourism has a prominent place. Nevertheless, the funding of tourism development has not yet been adequate and the European funds applicable for tourism and regional development have a complicated system of access and thus are under-utilized. This is one of the reasons why the Association for Support of Tourism Development in Slovakia that groups tourism entrepreneurs, has the feeling that in spite of the official declarations, tourism has a low priority with the government (http//:www.arcr.eu).

Regional development is regulated by Act No.503/2001. There have been unsuccessful efforts to update this Act. It formulates Regional Operational Programmes (ROP) that should give shape to strategies for regional development on the national, county and local levels. Regional development is supported also by the possibility of forming Euroregions in cases where there is a natural or cultural subject of mutual interest (like the karst areas in SE Slovakia and NE Hungary, the basalt outcrops in Cér Mts, that are situated in south Slovakia and N. Hungary, the High Tatra Mts that are in N. Slovakia and S. Poland, etc.). In some cases Euroregional associations help to develop and manage natural resources relevant to geotourism. The ROPs on the national, regional and local level (i.e. the levels of NUTS 1, NUTS 3 and NUTS 4) need to be harmonized in order to get financing. NUTS are statistical entities used by the European Union. The whole country is NUTS 1, regions (like E. Slovakia) represent NUTS 2, counties represent NUTS 3, etc.

Non Governmental Organizations (NGOs) in cooperation with other partners can have an important role in tourism development because they are more flexible. They can address EU grant programmes to help the local communities to get prepared for entrepreneurial activities as well as to gain the necessary skills needed for tourism development and management. In the case of mining heritage related activities, in some regions the state sector helps to develop the man-made industrial and historical objects into tourism attractions; in other areas the NGOs, regional associations, local government and Higher Education Institutions (HEIs) do that task, aided by EU grant resources.

In the case of cities like Banská Štiavnica (Schemnitz), where the Central Mine Archives, the Mineralogical Museum, Museum of Mining and a number of 5th Form (mining, chemical, forestry) colleges are clustered, development is shaped by these organizations. In the case of the city of Gelnica or the village of Dubník – places, where no state institutions are present to deal with mining heritage, local government with NGOs, local entrepreneurs and sometimes regional HEIs are the driving force behind tourism development.

Still, in spite of the efforts of the government and of the organizations noted above, Slovakia underperforms in tourism revenue and job creation. Whilst the EU average revenue from tourism is about 10 per cent of the GDP, in Slovakia it is only about 2.5 per cent. In spite of an 8 per cent increase in foreign tourist visits in 2008, the government predictions for 2009 (Table 13.1) have not materialized (Ministry of Economy, 2009), but that may be also be attributed to the economic crisis of 2009.

Geological features of Slovakia and geotourism

The relief of Slovakia features two distinct geological and geomorphologic formations: the Carpathian Arc and its adjacent lowlands. It was shaped by young tectonics, tertiary volcanism and glacial sculpturing as well as by periglacial, glacio-fluvial and fluvial processes.

In the Central and Inner Carpathians, erosive denudating processes have removed the cover of the younger sedimentary rocks as well as the upper pile of nappes to such a degree that the crystalline basement, made up of granitoid rocks or of various crystalline shales, outcropped on the surface. These include the highest mountain ranges of Slovakia – the High and the Low Tatra Mountains. Their relief has changed depending on the mechanical resistance of the rocks. The cores made of granitoid rocks represent a more resistant basement. They appear usually as rounded, smoothly modelled surface forms and rarely as steep rocky relief (with the exception of a glacial relief).

The shape of Mesozoic rocks is very diverse. Resistant Mesozoic sedimentary rocks, particularly the thicker limestone formations, dolomites and quartzites often outcrop in the form of rocky peaks and various sharp forms. They contrast with smoothly modelled, less resistant Mesozoic marls, slates and other sediments.

The volcanic mountain ranges of Slovakia have predominantly a strato-volcanic structure. It means an alternation of less resistant pyroclastic rocks with resistant massive andesite lava

Table 13.1 Tourism development indices given by the "Strategy of tourism development in the Slovak Republic" for the 2003–2013 period

Index/year		*2003*	*2006*	*2009*	*2013*
1.	Average no. of nights spent	3,6	4,0	4,3	4,8
2.	No. of overnight stays in thousands	3,373	3,600	3,900	4,300
2a.	No. of foreign tourists	1,386	1,450	1,570	1,710
3.	Foreign arrivals in thousands	24,984	25,900	26,500	27,200
4.	Total no. of overnight stays	12,058	14,400	16,770	20,640
4a.	No. of foreign tourists	4,964	5,800	6,751	8,208
5.	Average spending of tourists per person and day in USD	46,32	52	60	70

flows. Less frequent rhyolitic rocks have been preserved as extrusive domed bodies, short lava flows and tuff layers frequently altered into bentonite or zeolite. The youngest products are basalts. The basalt lava flowed originally through river valleys. Because of a relief inversion, they often form rocky peaks and elevated ridges. Formation of less resistant pyroclastic rocks (tuffs, tuffites) is usually found in smooth parts of slopes, while steeper to rocky parts of slopes are bound to outcrops of massive andesites and basalts.

In the interior part of the Carpathian range there are Neogene volcanites and neovolcanites. Some of the volcanic activities continued up until the Quarternary period. The oldest rocks are rhyolites and rhyodacites (Lower Miocene). The middle stage of the volcanic activity resulted in andesites and dacites. At that stage, the stratovolcans of Pol'ana, Javorie, Banská Štiavnica and Kremica were formed. The final stage of the volcanic activity had a basic, alcaline character. It occurred mostly in the area of B. Štiavnica, Nová Baňa and Cér Mountains. (Centre for Scientific Tourism in Slovakia: www.ecosystems.sk). Abundant mineralization has led to intensive mining activities for millennia. All these geological formations provide a rich ground for the dedicated geotourists, as well as for those who enjoy the manifestation of the creative power of both the inanimate and vegetal systems.

History of geotourism in Slovakia

Geotourism and sports related to geotourism (like trekking, mountaineering, rock climbing, cave exploration, etc.) have been practised here for a long time. The High Tatra Mountains are perhaps the oldest centres of summer and winter sports. Their special climate has made this area a well known climatic spa since the times of the Austro-Hungarian monarchy.

Presently, every year around 3 million visitors come to this area (SME, 2008). The earliest mention of caves in the territory of Slovakia comes from the work of Petrus Ransanus (AD 1490, Kordos *et al.*, 1980), who discovered a cave in the present day Sliač spa and the ice cave of Dreveník. The first speleological research here started around 1910 (Kordos *et al.*, 1980). The first cave that was developed into a tourist attraction was the Domica-Baradla cave. At present, there are 12 caves open to public. The first tourist and mountaineering clubs (Magyarországi Kárpát Egyesület and the Karpatenverein) were established in 1873 in the High Tatra region. They also introduced skialpinism to the region in 1875 (Luczy, 2008).

Mountain trekkers in the High Tatra Mountains were divided from the very beginning into two groups – those who enjoy the challenge of conquering the peaks and those who simply enjoy the beauty of the mountains. These latter were the first geotourists. From the very beginning trails were marked, descriptions of tours and maps published. At present, this development is further aided by the Slovak Tourist Club, which cares for the trail markings, new trail design and in general coordinates walking activities. There are about 12,000 km of marked tourist trails in Slovakia. Climbers are grouped mostly in JAMES – an organization first established in 1921 and re-established after 1989. JAMES is an acronym standing for idealismus, alpinizmus, morality, eugenics and solidarity – the motto of the JAMES founders. Cycling tourists are catered for by a number of national and regional organizations. Skiers (downhill or cross country) have been organized since the beginning of twentieth century within the aforementioned clubs. Professionals in geotourism and geotourists are grouped in the new International Association of Geotourism (see www.iageotour.com).

An outline of selected geotourist sites

Of the Slovakian geotourism sites, the High and Low Tatra Mountains are perhaps the most popular. The Tatra Mountains form a part of the Carpathian range. The High Tatra Mountains occupy about 260 m^2. They have more than 300 peaks and 35 valleys. There are 26 peaks higher than 2500 m. The highest lake is 2157 m above the sea level. Together with the White and Western Tatra Mountains, they form the Tatra National Park.

The morphologically most interesting parts are built by Carboniferous granitoides. The envelope of this range is built by Cretaceous sedimentary rocks. These also have interesting morphology and extensive cave systems. Some of the exquisite scenic features are due also to the effect of Quaternary glaciations (e.g. the impressive hanging valleys in the Bielovodská valley).

Geotourism in the High Tatra National Park is concentrated on trekking routes that usually lead through valleys and passes. There are many trekking routes that satisfy the needs of geotourism. The main value of these is the scenic quality of the majestic mountains, glacial lakes and valleys (Figure 13.1). Interestingly, in some parts of the High Tatra Mountains copper was mined in medieval times at Predné a Zadné Meďodoly.

The stratovolcano of Banská Štiavnica is geotouristically important due to its impressive dimensions, great variety of rocks and above all due to its mineralization (precious metals, polymetallic mineralization). There are also therapeutically important mineral waters – the Sklenné Teplice Spa, where there are 12 thermal springs (with water temperature of 39.8–52.3 °C).

Figure 13.1 Peaks around the Czech lake in the Tatra Mountains.

The Szabo's rock near Lehôtka pod Brehmi is named after the late Professor Szabo, the founder of the Hungarian School of Geology ('Szabó' in Wikipedia; see Biography of Professor Szabó József, 2009), who first described it. It was also the first protected site in the territory of Slovakia – it has been protected since 1907 by legislative measures. Szabo's rock is a cone of volcanic origin formed by cooling of acid, rhyolitic lava. The rhyolites have fluidal texture. In the nearby area, rhyolite was mined for the last 700 years. Close to the Szabo's rock (Anonymous: Geological History of Szabo's Rock, 2009) a deposit of perlite is still mined.

The Cér Mountains situated in the southern part of Central Slovakia bordering Hungary were formed during the final stages of the volcanic activity mentioned above. The region is a National Park. One of the interesting geotouristic phenomena is a basalt rock face with a fortress (Somoska) at the top of the mountain. At a locality called Hajnačka, rich palaeontological findings were made. At the Owl Fortress (Soví hrad) there is an archaeological site from the Stone and Bronze Age.

Near Brhlovce village, situated in the Ipel' mountains that are built by neovolcanites, there are outcrops of opalized wood (of Holzstein age) and in the pores of andesites, opal coatings can be found. Interesting man made artifacts are human dwellings from the sixteenth, seventeenth, nineteenth and twentieth centuries that were carved by locals into tuffs. In this area is the Novohrad – Nógrád binational Geopark, which includes the territory of 28 local communities in Slovakia and the territory of 36 communities in Hungary. It facilitates access to the youngest volcanic structures of the protected areas of Cér Mountains and Karancs Medves. (Štátna Ochrana Prírody, 2004; Cimermanová and Anderková, 2006).

In East Slovakia, one of the important geotourist sites in is the 'Herl'any Geyser'. Here, in 1872 drilling was done to secure mineral water for a spa. When the drill reached 330 m of depth, the mineral water started erupting to 112 m. Presently, the water erupts to 15 m within a 32 hour period.

The Spiš Fortress is one of the biggest fortresses in Central Europe and a UNESCO World Heritage Site since 1993. It is built on an impressive travertine cliff. The locality has been inhabited since the Paleolithic period (History of Spišský hrad, 2009). The fortress was established in the twelfth century; its habitable parts were mostly Romanesque and Gothic. A fire damaged the fortress in 1780. The fortress has been partially reconstructed. Next to the Fortress, there is the Drevenik 'Stone city' – an area of huge travertine pillars. The area is full of historical sites – like a Romenesque and Gothic cathedral in Spišské Podhradie, a Gothic cathedral in Spiške Vlachy, a gothic cathedral and a gothic church in Levoča, and a gothic church in Žehra, to name just a few (Jablonská *et al.*, 2007).

Slovak Karst is a region in the south east part of Slovakia. It contains karst plateaus, deep canyons and caves. A part of this karst region is shared with Hungary (Aggtelek). Both sides of the karst area are national parks. Perhaps the most famous cave here is the Domica-Baradla system that was first mentioned in 1786 (Kordos *et al.*, 1980: 5). In this binational area the Baradla part is in Hungary, the Domica part in Slovakia. For speleologists this karst is an excellent research ground, where the number of discovered caves increases every year. Four of the caves in this region – the Domica, Jasov, Gombasek and Silicka Ľadnica are accessible to public. Silická Ľadnica was first described by Matthias Bel in 1744 (Manikova and Popovics, 2009). The majority of these caves are only accessible to researchers.

Of the canyons of this region the most interesting are the Zádiel and Háj canyons. The Zádiel canyon was formed through the erosive activity of Blatná creek. On the Zádiel karst plateau there are also well preserved remains of a long Bronze Age fortification wall that made the then inhabited area better defendable.

Mining heritage related tourism

At present, mining heritage tourism in Slovakia is well developed only in Banská Štiavnica (Figure 13.2). The well-preserved technological and historical landmarks linked to the famous mining history and the spirit of identity of its contemporary inhabitants with their history contributed to the inscription of this town on the UNESCO World Heritage list in December 1993.

Other parts of Slovakia, which – still within the Hungarian Kingdom – used to be for centuries the European centre of precious metals production as well as copper and iron mining also provide opportunities for mining heritage tourism, but are less well developed in this respect. The reasons are many, but perhaps the slow growth of public awareness of the possible benefits of marketing the famous mining heritage sites, the cost of the reconstruction works as well as the costs of building the necessary infrastructure needed by tourism may be the most prominent ones.

Banská Štiavnica

The origin of the town and its development is linked to extraction of gold and silver in the area. In the Middle Ages, the Hungarian Kingdom became the richest country in Europe regarding the amount of mined precious metals, and Banská Štiavnica was credited for that wealth more than other parts of the Kingdom.

The first credible data on silver extraction at Banská Štiavnica come from AD 1217. The document of the Hungarian emperor Andrew II puts the quantities mined to as much as 600 kg of silver per year. In 1740, gold extraction reached 680 kg annually, and the yearly silver output was 26,000 kg. During the second half of the18th century mines in the territory

Figure 13.2 Banská Štiavnica – Centre of the Old Town.

of Slovakia produced four times more silver than Saxony, six times more than Transylvania, and nine times more than Bohemia (Rybár and Lichner, 2007).

Banská Štiavnica played a great role also in the field of mining education. In order to educate mining specialists, the School of Mines (Bergschule) was established in 1735, directed by the outstanding scholar Samuel Mikovini. After two years of studies, the graduates achieved both theoretical knowledge and practical skills in mining operations. In 1762, Austro-Hungarian Empress Maria Theresa founded the first ever mining college – the Mining Academy (Bergakademie) of Banská Štiavnica. Later, the Academy was transformed to the Mining and Forestry College in Banská Štiavnica. After the cessation of the Austro-Hungarian monarchy – after WWI, the College of Mining and Forestry was transferred to the post-war territory of Hungary. Due to disputes about who is the real successor of the famous College, in 2001, the Declaration of the Intellectual Successors of the former Mining Academy and College at Banská Štiavnica was endorsed in Košice. Signatories of the treaty recognized the shared rights of successors of the Mining Academy of Banská Štiavnica. These HEIs were granted these rights: Mining Faculty, Miskolc University (Hungary), Technical University Leoben (Austria), BERG Faculty, Technical University Košice (Slovakia), Faculty of Forestry Sopron (Hungary), Faculty of Mining and Geology Ostrava (Czech Republic), and Faculty of Forestry Zvolen (Slovakia).

From the mining heritage tourism point of view, it is important to know that in Banská Štiavnica all the original monuments and buildings where the Mining Academy and the College of Mining and Forestry existed have been preserved.

The new Guide to the Technical Monuments in and around Banská Štiavnica describes 79 mining technical objects, along three trails – A, B and C. They all start at the Holy Trinity square, right at the centre of Banská Štiavnica.

Route A leads to 24 landmarks through significant buildings of Banská Štiavnica, such as Berggericht, Belházy's house, Kretschmáry's house, the Knocker, etc. It offers a visit to the Open-air Mining Museum, as well as to the monumental historical inclines, including the hereditary Bieber incline. The trail ends at the Veľká a Malá Richnava and Počúvadlo mine water reservoirs.

Route B shows important buildings in the centre of the town, such as the Kammerhoff. As early as the thirteenth century this was the seat of the Mining Chamber where miners brought precious metals for assessment or exchange. A part of the extracted ore, the so called urbura was taken for the ruling monarch as tax. From 1598 on, the building became the residence of the council of the Chief Kammergrafs. Since 1966, the building houses the Slovak Mining Museum. Another point of interest on the route B is the Glanzenberg hereditary incline. The oldest documented reference to the incline comes from 1560. Early on, lead and gold and silver ores were excavated there. The incline was also used for the students of the Mining Academy to get first-time mining experience. After taking a ride in the incline, during a ceremony called Schachtag, the students were admitted into the mining profession through the ritual called 'a jump over the leather stump'. The incline is accessible for 450 m of its length and ends at the Emperor's staircase with an open space under it. The Emperor's stairs were constructed in 1751, on the occasion of the visit paid by Francis Stephen, the husband of Empress Maria-Theresa. A memorial plaque to commemorate the visit was set into the wall of the incline. Later, more and more plaques were placed in the Glanzenberg incline, to honour the visits of princes, emperors and presidents. The latest plaque embedded in the wall was honouring the deans and rectors of colleges and universities, who visited the incline along with Mayor Lichner when signing the declaration on Intellectual Heritage of the Mining Academy at Banská Štiavnica, see Rybár, Hvizdák (forthcoming) and also Anonymous (2008).

Route C offers sightseeing around historical buildings of the Mining and Forestry College and the Mining Academy compound. The trail then leads to landmarks related to surface mining, as well as to the system of artificial water reservoirs, created to supply water to mining processes – 'teichs'. The expositions and trails were visited by 103,983 visitors in 2008. The peak period for visits is in spring and summer (Krno, 2009).

Banská Hodruša

Here, mining activities have been known to exist since the thirteenth century. Construction operations of the open-air museum started a few years ago. They are financed by Slovenská Banská Ltd, which still operates both copper and gold mines in the Hodruša ore district. The underground parts of the reconstruction works are almost complete with buildings on the surface area in the process of restoration since 2005.

The underground parts, with 1,500 m of accessible walkways document the mining development of the past 700 years. Tourists can see inclines carved manually in stone by a small hammer and iron graver (cutting tool). Later, inclines and extracts were excavated using gunpowder or still later, using modern mining explosives.Visitors can see primary and secondary minerals on the walls of the galleries of the Old All-Saint Incline, which have been hand chiselled. There is also a wooden statue of Saint Kliment, the patron of all miners, and the image of a little hammer and iron graver cut into the ore. The heritage value of the incline is highlighted by historical visits paid by his Royal Highness Francis Stephen, husband of Maria Theresa, in 1751, and their sons Joseph II and Leopold II, who visited the place in 1764.

The attractions include the newly reconstructed portal of the Lill shaft, where, over the years 1884–1966, a water pump was used – a product of the Kachelmann Company, established in the close by village Vyhne. The pump is the only functioning device of that kind, and is also exhibited in the Slovak Mining Museum at Banská Štiavnica.

In the open-air museum, tourists can enjoy the amazing countryside with typical mining architecture, a knocker, and the miners' cemetery. The visitors can see a gold panning show and can take a chance at trying it themselves in the museum.

Kremnica

Kremnica (then called Cremnychbana) was granted royal mining and minting privileges by Charles Robert of Anjou in 1328. It had a wealth of gold and silver mines and it has a mint that has functioned as the State Mint up to now. The greatest gold yield was achieved in the fourteenth century. Presently, Kremnica does not offer underground mine work visits for the tourists, but apart from the famous Mint and Mint Museum, the city offers well preserved city fortifications, mediaeval merchant houses, a Franciscan monastery built in 1653, Gothic, Renaissance and Baroque churches as well as miners' houses from nineteenth century. The Mint established in 1329, produced copper, silver and famous golden coins. It is perhaps the only mint continuously in operation for more than half a millennium.

Gelnica

This town (first mentioned in 1246) is in the Slovak Ore Mountains (Slovenské rudohorie). It is difficult to determine when mining activities started here. According to historian J. Kořan, there are copper artifact findings in Asia Minor from the second millennium BC, that seem to have been manufactured using copper from the deposits in the Slovak Ore Mountains, more precisely the area of Gelnica. Another indication of prehistoric mining in the area is a fireplace found alongside pieces of iron at Jasov. It comes from the Halstadt period (800–400 BC). One assumption is that these iron fragments were made from local ores (Team, 2002). In the first century AD, historians G. Plinius and C. Tacitus made a comment about the Kvads a Kelts living in the territory of contemporary Slovakia, who used to produce iron linked to local iron ore mining activities. Slavs, who replaced the above mentioned tribes, appear to have continued with iron production in cooperation with the descendants of those tribes, which remained in this area. Documents giving evidence regarding mining were preserved only from the times after the Tartar invasion in the years AD 1241–1242, i.e. from the period when mining settlements were established by invited craftsmen who were mainly German.

Due to its natural riches and abundance of ore mining, the town was for centuries an important supporter of the prosperous Hungarian kingdom. This is documented in numerous written and pictorial reports as well as maps. Unfortunately, there is a lack of accessible historical mine-works of a length suitable for showing the mining history of the region.

The Institute of Geotourism FBERG TU Košice is considering the development of a full scale 3D virtual model of the historical underground mining works located in simulated 3D ore veins in a virtual underground for an animated 'stroll' around the historical mining zone by means of a special simulation 'vehicle'. In this way it would be possible for a tourist to see the mine as a medieval miner saw it.

Dubník

At Dubník, there is a precious opal mining site. The first mention of this deposit is from 1597 (Dubnické opálové bane). The mine was decommissioned in 1922, for economic reasons – the competition of Australian precious opals made the mining operations non-feasible. Thus its opal deposits were not exhausted and today they still contain a significant amount of opals. Efforts to reopen the mine have failed up to now. The largest opal from this mine, weighing 3035 ct (607g) called Harlequin ('Vienna Imperial Opal') was a part of the Austro-Hungarian Imperial Jewels. The mine is now a site of mining tourism, where visitors can see some of the old mine works. It is operated by the Dubnické Opálové Bane Company, which intends also to reopen the mine for opal production. The visitors are offered a visit to a part of the Viliam drive that was converted to an underground tourist trail. It is visited mostly in summer. About 3,500 people visit the mines every year. There are also summer camps for young people who help to renovate the mining site (R.S. media, 2007).

Solivar

This salt mine is situated near Prešov. The first mention of the salt deposit is from 1223. Industrial mining started in 1572, when the Leopold shaft was constructed. After the mine waters flooded the mine in 1752, only the brine was mined and processed for kitchen salt. The old mine and processing plant form the Solivar Open-air Museum. There are a number of very well preserved old brick and wood industrial buildings related to salt mining and processing that are open to tourists (Slovak Technical Museum and Solivar web pages).The salt brine has been extracted through boreholes since 1945 and the processing plant was operational until 2009.

Challenges of geotourism development

Geotourism – in spite of its great potential – depends on the overall development of tourism in Slovakia. Tourism development in Slovakia is envisioned in the 'New strategy of tourism development of Slovak Republic' defined by the Ministry of Economy (2007). This is supplemented by the Act on regional development as well as by the 'Regionalisation of Tourism in the Slovak Republic' (Weiss *et al.*, 2005).

Unfortunately, real development lags behind the expectations. Some of the reasons behind this performance gap may be the low accessibility of funds and grants available for tourism development, infrastructural problems (e.g. transport accessibility), low interest of the entrepreneurs in tourism (slow return of investments) and low motivation of a number of local communities to enter tourism business as a profession.

The sustainability of geotourism is addressed by EU materials on sustainable tourism from 2007 as well as by the Slovakian ministerial materials. The materials, however do not detail the way sustainability is to be quantified. The variables, which have to be considered include the number of visitors, their impact on the environment (e.g. waste material, destruction of vegetation, changes of flora and fauna, impact on rocks and mineral, mechanical vibrations, noise, traffic pollution), induced changes to the social and cultural environment of the local communities and the impact of a change of architectural style in the area (family hotels, hotels, restaurants, shops, parking lots etc.). An example of an analysis of the impact of tourism on the environment is given in Jablonská *et al.* (2008). The practical implementations of the results of this study as well as of other studies performed by the Slovak Environmental Agency are still to be seen.

One of the critical variables – number of visitors – is usually assessed only in an indirect manner, as in areas where the entry is not ticketed, no direct data are available. Thus for the High Tatra area up to 3,000,000 visitors per annum are estimated. In case of the Zádiel Canyon about 150,000–200,000 visitors are estimated per annum. Numbers are more exact in the case of caves, as here entrance tickets are issued. The East Slovakian caves have 12,000–30,000 visitors annually. On the basis of the visible impact of tourists (e.g. amount of litter generated, noise related discomfort on a trail) it seems that these numbers are higher than sustainable, but if more caves became accessible to tourists and with a better tourist management system, these factors could be eliminated.

There is also a conflict of interests to be resolved – the tourism enterprises are keen to maximize the number of tourists (Brida *et al.*, 2008), whereas the environmentalists would like to minimize it. While the tourism industry seems to prefer short term advantages, the concept of sustainability is based on long term advantages. In principle, the Ministry of Environment evaluates policy papers and Acts relating to tourism and regional development from the point of view of possible impact on the environment, and makes suggestions for proper implementation.

At present the protection of the environment on the local level is inadequate. Thus minerals, fossils or rocks are plundered by souvenir collectors, as there is no organized protection of most of the geotourism sites. This tends to persist even if similar items are on sale at the nearest souvenir outlets.

The aim of the government policies on tourism is to prolong the stay of tourists in the country. However, in the majority of geotourism sites, due to a lack of extended services, tourists tend to spend less than a day. This results in low local income from tourism, which again demotivates local entrepreneurs. The final result is a lag behind the government set targets for tourism development.

One could conclude that Slovakia has an excellent background for geotourism and mining heritage tourism. The Ministry of Economy and Ministry of Environment have defined visions for tourism development, but a coherent strategic plan for tourism development is lacking. There are EU grant schemes that in principle make funds available for investment in tourism, but a small country like Slovakia needs a tourism master plan to get the investments into areas that would make the country as such more attractive and the services more complete. In areas where an interest of the government manifested (like in Banská Štiavnica), coordinated development effort is more visible. In other localities, where the initiative comes from the regional or local level, it is more difficult to get grants for development. On the County level, strategic materials are prepared, but often are only partially carried out due to financial restrictions. Local initiatives – if driven by investors – result in developments that are preferred by the investor. This however, is not necessarily a part of an integrated development strategy and many of them result in failures or result in tourist influx into areas that should have greater protection. Simultaneously, historical objects that would badly need repair are left deteriorating. Nature resorts that would need cleanup, tourist flow management and protection, may be left unprotected. Therefore, a wider strategy depending on a tourism master plan that defines the priorities and action sequences in accord with local interests is needed. Such a strategy needs up to date field information. The Institute of Geotourism of the Technical University Košice maps the local, tourism related resources of all East Slovakian villages, cities and their surroundings in order to generate reliable data for strategic development projects. This work is done in cooperation with our undergraduate and graduate students.

The process of changing Slovakia into an attractive geotourism and mining heritage destination would be greatly aided also by the Section of Tourism of the Ministry of Economy gaining wider competencies, greater resources and a wider pool of tourism development specialists.

References

Anonymous: Geological History of Szabo's Rock. Available online at: http://www.poznajachran.sk/pages/sk_vulkanizmus_zk_mlady.html (accessed 13 July 2009).

Anonymous Information from the Centre for Scientific Tourism in Slovakia. Available online at: http://www.ecosystems.sk/pages/geo.html (accessed 13 July 2009).

Anonymous: Solivar. Available online at: http://www.preshowminerals.szm.sk/lomy/slanskevrchy/solivar/solivar.htm (accessed 13 July 2009).

Anonymous: Sprievodca po technických pamiatkach Banskej Štiavnice a okolia, B. Štiavnica, 2008.

Biography of Professor Szabó József. Available online at: http://hu.wikipedia.org/wiki/Szab%C3%B3_J%C3%B3zsef_(geol%C3%B3gus) (accessed 13 July 2009).

Brida, J., Carrera, E. and Accinelli, E. (2008) 'Preservation of Environmental Quality and Tourist Investment', *European Journal of Tourism Research*, 1(1): 10–35.

Cimermanová, I. and Anderková, E. (2006) Getting Through the Country of Youngest Volcanoes, The Novohrad – Nógrád geopark, Proc. GEOTOUR 06, BERG Faculty, Košice, 183–91.

Dubnické opálové bane. Available online at: http://www.opalovebane.sk/index.php?page=20&jazyk=1 (accessed 29 June 2009).

History of Spišský hrad. Available online at: http://sk.wikipedia.org/wiki/Spi%C5%A1sk%C3%BD_hrad (accessed 29 June 2009).

Jablonská, J., Timčák, G.M. and Špičuk, J. (2007) Correlations between tourism and environmental load in the Slovak Paradise – Slovenský Raj, E.Slovakia, *Acta Montanistica Slovaca*, 12(4): 334–40.

Kordos, L. *et al.* (1980) Magyarország barlangjai. Available online at: http://mek.niif.hu/00500/00575/00575.pdf (accessed 29 June 2009).

Krno, L. (2009) Vyhodnotenie návštevnosti a tržieb zo vstupného za r. 2008. Available online at: http://www.muzeumbs.sk/wp-content/obrazky//2009/03/sprava-o-navstevnosti-a-trzbach-za-r2008.doc (accessed 29 June 2009).

Luczy, A. (2008) Lyžovanie v Tatrách a Karpatenverein. Available online at: http://www.hory.sk/index.php?site=pph/luczy_5/index.htm (accessed 29 June 2009).

Magaraországi Kárpát Egyesület. Available online at: http://hu.wikipedia.org/wiki/Magyarorsz%C3%A1gi_K%C3%A1rp%C3%A1t_Egyes%C3%BClet (accessed 29 June 2009).

Manikova, M. and Popovics, J. (2009) Národný park Slovenský Kras. Available online at: http://www.sopsr.sk/slovkras/ (accessed 29 June 2009).

Ministry of Economy (2007) Nová stratégia rozvoja cestovného ruchu Slovenskej republiky do r. 2013, 32pp. Available online at: http://www.economy.gov (accessed 29 June 2009).

Ministry of Economy (2009) Latest statistical data on tourism in Slovakia. Available online at: http://www.economy.gov (accessed 29 June 2009).

Preshow Minerals. Available online at: http://www.preshowminerals.szm.sk/lomy/slanskevrchy/solivar/solivar.htm (accessed 29 June 2009).

R.S. media (2007) Opálové bane. Available online at: http://www.pis.sk/portal/?c=12&id=2897 (accessed 29 June 2009).

Rybár, P. and Hvizdák, L. (forthcoming) Využitie starých banských a geologických máp pre montánny turizmus – prípadová štúdia Gelnica a okolie. Proceedings, Z dejín vedy a techniky v stredoslovenskom regióne, Univerzita Mateja Béla Banská Bystrica.

Rybár, P. and Lichner, M. (2007) Banská Štiavnica the Crown Jewel among the Slovak Mining Cities, in *Geoturystyka*, 2(9): 35–46.

Slovak Environmental Agency. Available online at: http://www.sazp.sk (accessed 29 June 2009).

Slovak Technical Museum – Solivar pri Prešove. Available online at: http://www.stm-ke.sk/vysunute/presov.htm (accessed 29 June 2009).

Slovakian Travel Portal. Available online at: http://www.slovakia.travel/portal.aspx?l=2&smi=108007&ami=108007 (accessed 29 June 2009).

SME (2008) Available online at: http://pocitace.sme.sk/c/3911864/V-Tatrach-zrata-turistov-automat.html (accessed 29 June 2009).

Štátna Ochrana Prírody (2004): Geologická náučná trasa v CHKO Cerová vrchovina a Karancs-Medves, CHKO Cerová vrchovina, ŠOP SR, 4–12.

Team of authors (2002) Dejiny baníctva na Slovensku 1. diel, Banská agentúra, Košice.

Turanová, L., Bizubová M. and Minka J. (2005) Minerály a horniny Slovenska. Available online at: http://www.mineraly.sk/files/lok/100-200/126_exkurz_stiav_vrchy_6_cast_szabova.htm (accessed 29 June 2009).

Weiss, P. *et al.* (2005) Regionalizácia cestovného ruchu v Slovenskej republike, Ministerstvo hospodárstva SR, odbor cestovného ruchu, Bratislava, 114.

14 Mining tourism in the Spanish province of Huelva

The case of the Riotinto Mining Park

Alfonso Vargas-Sánchez, Nuria Porras-Bueno and Mª Ángeles Plaza-Mejía

Introduction

An integrated and well-balanced tourism development for any territory must take full advantage of all the resources that are available there to be exploited and, at the same time, must treat them with proper respect. Among such resources, the history and culture of the place are inherited assets that form part of our collective heritage; the visitor should be able to access these resources in order to get to know the place and, through that knowledge, to appreciate and value it (Gunn, 1994; Hall, 2000; Inskeep, 1988).

There can be no doubt that a part of the past, and even of the present, of the province of Huelva (Spain) is bound up with the industrial exploitation of its natural resources. One of the most striking manifestations of this exploitation is the mining activity that has been taking place here for thousands of years. That legacy is worth preserving in its entirety so that future generations may learn about it and take pleasure from it. Tourism can become an effective instrument for the regeneration of that heritage, some of it marginalized, under threat and in progressive deterioration. In a recently-conducted study on this industrial tourism segment (Vargas *et al.*, 2007) the authors have identified, with the help of experts in the subject (both academics and professionals active in the sector), the segments with the most potential in the province, among which mining tourism stands out.

The consumers of tourist products are becoming increasingly sophisticated and demanding; they require an offer that complements the traditional Beach and Sunshine model – which has been the case of this province, with products, for example, linked to local history and culture. In addition, the tourism market is increasingly heterogeneous, with a multiplicity of segments or niches in which suppliers need to position themselves and then strengthen their competitive positions; one of these is the sector known as 'industrial tourism'.

Industrial tourism is an activity that has recently been added to the tourism offer, and which broadly involves making visits to industrial centers (factories, workshops, stores etc.) to extend the cultural experience enjoyed by the tourist, and as an interesting way of learning about the economic activity of other peoples, past and present. Industrial tourism in general, and mining tourism in particular, is a phenomenon already being exploited in other regions and countries, where there are numerous mines and industrial installations linked to extractive activities which have been adapted for purposes of tourism. This segment is closely related to a type of consumer (tourist) who seeks interesting new experiences or emotions, who seeks to know better and in greater depth the places being visited. (Edwards and Llurdés i Coit, 1996; Light, 1989; Richards, 1996, 2001; Uriely *et al.*, 2002; Van Der Arka and Richards, 2006; Vargas *et al.*, 2007a, Vargas *et al.*, 2009; West, 1988; Wynne, 1992).

Despite the notable experience of the Mining Park of the locality of Minas Riotinto,[1] this phenomenon is still in its very early stages in the province of Huelva. It does however, offer possibilities for growth and for generating complementary revenues in territories that are very much in need of them, following the crisis subsequent to the abandonment of extractive activity. See the map below for their geographic location (Figure 14.1).

Thus, the basic objective of this study is to explore the potential of industrial tourism linked to mining in the province of Huelva, reinforcing its character as a tourist destination. Concerning the actions to be undertaken, it is necessary to ensure full compatibility with the overall Strategic Plan for Tourism Development of that province, drawn up by the provincial government, and with the development of its tourism brand: 'Huelva, la luz' ('Huelva, the light'). More recently, the Plan for Sustainable Tourism of the Pyrites Belt, which also extends to municipalities of the province of Seville[2], is another work to be taken fully into account, so that efforts under these plans converge rather than diverge.

Methodology

The work presented in this chapter is framed within a wider research project analysing the possibilities for the development of industrial tourism in the province of Huelva (Spain). There is no doubt that the most interesting experience of this kind at the provincial level is the Riotinto Mining Park. Therefore this has been taken as a focus for gaining more knowledge of the features of this type of tourism, but on the clear understanding that the field of action is far wider than the municipality of Minas de Riotinto itself, and extends at least to the whole area.

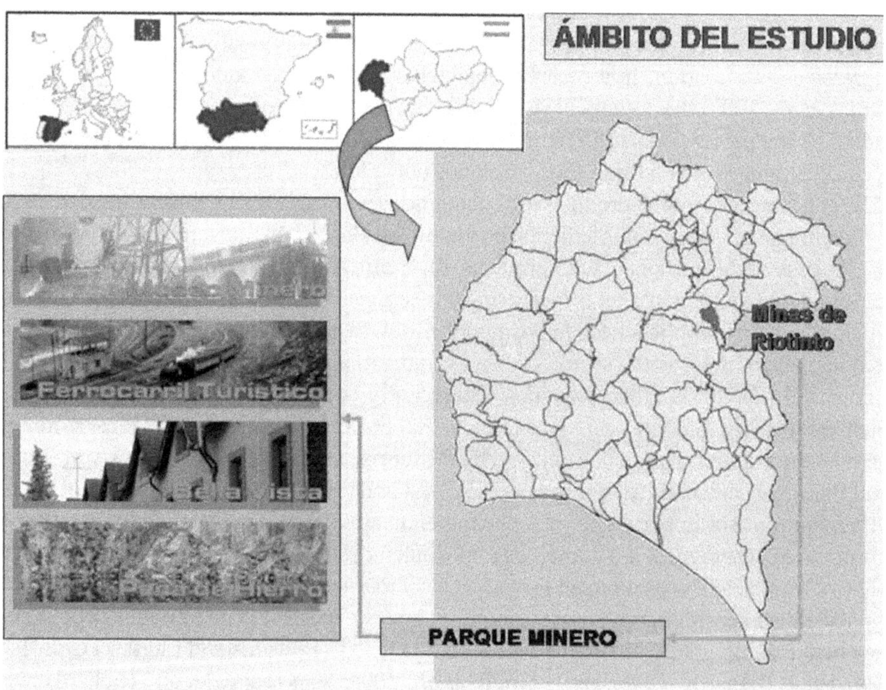

Figure 14.1 Scope of the study.

In this chapter we summarize the results of the empirical study carried out to discover the socio-demographic profile, attitudes, opinions and behavior of the tourists who visit this mining locality, with the object of studying in greater depth the possibilities for mining tourism in all the municipalities of the area with resources of this type, with the assistance of a multidisciplinary group of experts.

Scope of the study

For centuries, the town of Minas de Riotinto has been almost totally dependent on one single economic activity, open-cast mining. Among many other consequences, this activity has given rise to a landscape that, for its unique and spectacular appearance, is today the principal attraction of the zone. The tabular shapes, the variety of coloring in shades of violets, oranges, reds, ochres, presented by the scenery on a vast scale, confront the spectator with images of beautiful contrasts. From another perspective, the urban configuration of the municipality still retains a marked colonial character, inherited more recently from the English managers of the mines who lived there with their families for more than eighty years (1873–1954). Both the importance of its history and the special features of its natural and urban environment make this enclave an ideal place for the development of industrial cultural tourism. The Rio Tinto Foundation was created with this objective.

In respect of the Mining Park specifically, its cultural tourism offer comprises the following main elements: the Mining Museum; the old mining installations of 'Peña del Hierro'; the English residential District of 'Bella Vista', dating from the Victorian era; the Roman Necropolis of 'La Dehesa' (second century, AD); and a tourist railway using part of the former mine's track. This municipality also contains the 'Corta Atalaya', Europe's largest open-cast mining site, and the following image gives some idea of its spectacular size (Figure 14.2).

Confirming the current significance of this example of Huelva's industrial archaeology, the Riotinto Mining Park is placed third in the national ranking of places with mining heritage

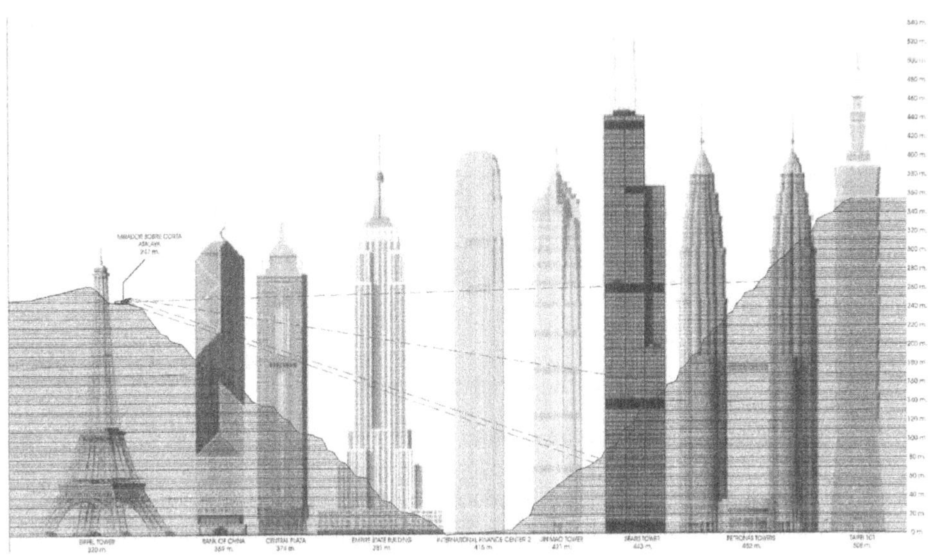

Figure 14.2 Depth of 'Corta Atalaya' (Source: Jesús Navarro Goy, Riotinto Tour).

that can be visited, among the approximately 100 sites of this type that exist in Spain. Another example of its relevance is the fact that the Mining Museum of Riotinto is the provincial museum that receives the most visits each year. Specifically, in 2008, the Mining Park of Riotinto received a total of 78,683 visits; this total has shown a growth trend over several years. The time period selected for conducting the surveys was the months of July and August, 2008 (summer holidays).

From a random sampling, the number of observations obtained was 400; therefore, with reference to a population total of 73,900 visitors, the resulting margin of error was 4.73 per cent (smaller than the usual five per cent) and the level of confidence was 95 per cent, with a population variance of 50 per cent. The questionnaire was given to the visitors for completion during the return trip on the visitor railway of the Riotinto Mining Park; three survey assistants were always on hand to support the process, helping the respondents by clarifying the questions.

The questionnaire contained a total of 23 questions structured around eight major dimensions to be studied: socio-demographic profile (gender, age, occupational situation, professional category, educational level achieved, and place of residence); motivation of the visit; interest in industrial tourism; qualitative behavioral aspects (the type of trip/journey within which the visit to the Mining Park is made, the length of stay in Huelva anticipated by the traveler, the principal reason for the visit, the people with whom the trip is being made, and which mining sites are expected to be visited in the locality); quantitative behavioral aspects (the average daily expenditure by the tourist and the number of persons covered by the amount of expenditure reported); knowledge of the destination and of the world of mining; satisfaction; and loyalty (previous visits).

The questionnaire was designed in accordance with the structure utilized in previous studies published by various authors (Beeho and Prentice, 1997; Chandler and Costello, 2002; Confer and Kerstetter, 2000; Davis and Prentice, 1995; Keng and Cheng, 1999; Kerstetter *et al.*, 2001; Laws, 1998; Moscardo, 1996; Taylor *et al.*, 1993; Vargas *et al.*, 2007a; Vargas *et al.*, 2007b; Vargas *et al.*, 2007c; Vargas *et al.*, 2008a; Vargas *et al.*, 2008b; Vargas *et al.*, 2009).

The visitor survey data were submitted to two kinds of analysis: first, a univariate analysis of descriptive nature was performed and, second, a bivariate analysis was performed by crossing variables, with the object of analyzing and identifying significant correlations and relationships of dependence between the variables. For this second type of analysis, use was made of the most appropriate test of differences of means in function of the characteristics of each variable.

With respect to the focus group session on the future of mining tourism in the province of Huelva, it was conducted with the participation of six academic experts of the University of Huelva and eight professionals, local public and private agents involved in the development of this tourism segment. This session was carried out in December 2008, and headed by the authors of this chapter.

Behaviour and satisfaction of the tourist: results

The socio-demographic profile of the average visitor to the Mining Park is that of a person resident in Andalusia, of between 30–44 years of age, employed or self-employed, and who has a High School or University level of education (the proportion of those without educational qualification is very small). Three provinces, Huelva, Madrid and Seville, account for half of the visits by Spanish nationals. In the case of those living in Huelva, 80 per cent of the visits are by those living in the southern third of the province; there are very few visits by those living in the mining district itself.

With respect to motivation, the reason that tourists visit the Mining Park is the experience of learning; no evidence is found of the effect of the other motivations considered – entertainment, circumstances, and even less emotion. In reality, the motivation shown by the visitor to the Mining Park is fairly weak.

The people who visit the place do so mainly 'to get to know' the landscape, the world of the mine and its history. In all cases, this interest in learning more appears to result from the curiosity of the layman rather than from any desire for deeper knowledge by the more erudite visitor, a finding that is corroborated by the low degree of existing personal knowledge about the world of mining and its history reported by the average visitor.

In any case it was found that, among visitors of the older age group, retired persons and pensioners, the main motivating interests are those of entertainment – simply spending a day out, and relaxing – and circumstantial factors – the price of the visit being 'reasonable' or the inclusion of the visit in the program for their trip. Learning as the main motivation reported for the visit is highest among those of higher professional category.

The tourists surveyed showed a moderate degree of interest in other experiences of industrial tourism available in the province, principally those related to mining tourism and the agro-food industries, such as wine-making, Iberian ham and olive oil production.

At the opposite end of the scale, the industries that evoked less interest among the visitors to the Mining Park are those of the chemical and basic industries, such as the functioning of a combined-cycle electricity generating station, the refining of petroleum or the manufacture of paper; an exception to this is the relatively high degree of interest reported in the casting of copper, probably because it is closely linked to the mining industry.

The degree of interest expressed in almost all the other suggested experiences of industrial tourism in Huelva has a direct relationship with the visitor's motivation of learning; it is those visitors who rate more highly this type of motivation for visiting the Mining Park who also declare greater interest in experiencing other forms of industrial tourism.

With regard to tourist behavior, half of the visitors to the Mining Park travel to the locality of Minas de Riotinto principally to visit its mining installations; this confirms the importance and tourist appeal for the locality deriving from its mining heritage and English legacy. One third of those visiting the Park are day-trippers, and the rest are tourists staying for longer; these latter report an average length of stay of almost ten days away from their place of residence; in this respect it is notable that 40 per cent of total visitors are making a visit of several days to the province of Huelva.

Almost 70 per cent of the visitors to the Mining Park of Riotinto planned to spend the whole day in the locality; typically this type of visitor is accompanied on the holiday or day-trip by a partner or other family members; those traveling alone, with work colleagues or in an organized group account for only small proportions. One-third of the tourists have not visited and do not plan to visit any other mining site in the area; one out of every five visitors is thinking about visiting another mining site in the province; and one out of every six may visit two more sites. Apart from the Mining Park, the rest of the mining resources of Huelva are found to be less well known, and relatively few of the tourists surveyed intend to visit them.

The average daily expenditure per person of the visitors to the Mining Park is €41.5, which is similar to the amount that characterizes Huelva's cultural tourism demand (€41.08). This variable diminishes the greater the emotional motivation of the visitor, and reaches minimum values among the groups of older age, retired persons and pensioners, and those making the trip with their family, those visiting family or friends, and those coming specially to visit the mines; on the contrary, those visiting the locality for reasons of business or work, and those traveling alone or with work colleagues, report a high average daily expenditure.

It is estimated that the costs of entry tickets to the Mining Park represent between 7–41 per cent of the visitor's average daily expenditure. In the light of these figures, it is evident that the revenues generated by mining tourism go far beyond the price of access to the attractions specifically offered by it. It is therefore important for these localities to provide adequate and appropriate infrastructures and offers for tourists, particularly hotels, restaurants, cafeterias and shops, to help ensure that most of the revenues are received by the actual inhabitants of the mining municipalities and thus contribute to the generation of wealth and employment, in zones that are endemically depressed but have tremendous tourism potential.

In fact, extrapolating the average daily expenditure of this sample to the 'population' of 73,900 persons who visited the Mining Park in the past year, a total tourist expenditure of more than €3,000,000 per annum is obtained[3].

Almost 30 per cent of the visitors to the Mining Park reported acquiring some knowledge of the locality of Minas de Riotinto and its Mining Park as a result of the recommendation of friends or family. This gives an idea of the importance of the informal 'word of mouth' channel as a means of promoting this mining tourism experience. Overall, two other formal channels, the Internet and tourist leaflets and guides, represent two media for formal promotion that are also very important. A comparison of these results with those obtained in a similar study two years ago reveals the increased effectiveness of the formal promotion campaign of the Mining Park; this is also demonstrated by a progressive increase in the annual number of visitors.

In general, the visitor to the Mining Park has little personal knowledge of the history and the mining heritage: three out of every four visitors accept that they have little or no knowledge about the history and mining resources; only a small proportion describe themselves as 'experts' or persons with good knowledge of the subject. There is a linear and direct relationship between knowledge of the world of mining and interest in industrial tourism in general, and interest in visiting mines and other mining resources in particular.

The degree of satisfaction of the visitor both with the visit itself and with the overall tourism offer of Minas de Riotinto is fairly high: average scores for these two aspects were 7.6 and 7.3, respectively, on a scale from one to ten. It is not surprising, therefore, that 92 per cent of the tourists would recommend a visit to this place; this percentage is higher for respondents of older age, those with more interest in industrial tourism, those with more knowledge of the world of mining, and those expressing more satisfaction with their visit. The degree of satisfaction is increased also for those respondents whose motivation for the visit is to learn, those with more knowledge of the world of mining, and those more interested in industrial tourism.

In global terms, industrial tourism is the factor of the locality's tourism offer that is rated most highly by the visitors to the Mining Park, at 7.7 out of ten; next is the tourism environment, rated at 7.2; and lastly the quality/price relationship, at 6.9. The aspects most highly valued by the visitor are the natural and human environment of the locality, whereas the least valued aspects, in all cases with an average close to notable, are the communications, signage and information.

The visitors with reduced motivation, of any type, express the lowest levels of satisfaction with the industrial tourism and the tourism environment offered by Minas de Riotinto, with the quality/price relationship of its tourism offer, and globally with the entire tourism offer.

A little more than a quarter of the visitors to the Mining Park state that they had visited the locality of Minas de Riotinto previously, although it is also surprising that six per cent of the tourists state that they had previously visited the municipality on more than five occasions.

The degree of loyalty is higher among the day-trippers, those traveling alone or with work colleagues, those who are in the locality principally for reasons of business, work, a meeting or to visit friends, and among those with greater personal knowledge of the world of mining.

Focus group session: results

The focus group session was centered on two aspects with regard to the future of mining tourism in the province of Huelva: ideas for increasing tourism in the zone and barriers and difficulties to be overcome.

It is evident that mining tourism should not be limited to the Riotinto area. An effort must be made to generate a unified picture of this tourism segment in the whole province and even beyond, to link together the various different points of reference and to seek complementarity between them.

Mining tourism must not be treated in isolation, separate from other forms of tourism; instead it should be combined and integrated within a broader concept of experience for the tourist that includes other attractions (natural features, gastronomy, folklore, heritage, etc.). At the same time, the content of the mining tourism experience offered should have more depth and substance, in order to meet the needs of the specialized markets.

It should be the better part of a broader project to stimulate the growth of tourism in the mining basin, with the creation of a mining tourism network at the provincial level that will work in collaboration with other initiatives at the regional level, and with neighboring regions of Portugal.

The investment in basic infrastructure for tourism, such as accommodation, would help to convert simple day trips into overnight stays. Similarly, the improvement of communications would facilitate accessibility and hence the flows of tourists, especially those coming from the coast. It would be also interesting to see the creation of innovative and attractive holiday packages, which would facilitate sales and distribution.

There are some also some barriers and difficulties to be overcome. In the first place there appears to be an apparent lack of consistency in the policies of the competent public authorities and the absence of a global and territorially coordinated project.

But there are other aspects noted in the focus group research to take in account:

- Many of the principal heritage attractions are not accessible for tourism.
- Loss of a large part of the heritage (some of which has even been sold as scrap).
- Limited understanding of the agents offering services/products regarding the operating 'rules' of the tourism industry and its promotion and sales channels, particularly the Internet. For instance, if the suppliers of products would allow reservations and sales of tickets on-line, this would increase sales spectacularly.
- Limited understanding of these agents about the economics of tourism intermediaries, and lack of trust in them. If commercially attractive conditions were offered, this would greatly facilitate agreements with wholesalers and retailers at the national and international level.
- In the local community, a high degree of skepticism regarding of the future of this activity can still be detected; to this must be added a long history of dependence on the public sector and a very weak entrepreneurial tradition.
- Lack of basic tourist services, such as the supply of accommodation, which is very scarce.
- Insufficient investment, both private and public.
- The area lacks a tourist information point.

- Need for better professional qualification of personnel, to improve the quality of service.
- Lack of a long-term strategic plan.

Summary

The province of Huelva meets several conditions for the success of mining tourism. In particular and first, its wealth of resources for mining tourism, covered by various measures of public protection. In second place the uniqueness, in many respects, of what can be offered. There is also an evident demand for products of this type. And finally there are other complementary attractions and reasons for traveling throughout the province.

As a starting point, it would seem that the opportunities and strengths already referred to are much greater than the weaknesses and resistance identified in this analysis, which could be summarized mainly as the lack of an overall, unified and long-term plan orientated to the sustainable development of the territory and insufficient mutual trust among the stakeholders. The need of investment in tourism infrastructures by both public and private agents has also been identified. No less important is the need to take full advantage of Internet technologies for marketing tourism products.

These latter negative points could be considerably reduced with a change of attitude on the part of all the interested agents, working more collaboratively and in a more coordinated way, making possible an atmosphere of greater cooperation. Additionally, greater doses of creativity and support for tourist services are required, as well as innovatory products orientated towards providing a unique, different experience, which would also serve to retain the visitor for longer and overnight stays. Included in this would be all types of organized events (fairs, courses, exhibitions, etc.) that would serve to attract greater numbers of visits.

In the view of the authors the information gathered from visitors to the Riotinto Mining Park gives market support to the expert's conclusions, and it is a valuable asset for developing and implementing more specific proposals by both public and private agents.

Notes

1 http://www.parquemineroderiotinto.com

2 The Interprovincial Association of Municipalities for the Development of Sustainable Tourism (ADTS) of the Pyrites Belt was created in the year 2007 as an entity to unify the territory of the Districts of Corredor de la Plata in the province of Seville and the Mining Basin in the province of Huelva. Its main objectives are to encourage the dynamic development and promotion of this area with significant mining past, and to make developers and promoters more sensitive to the need for sustainability. This organization acts as the Promoter of the 'Plan for Sustainable Tourism of the Pyrites Belt of the Southwest of the Peninsula: Puerta de Sierra Morena 2007–2011'.

3 73,900 visitors × 41.5€ = 3,066,850€.

References

Beeho, A. and Prentice, R. (1997) 'Conceptualizing the Experiences of Heritage Tourists', *Tourism Management*, 18(2): 75–87.

Chandler, J.A. and Costello, C. A. (2002) 'A Profile of Visitors at Heritage Tourism Destinations in East Tennessee According to Plog's Lifestyle and Activity Level Preferences Model', *Journal of Travel Research*, 41: 161–66.

Confer, J. and Kerstetter, D. (2000) 'Past Perfect: Explorations of Heritage Tourism', *Parks & Recreation*, February: 35–38.

Davis, A. and Prentice, R. (1995) 'Conceptualizing the Latent Visitor to Heritage Attractions', *Tourism Management*, 16(7): 491–500.

Edwards, J.A. and Llurdés i Coit, J.C. (1996) 'Mines and Quarries Industrial Heritage Tourism', *Annals of Tourism Research*, 23(2): 341–63.

Gunn, C. (1994) *Tourism Planning: Basics, Concepts and Cases*, New York: Taylor and Francis.

Hall, C.M. (2000) *Tourism Planning: Policies, Processes and Relationships*, New York: Prentice Hall.

Inskeep, E. (1988) 'Tourism Planning: An Emerging Specialization', *Journal of the American Planning Association*, 54(3): 360–72.

Keng, K. and Cheng, J. (1999) 'Determining Tourist Role Typologies: An Exploratory Study of Singapur Vacationers', *Journal of Travel Research*, 37(May): 382–90.

Kerstetter, D., Confer, J. and Graefe, A. (2001) 'An Exploration of the Specialization Concept Within the Context of Heritage Tourism', *Journal of Travel Research*, 39(3): 267–74.

Laws, E. (1998) 'Conceptualizing Visitor Satisfaction Management in Heritage Settings: An Exploratory Blueprinting Analysis of Leeds Castle, Kent', *Tourism Management*, 19(6): 545–54.

Light, D. (1989) 'The Contribution of the Geographer to the Study of Heritage', *Cambria*, 15: 127–36.

Moscardo, G. (1996) Mindful visitors: Heritage and Tourism, *Annals of Tourism Research* 23 (2): 376–97.

Richards, G. (1996) *Cultural Tourism in Europe*, Wallingford, UK: CAB International.

Richards, G. (2001) *Cultural Attractions and European Tourism*, Wallingford, UK: CAB International.

Taylor, D., Fletcher, R. and Calabaugh, T. (1993) 'A Comparison of Characteristics, Regional Expenditures, and Economic Impact of Visitors to Historical Sites with other Recreational Visitors', *Journal of Travel Research*, Summer: 30–35.

Uriely, N., Israeli, A. and Reichel, A. (2002) 'Heritage Proximity and Resident Attitudes Toward Tourism Development', *Annals of Tourism Research*, 29(3): 859–61.

Van Der Arka, L.A. and Richards, G. (2006) 'Attractiveness of Cultural Activities in European Cities: A Latent Class Approach', *Tourism Management*, 27: 1408–13.

Vargas, A., Plaza, M.A., and Porras, N. (2007a) 'Industrial Tourism Development and Local Community Perceptions: The Case of Minas de Riotinto (Spain)', *Advances in Tourism Economics*, Portugal: Instituto Piaget.

Vargas, A., Porras, N. and Plaza, A. (2007b) *Turismo industrial en la provincial de Huelva. Presente y futuro*, Cátedra Cepsa de la Universidad de Huelva: Huelva (Spain).

Vargas, A., Porras, N., and Plaza, M.A. (2007c) 'Tourism Development and Local Interests of the Residents', *NARODNOSTOPANSKI ARHIV*, LX(4): 25–32.

Vargas, A., Porras, N., and Plaza, M.A. (2008a) 'Residents' Attitudes towards Tourism Development in the Spanish Province of Huelva', *NARODNOSTOPANSKI ARHIV*, LXI(1): 102–22.

Vargas, A., Porras, N., and Plaza, M.A. (2008b) 'Residents' Attitudes towards Tourism Development and its Inductors: Empirical Evidence', International Meeting on Tourism Management. EIGTUR 2008, Universidade Federal de Ouro Preto, Ouro Preto, Brazil.

Vargas, A., Plaza, M.A., and Porras, N. (2009) 'Understanding Residents' Attitudes toward the Development of Industrial Tourism in a Former Mining Community', *Journal of Travel Research*, 47(3): 373–87.

West, B. (1988) 'The Making of the English Working Past: A Critical Review of the Ironbridge Gorge Museum', in R. Lumley (ed) *The Museum Time Machine: Putting Cultures on Display*, London: Routledge.

Wynne, D. (1992) *The Culture Industry: The Arts in Urban Regeneration*, Avebury: Aldershot.

PART V

Globalization and the future of mining attraction destinations

15 Planning for the future

Tourism options for an open pit coal mine at Ha Long Bay, Vietnam

Cuong Nguyen Duc Hoa, Nancy Chesworth and Lee Jolliffe

Introduction

The Ha Tu Mine has a long history and has been enlarged several times. Owned and operated by the Vietnam National Coal Corporation (Vinacoal) the mine is facing closure in 2015 when it is estimated the existing coal deposits will be exhausted. Situated a few kilometers north of one of Vietnam's largest tourist sites, Ha Long Bay, the mine has potential to become a tourist attraction.

Ha Long Bay (the name literally means descending dragon bay) is located in the coastal province of Quang Ninh, 180 km northeast of Hanoi. In 1994, UNESCO designated Ha Long Bay as a World Heritage Site laying a foundation for a strong growth of the tourism industry there. It is now a popular tourist destination in the North of Vietnam. In 2007, Ha Long was given city status by the government and recognized as an urban tourist destination.

This chapter discusses a variety of options for future use of the nearby Ha Tu Mine which to date has only been accessible to the public on a limited basis. This review is based on the mine history from relevant documents and publications, a field visit to the mine, and follow up telephone interviews with key informants in Quang Ninh province. In addition in the discussion part of the chapter the site is compared to a model project for reclamation of a mine site, the Eden Project in the UK.

The Ha Tu Mine

The mine is situated in the Ha Tu commune in the eastern part of Ha Long City. The history of the mine can be divided into two periods: before and after 1955, the latter being an important milestone of change in mine ownership. In the late eighteenth century the Nguyen Dynasty signed a lease with a French investor for exploitation of natural resources in the region for a period of 100 years. This region had originally been a vast mangrove forest with a grand sea canal from the Luon Cave to Cai Seaport. During high tide, ships could pass through easily and during low tide, local residents came to collect seafood. In French colonial times, solid waste from coal mining filled up the mangrove forest and turned the region into a barren land. As a consequence, local residents and mine workers moved to live on the land (Tu *et al.*, 2000).

The French colonial government in Indochina certified the ownership of the mining area to a French investor, Mr. Bavier Chaufour, in March 1887. On 24 April 1888, the French Northern Coal Company (*Société Française des Charbonnages du Tonkin*) defined a mining area from Bai Chay town to Mong Duong town, including Ha Tu. At the time, mining was labor intensive with 3,000 workers using simple working tools and equipment. Conditions for the workers were poor and worsened during times of economic crisis, for example during the Second World War (1939–1945). The situation became critical when the Japanese military force attacked Indochina (1940–1945), resulting in mass unemployment and hunger, and deaths. After the victory of the Dien Bien Phu Battle of Viet Minh communist revolutionaries, the First Indochina War ended with the 1954 Geneva Accord, under which France withdrew from the Indochina colonies. On 24 April 1955, the last French colonial troops who had occupied and governed the mine withdrew, handing ownership to the Government of Vietnam.

The second period of the mine's history began in May 1955, the Ha Tu mine was expanded and production increased. At that time, the mine consisted of two underground mines and one open pit mine. During the Vietnam War, known to the Vietnamese as the American War, the mine was twice the target of a series of attacks by the American military. From 1966 to 1968, the mine maintained operations during the first series of attacks by the American Air Force. In 1967, the American Air Force attacked the mine 13 times. In November 1968, limited American bomb attacks on North of Vietnam allowed the mine to recover its production in the period 1969–1971. However, the Americans initiated a second series of attacks in the North of Vietnam in May 1972, until their defeat in late December 1972. On 27 January 1973, the Paris Accord was signed, bringing the war to an end.

Since the end of the American War, the Ha Tu Mine has boosted output. The production line has been synchronized and improved. Living conditions and professional skills of workers have been strengthened to support the increased mining production of 1,294,064 tons in 1975 and 1,816,547 tons in 1980. However a decade later, the high post war production had declined to 606,471 tons, in part due to the diminishing coal resource. By 2000 production had increased again to 978,055 tons.

The mine and the community

Quang Ninh province is located in north-eastern Vietnam with a total area of about 600,000 hectares, 245,000 hectares of agricultural land and 6,800 hectares of residential land. Mountains and hills cover 80 per cent of the total area of the province, including more than 2,000 islands (two-thirds of the islands in the country). To the North, Quang Ninh borders China for a 170 km stretch; in the south, it borders Hai Phong province, and Lang Son province to the west. The coastline on the East Sea is 250 km long and the average temperature is 25°C (77°F). The population is 1.44 million people, half of whom reside in urban areas (http://www.quangninh.gov.vn).

Alongside tourism, the area's coal resources are a steady contributor to the national economy, providing 90 per cent of Vietnam's coal production. Compared with the rest of the country, a relatively high proportion (about 21 per cent) of the total working population is employed in industry, with the coal industry employing some 65,000 people. Some 32,000 other workers are spread out across a host of ancillary businesses (including training centers, health facilities, schools, hotels, and construction companies) (Soai, T., personal communication, 16 September 2009). In the present economic structure of Quang Ninh (2008), industry accounts for about 50 per cent of the gross provincial product while agriculture accounts for only 5 per cent. The total coal reserve (mainly anthracite) is estimated at 3.6 billion tons, primarily in three areas of Ha Long, Cam Pha and Uong Bi – Dong Trieu. In 2006, production reached 37 million tons which of approximately 20 million tons were exported (Quang Ninh e-newspaper, 2008; Quang Ninh Department of Planning and Investment website, 2009; Quang Ninh Province website, 2008).

There are three state run coal companies and hundreds of small, private, unlicensed coal mining operators reported to be working in the province. The state companies produce the majority of the coal output contributing to numerous employment and income opportunities for the locals, a rise in the local gross provincial product and urbanization in the province. However, illegal coal mines pose threats with regard to local economic leakage as well as social and environmental problems. Social problems stem from illegal coal production and trading (i.e. exporting to China) and violence among illegal miners at local harbors. Operators of dozens of these mines use explosives contributing to an increased rate of accidents and causing an insecure living environment in the city.

Thuren (1997) pointed out that coal mining activities, whether licensed or not, are polluting Quang Ninh's air, water and land resources. Chinh *et al.* (2006) identify the main environmental impacts of coal mining in Vietnam as being dust pollution at coal transfer points or from piles in storage; noise pollution resulting from heavy equipment usage; and high volume of discharged acid and turbid mine water. A direct impact on tourism here is that polluted water has affected Bai Chay Beach, a major tourist area. Environmental issues are only one of the many conflicts here between the coal mining industry and tourism (Anna, 1999).

Tourism development in the province

Alongside occupation and the exploitation of local coal mines in the period 1883–1955, the French did not forget to explore the tourism potential of Ha Long Bay, ranking it among the top three wonders in Indochina together with Hue Citadel and Angkor Wat Temple. Starting with unplanned short excursions, by the 1920s there were package tours, marketing activities and then establishment of local tourism service suppliers in Ha Long and travel service businesses in Hanoi and Hai Phong province. Ha Long Bay was marketed globally, attracting tourists from Asian and European countries. Bai Chay Nautical Club and the Mine Hotel situated in Hon Gai provided comfortable amenities and services for visitors to the bay. In the Bai Chay area a Japanese businessman operated "Ha Long Bay Lodge". Some cruise boats provided luxury overnight services on the bay. The main destinations included Sung Sot Cave or Ba Ham Lake in Ha Long Bay. In the 1930s, excursions were expanded to the vicinity of the bay, establishing tourism routes from Hon Gai to Van Hoa Bay (on Cai Bau Island in Bai Tu Long Bay) or even as far as Tra Co Beach in Mong Cai Town.

These historic tourist activities in Ha Long Bay were also diversified, offering tours to local coal mines in Ha Tu, Cam Pha and Cua Ong. In his book on sightseeing tours in Ha Long Bay (pp. 20–28), the author Juan Renaud wrote:

> Our ship came to Hon Gai harbor. This is an important mining center in the north of Vietnam. The mine annually produces 600 thousand tons of coal. Coals produced here are best sold in the Far East markets. An investment of 4 million French Francs has been made in the Northern Coal Company to operate the coal mine business. The company hires thousands of workers from Nam Dinh and surrounding provinces. Hon Gai harbor is located in a spectacular bay which has frequently received ships from many countries, especially the UK and Japan.

Renaud also said Tra Co beach was as appealing as a famous beach in the northwest of France. Tourist activities and services in the region kept growing until 1953–1954 when they stagnated as a consequence of the war (http://datmo.halong.com.vn/datvanguoi/details.asp?id=2414).

Not until the late 1970s was tourism development revived in the Ha Long Bay area when people from Hanoi, Hai Phong and other cities in the North went to Bai Chay Beach for summer holidays. Typically these vacations were a form of social tourism; annual incentive trips provided and conducted by the trade unions for selected employees in government organizations and state owned companies. At that time, an incentive beach holiday was valued as a reward to workers in Vietnam. Sightseeing attractions offered were limited to Bai Chay Beach and some caves in the bay.

After 2004 when Ha Long was recognized as a UNESCO World Heritage Site, the city saw an enormous growth in tourism, making it the fastest growing tourist destination in Vietnam in terms of tourist arrival numbers. Ha Long received only 236,000 tourist arrivals in 1996. As of 2001 the city had approximately 300 hotels ranging from one to five star standards, providing 4,500 rooms (Binh *et al.*, 2001). By 2008, tourism arrivals had increased to 2.64 million. The city attracts one-third of the country's international tourists and has contributed to Quang Ninh province's 37 percent annual growth in tourism since 1990.

In the past, conflicting interests between coal mining and tourism have been acknowledged (Sinh, 1995). However, recently new tourist activities combining both sectors have been introduced. In the summer of 2006, the Tourism Department of Quang Ninh developed a pilot coal mine tour in collaboration with Ha Tu Coal Mine and tour operators from Quang Ninh and Hanoi. The goal was to educate tourists on the coal production processes, mining technologies and the economic value of the coal mining industry. Participating in this tour, the tourist would experience the life of mine workers and their hardships, as well as their pride in producing 'black gold' for the country. Two selected coal companies participating in this project were Ha Tu and Nui Beo. Different from what prospective visitors may think, visiting sites in these mines is a very clean and peaceful experience. Tourists put on a mining worker uniform, and were transported to visit open-pit mines. They visited the mine tradition house which exhibits a film, photographs and items about the history of the coal industry since the French colonial period. While at the tradition house, they experienced a performance of song, dance and poetry about mining by mine workers. Visitors were also able to purchase coal related souvenirs (for example, sculptures in coal of boats and islands), like those available at the handicraft village in nearby Hon Gai. At the mine a meal was provided in the form of a buffet with the workers, followed by a visit to the Quang Hanh hot mineral water resort to bathe in the waters of the mineral springs and drink mineral water.

Ha Tu Mine visitor facilities

This section reviews the existing facilities for visitors to the Ha Tu Mine. The mine traditional house where visitors are welcomed is located on the access road to the mine near the entrance. It is a one-storey concrete building. One end of the building is used for award presentations, addresses by dignitaries and meetings of the mine workers and staff. The other end houses exhibits and models of mining equipment. Exhibits are housed in glass topped cases, approximately three by four meters. Topographically detailed, including grass, trees and one or more lakes, three display cases illustrate the development of the mine from its inception. The first display case provides a view of the original mine, complete with buildings, tools, large equipment and machinery. The second shows the growth of the mine over time, to the middle of the last century.

The expansion of the mine is clearly seen in the third case which includes scale models of machinery currently in use, new buildings and facilities for workers and two much larger lakes. While interpretation is not provided in written form, a guide can provide narrative and answer questions. The models on display are large enough in scale for observers to be able to determine the purpose of a given machine and how it was used. Exact in all details, the models illustrate advances in mining technology and efficiency, some are in working order, however as they are primarily exhibits, they are for visual examination only. The oldest models are fragile and not appropriate for handling in any case.

The static displays here, a form common in museums in earlier decades (Hudson, 1975) do not invite more than a short examination by the viewer. Their large format however, would easily lend themselves to updating through addition of lights, action figures and working miniature models. The delicate materials used in constructing the displays, makes containment in a case necessary. However, animation would make them attractive to all generations. This applies particularly to children who are accustomed to visually engaging media.

This building is open for pre-arranged tours, but is not open to the public for casual visits. School and other educational groups also visit the building to learn about the mine, the achievements of the workers and the role the mine plays in the local and national economies. There is no souvenir shop or other facilities for visitors.

On-site there are two main open-pit mining grounds (Nui Beo and Ha Tu), the Lo Phong waste dumping ground, and a number of mining facilities (e.g. truck and bus terminal, rail terminal, electric and vehicle maintenance workshops, coal sorter belts, viewing tower and operation control office, cafeteria, etc.) with a total area of over 700 hectares. The majority of these facilities are located in the administrative zone behind the main entrance gate. Nui Beo mining ground located near the administrative zone had been in operation since the French times. Operation at that mine ceased some years ago, making Ha Tu the main production site. Both mines are hundreds of meters in depth with coal beds and roads running down to the bottom, creating a superb view to the entire mining grounds. On the top of a hill behind Ha Tu mining ground, there is a viewing tower where the mining control office is situated.

Discussion

As noted above Ha Long Bay with its unique limestone (karst) formations and hundreds of islands and islets is one of the most visited places in Vietnam. While there are a number of diversions available to visitors, the addition of another tourism facility would be appropriate and could help to relieve the pressure of too many tourists concentrated in Ha Long Bay itself (Galla, 2002).

This part of the chapter therefore addresses the potential optional uses of the Ha Tu Mine. The observations, comments and ideas discussed in this section arose from an escorted tour of the mine in March 2009. Mr. Tran Soai, narrator at the company tradition house acted as our guide. Information provided by the guide and the tour of the mine itself, led to the generation of the ideas which comprise the remainder of this chapter. Three options for development of the existing mine, discussed below are: an interpretive center, a recreational area, and an attraction (adaptation of the Eden Project, UK).

An interpretive center

The existing mine tradition house, housing exhibits and models of mining equipment used over the life of the mine, would be an appropriate venue for an expanded interpretive center. Already equipped with exhibits, with the addition of photographs and mining tools to augment the existing collection of models, it would make a ready-made interpretive venue. Conversion of the existing building into an interpretive center is an obvious first choice in terms of cost savings. Minimal expense would be required to augment existing facilities and exhibits and to build a parking area. Space is available in the building for the addition of a video viewing area (*mini-theatre*) and additional models and exhibits. In addition, the community around the interpretive center would benefit from the introduction of a tourism generating facility. One of the advantages of building conversion, other than the creation of a parking area, is minimal impact on the immediate environment. Indeed, developing an interpretive center is an opportunity to improve the exterior and interior appearance of the building, the area around the building and nearby community, as well as generating new income for the local area.

The primary goal of most interpretive centers is to inform or educate the public (Moscardo, 1999). As such, they can also serve as a surrogate for the actual experience of a sensitive or dangerous area. Inskeep (1991: 274) stated that such a facility complex would typically include the reception/lounge area, an information counter (often combined with the reception area), a shop or shops selling items related to the attraction feature, a snack bar or restaurant (or both), an exhibit area or small museum about the attraction, perhaps a small lecture and audio-visual presentation hall, an emergency medical service, toilets, and related landscaping and parking for cars and tour buses. The attraction management office could be also located here.

At the Ha Tu Mine, the opportunity exists to educate the public regarding the mining business and processes. A carefully planned interpretation center could help visitors develop an appreciation of the role the mine and miners have played in the history of Vietnam and the contribution of the mine to the national and local economies. The Ha Tu Mine is an open pit mine and the impact of mining operations carried on over many generations have left their mark on the land. Thus, the opportunity exists for an interpretive center to demonstrate the history of the mine, consequences of mining, the tradeoffs between conservation and economic gain, and the need to develop sustainable options for a planet no longer able to sustain life if fossil fuels continue to be used. It may also be possible to use the site to interpret the role of the mine in the wartime heritage of Vietnam (Henderson, 2000).

The addition of a video of the mine as it exists now, showing the mining process, would bring the mine to life. This could communicate the experience of mining more directly than viewing an exhibit or handling tools. In the case of the Ha Tu Mine, direct use of mining tools would be unadvisable, as the tools are enormous pieces of equipment requiring special training to operate. Further, the general appeal of moving images could contribute to the experience. A well designed video could capture the viewer's imagination and possibly add impact, making the visit more memorable as well as adding a "pull" factor to a visit to the interpretive center.

Families could be a prime target market for this kind of facility because it would provide another option for family centered experiences. As noted, the Ha Long Bay region is already a popular weekend destination with numerous tourists attracted to nearby Ha Long Bay due to its historical importance to the nation and the culture of Vietnam. The facility could attract the growing numbers of domestic and international tourists in Vietnam, taking some pressure off the limited number of attractions in the area (Galla, 2002). It could provide a rare glimpse into the history of industry in Vietnam. As such it could also serve as a focal point for company groups looking for outings to build morale and team work among employees. Timothy and Boyd (2003: 37) note that many visitors are interested in this type of industrial heritage site where they can experience industry in an outdoor setting. The fact that pilot tours have already been conducted is a positive sign for continued development.

A recreational area

before any land redevelopment takes place, three studies are recommended. The first recommendation is for a study of the viability of the use of the site for recreational and leisure purposes for both residents of the surrounding area, as well as visitors. The second, an environmental assessment, is necessary before site plans are developed. A detailed study of possible toxins arising from mining operations needs to be part of the environmental assessment, to avoid problems that could arise from ignoring the issue of toxins related to mining and the concomitant development of a negative perception on the part of potential visitors. Ignoring the potential issue of toxic contamination could invite negative media exposure on a global level. Research into the possibility of toxins and methods to reduce their impact would be the responsible path and could serve as a positive example to the international community. Third, a study of the financial feasibility of recreational or other tourist development would either be part of the viability study or could be undertaken following the completion of the studies recommended above.

Potential activities depend on the safety of the mine after closure. Tours, for example would ideally be offered both as a guided and an independent experience. Most mining operations in other countries include some kind of mine tour, for example, as in Norway (Swarbrooke, 2002: 357). Interpretive panels explaining various features of the land and mine could enhance the visitors' experience. A trail from the interpretive center to a viewing area where visitors could see some of the former mining site would enable visitors to appreciate the enormous scale of the mining operation. The trail experience could be augmented by the addition of fitness stops at fixed intervals. Various exercises or stretches could be outlined on signs designed for the purpose of encouraging physical activity and to appeal to a wide audience. The existing lake on site also has recreational potential (Figure 15.1).

Figure 15.1 Ha Tu Mine Lake.

Source: Nancy Chesworth.

Given the size of the mine, a gazebo currently located near the main office, on top of the edge of the existing pit, could be used as a viewing platform or look off point (Figure 15.2). As it is a small structure, it could either be enhanced to serve as a viewing point or rest stop, or moved closer to the main entrance to mark the beginning of the trail.

An attraction

When visiting the mine, the similarity of the site to the Eden Project came to mind. The Eden Project utilizes a former kaolin, or China clay mine, of 35 acres, which presumably had minimal, if any toxic potential. On the presumption that the environment holds no threat to area residents or visitors, or that any threat can be dealt with in an effective and low cost manner, the third option of developing the mine site as an adaptation of the Eden Project is offered for consideration.

The Eden Project opened officially in March 2001 following two and a half years of construction. It consists of several geodesic domes or biomes, which serve as greenhouses centered around three themes. The Eden Project's website defines a biome as: "a climatically and geographically defined area of ecologically similar communities of plants, animals, and soil organisms, often referred to as ecosystems" (Eden Project, 2009).

The project consists of four components. The first, a rainforest biome, the world's largest greenhouse, contains over 1,129 plant species. A slightly smaller group of domes house the Mediterranean biome. The outdoor biome or gardens focus on native plant life and the educational center called the "Core", contains exhibits, a restaurant, gift shop and space for meetings and workshops. The project is more than just a collection of greenhouses. Its main aim is to provide an educational experience, to "connect people to the natural world" and "give people information so they can make informed choices" (Blewitt, 2004). Blewitt further states that "Eden, through its messaging, communication and informal learning environments, can change people", by which is meant that the visitors' exposure to information will enable them to make better informed choices with regard to how they live, including making ecologically enlightened choices in purchasing and in living in their home environment (Blewitt, 2004). The project hopes to accomplish this goal through direct learning, interpretation and edutainment. To make this goal a reality, Eden uses a variety of strategies starting with the visually engaging architecture of the biomes. In addition, considerable thought has gone into

Figure 15.2 View from existing viewing point at the Ha Tu Mine.

constructing an environment where visitors can engage in learning in as much depth as they wish. Guides found at mobile interpretation stations throughout the project, interact with the public casually on an as-desired basis, with the emphasis on authentic experiences.

Options for redevelopment

As an adaptation, the Ha Tu Mine property could serve a variety of purposes as indicated by Figure 15.3 which illustrates a possible concept. It should be noted that the Ha Tu Mine covers 700 hectares, or approximately 1400 acres, an area much larger than the Eden Project. The first step would a land reclamation project. Reclaiming the land would be a major effort, possibly taking several years. Creative use of composting and finding uses for products that break down quickly would be a natural path to reclaiming the land for other uses. In considering the concepts underlying the Eden Project, a focus on the natural world could also be employed at Ha Tu in several ways, for example:

- a center for ecological education
- a research facility examining tropical plants for medical uses
- a research center for food crops and adaptability of various crops under changing climate conditions
- a fitness, hiking and environmental exploration center
- an entertainment venue focused on responsible consumption
- a waste neutral recycling demonstration center
- a center for the study of adaptation, mitigation and relocation possibilities for climate change.

All of these options would require considerable financing. The Eden Project, for example, received a grant of £43 million, and was completed at a cost of £86 million (Blewitt, 2004). The cost, however, is counter balanced by the financial benefits. From its opening to 2004 the Eden Project brought over £430 million to Cornwall (Blewitt, 2004).

Vietnam has an opportunity to develop the Ha Tu site into a world leader in research in several fields. As an example, we suggest the construction of a facility to research the impact of climate change on food production, in Vietnam and the rest of South-east Asia. Vietnam is the world's second largest producer of rice and the third largest producer of coffee, and thus has a vested interest in planning ahead in the face of the changing global climate and changing global economies. The International Rice Research Institute (IRRI), located in the Philippines currently is charged with research into rice for the Asian community. Given the current rice crisis, and the likelihood of the loss of rice growing areas, it seems reasonable to suggest a research site in Vietnam that would be protected from flooding. The Ha Tu site would potentially offer this protection, barring super typhoons striking northern Vietnam. Thus, the first priority suggested is the construction of a sustainable food research facility.

By combining the first and second options discussed above, the site could pay for itself by offering a variety of educational and activity experiences as well as research centers, becoming self-supporting and earning a profit over the long term. Careful planning is of course necessary to the success of such a large development project. It is hoped that the opportunities offered by the potential transformation of the Ha Tu Mine will be considered and acted upon.

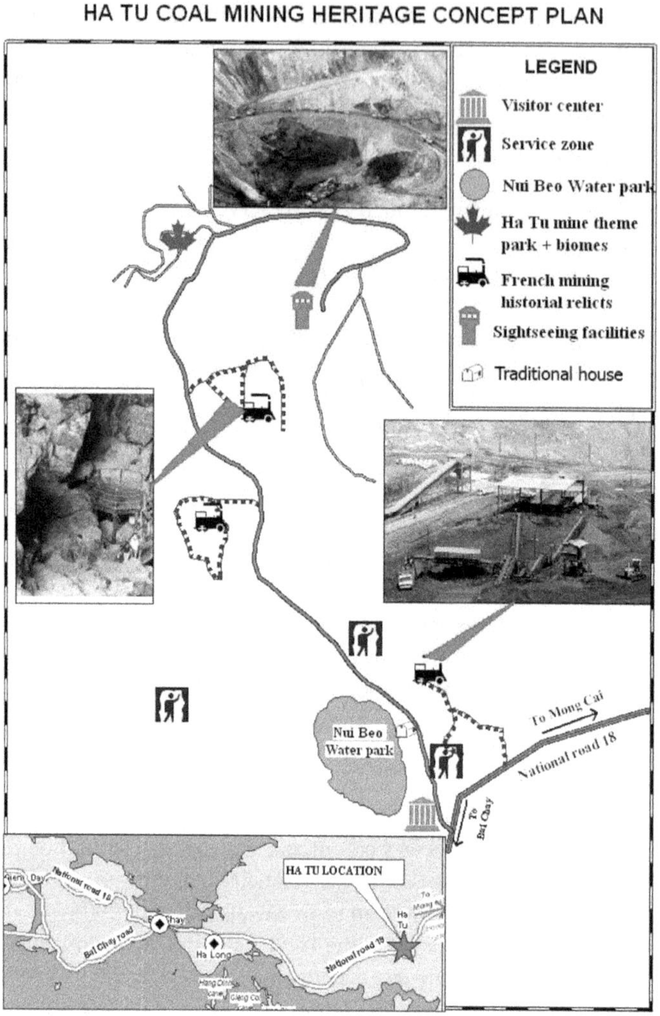

Figure 15.3 Ha Tu Coal Mining Heritage concept plan.

Conclusion

Ha Tu Mine has a rich social and industrial history as the premier coal mine in the north of Vietnam. When the mine closes in 2015 there is the opportunity for it to be adapted for a new use. Given that tourism is the first priority in this province and that there is an urgent need to disperse visitors away from the congestion at Ha Long Bay World Heritage Site conversion to some type of mining attraction is logical. As discussed in this chapter, proposed development initiatives for the mine as a tourist attraction site include an interpretive centre, a recreational area, a visitor attraction or even a combination of these three options. These options offer the opportunity to revitalize and renew this historic mining site contributing to local tourism development, generating positive economic and employment impacts while revitalizing local natural resources and the environment.

References

Anna, S. (1999) 'Conflict of Interest between Coal Mining Industry and Tourism' in *Ha Long Bay: A Cost Benefit Analysis*. Umea: The Umea University.

Binh, V.T. *et al.* (2001) *Vietnam Tourist Guidebook*. Hanoi: Vietnam National Administration of Tourism.

Blewitt, J. (2004) 'The Eden Project – Making a Connection', *Museum and Society*, 2(3): 175–89.

Chinh, L.D., Gheewala, S.H. and Bonnet, S. (2007) 'Integrated Environmental Assessment and Pollution Prevention in Vietnam: The Case of Anthracite Production', *Journal of Cleaner Production*, 15(18): 1768–77.

Eden Project (2009) 'Rainforest Biome: What is a Biome?' Available online at: http://www.edenproject.com/come-and-visit/whats-here/rainforest-biome/index.php (accessed 2 September 2009).

Galla, A. (2002) 'Culture and Heritage in Development: Ha Long Ecomuseum, A Case Study', *Humanities Research,* IX(1).

Henderson, J.C. (2000) 'War as Tourist Attraction: The Case of Vietnam', *International Journal of Tourism Research*, 2(4): 269–80.

Hudson, S. (1975) *A Social History of Museums*. London: Macmillan.

Inskeep, E. (1991) *Tourism Planning – An Integrated and Sustainable Development Approach*. New York: Van Nostrand Reinhold.

Moscardo, G. (1999) *Making Visitors Mindful: Principles for Creating Sustainable Visitor Experiences through Effective Communication*. Champaign, Illinois: Sagamore.

Quang Ninh Department of Planning and Investment (2009) The Official Website. Available online at: http://www.quangninhdpi.gov.vn/comment.asp?id=343&CatId=151 (accessed 19 September 2009).

Quang Ninh e-newspaper (21 October 2008)'Forty-five Year Review of Quang Ninh Province (1963–2008)', Available online at: http://www.baoquangninh.com.vn/print.asp?id=29830 (accessed 19 September 2009).

Quang Ninh Province Official Website (19 August 2008) Available online at: http://www.quangninh.gov.vn/Trangchu.aspx (accessed 18 September 2009).

Sinh, B.T. (1995) *Policy in Vietnam: Conflicting Interests in Quang Ninh Province – Coal Mining, Tourism*. Ministry of Science, Technology and Environment: Hanoi.

Swarbrooke, J. (2002) *The Development and Management of Visitor Attractions*. Oxford: Butterworth-Heinemann.

Thuren, A. (1997) *Environmental Protection in Open-Pit Coal Mining in Quang Ninh Province*. The proceedings of the first EIA workshop of the EU project VNM/B7-6200/IB/96/05.

Timothy, D. J. and Boyd, S. W. (2003) *Heritage Tourism*. Harlow, Essex: Pearson Education Limited.

Tu, T.V., Uy, N.N., Gang, V.V. and Soai, T. (2000) *History of Ha Tu Mine: 40 Years of Establishment and Development*. Quang Ninh: Quang Ninh Printing Company.

16 Mining tourism in Ouro Preto, Brazil

Opportunties and challenges

Ângela Cabral Flecha, Marcos Eduardo Carvalho Gonçalves Knupp, Gui Lohmann, Antonio Liccardo

Introduction

The eighteenth century, during which Brazil was a colony of Portugal, was a culturally and economically remarkable period in history. In Brazil, that century is known as the Gold Cycle, or the Mining Cycle, because gold and other precious minerals started to be mined at that time. The most important urban center related to mining was Villa Rica, in the state of Minas Gerais, later renamed Ouro Preto (*Black Gold* in English), the subject of this chapter.

Mining is a traditional economic activity in Brazil. The country's underground is host to important mineral deposits. In the year 2000, Brazil held 92 per cent of the global market share of niobium, 20 per cent of iron ore, 22 per cent of tantalite, 19 per cent of manganese, 11 per cent of aluminium, and 19 per cent of asbestos and graphite, amongst others with an average 5 per cent global participation, such as ornamental rocks, talc and vermiculite. Particularly important is the south-eastern state of Minas Gerais, one of the leading economies in the country, with extensive mining activity since the early colonization period in Brazil. As a result, the potential for mining tourism, a form of geotourism that can be developed in mining locations as well as in urban areas affected by mining activities, in Minas Gerais is remarkable.

Apart from being a famous part of Brazilian history, Ouro Preto is also a place of mineral trade and visitation to former mining sites. For centuries, the city enjoyed the prestige and wealth provided by intensive mining activities. During the second half of the eighteenth century, Ouro Preto experienced a burst of architectural activity, being a nucleus of intellectual and artistic activities as well as the breeding ground of the 'Inconfidência' revolutionary movement which eventually led towards the independence of Brazil. As mining activities declined by the end of the century, so did the city's social life. In 1895, when the state's seat of government was transferred to Belo Horizonte, Ouro Preto entered a period of isolation.

This chapter identifies and presents major opportunities and challenges for mining related tourism in Ouro Preto. The cultural and mining heritage of the city constitutes a fascinating product when presented as tourism attractions, particularly considering that tourism is currently the heartbeat of the local economy. The chapter is divided into four sections. The first addresses some conceptual and general aspects of geoparks, geotourism and mining heritage tourism. The area in which Ouro Preto and a UNESCO related geopark project are located, the 'Quadrilátero Ferrífero' (*Iron Quadrangle*, in English) is contextualized in the subsequent section. The third part discusses tourism in Ouro Preto, where many attractions are related to the mineralogical setting. Finally, the last section presents and discusses opportunities and challenges for Ouro Preto as a mining tourism destination, along with some recommendations on the interpretation and use of mining heritage resources for tourism.

Geoparks, geotourism and mining heritage tourism

Associated with mining heritage tourism are the concepts of geoparks and geotourism. According to UNESCO (2008), a *geopark* is 'a nationally protected area containing a number of geological heritage sites of particular importance, rarity or aesthetic appeal'. These sites are usually part of an integrated concept of protection, education and sustainable development. Eder and Patzak (2004) state that the geopark concept was developed mainly in Europe in cooperation with UNESCO, with the purpose of preserving geological heritage sites for future generations (conservation), educating the general public about the geological landscape and environmental matters (education), providing research facilities for geoscientists, and ensuring sustainable development (tourism). The size and function of geoparks are usually large enough to allow for economic activities and include a sufficient number of geosites with important geological features able to stimulate economic development. Geoparks may also include archaeological, ecological, historical or cultural features.

The regional impact of a geopark is that it usually improves human living conditions and the rural environment, strengthens identification of the population with their neighborhoods and triggers cultural renaissance, stimulating the creation of innovative local enterprises, small businesses, cottage industries and new jobs that generate new sources of revenue and geotourism products (Eder and Patzak, 2004). Within a geopark, sites of greater relevance known as geosites may include, for example, cities and their surroundings. Geosites must, however, be encompassed by the concept of geological heritage.

The modern concept of museology (or museum studies) promotes the facilitation of experience, attention being focused on the importance of scenario as the main facilitator since, in practice, this is the only variable that managers can control. In tourism, scenarios are often destinations or attractions in themselves, and the main goal is the benefit of the trip. Prentice *et al.* (1998) investigate which experiences are to be included in an attraction, by identifying types of tourists and how their experience can be facilitated. Hewison (1987) sees museums as popular components of attraction-based tourism and resources of the destination areas. Coupled with an increasing interest in history, the supply of museums led to the growth of the so-called heritage industry in the 1980s.

Pretes (2002) realized that the conjunction of geoparks, museums and heritage sites fosters the establishment of mining tourism as a cultural element. It is the transformation of mining areas into spaces where tourists can also appreciate the historical and cultural heritage. According to the same author, in a region intensively mined in the past there is usually a 'mining community' and it is the local population that identifies itself with mining tourism. The economy of these areas used to be, and often still is, based on mining, which means that part of the population is directly related to that activity. In the case of Brazil, and especially the State of Minas Gerais, one of the largest mineralogical provinces on Earth, its mining heritage is of great importance to the development of geotourism.

The Quadrilátero Ferrífero as a geopark

Since the beginning of mining activity in Brazil, mineral production has come in large part from the region now called Quadrilátero Ferrífero, a four-sided block of Precambrian structures exposed by differential erosion in central-southeastern Minas Gerais (Barbosa and Rodrigues, 1967). Its area covers approximately 7,000 km^2 and it is surrounded by 1,300–1,600 meter-high crests that belong to four different mountain ranges (Figure 16.1).

Map 16.1 The location of Quadrilátero Ferrífero and Ouro Preto.

Given its history and richness, the Quadrilátero Ferrífero can be considered the most important mineralogical province in Brazil. Rich in mineral resources, scenic landscapes, and geological and human history, the region is under consideration to become a geopark to be operated by a forum of several institutions (Ruchkys, 2007). More recently, various tourism-related activities have grown significantly in importance, which will certainly contribute to the proposed geopark establishment.

Ouro Preto and its attractions are the main geosites. One of the issues regarding Ouro Preto is the conservation of physical and non-physical heritage elements, including their mining culture aspects and their visitation. In this context, it is essential to discuss the relative importance of the mining industry and the efforts to protect the environment and the scenic beauty of the Quadrilátero Ferrifero landscape. According to Mantesso-Neto *et al.* (2009), given the local economic situation, there is no scope for a dichotomy between mining and preserving the mines. What should be discussed instead is the concept of sustainability and its applicability.

Imperial topaz, soapstone and stone masonry as tourism attractions

Two minerals found in Ouro Preto stand out not only for their historical, but also current economic importance: the imperial topaz and soapstone. The first is found in Ouro Preto and is unique. The second provides the material for handicrafts and architectural projects that form most of the monuments in the historic center of Ouro Preto. The importance of these two mineral resources is discussed in this section.

The imperial topaz is one of the rarest gemstones in the world, whose only occurrences of gemological value are located in Ouro Preto. This variety of topaz (Figure 16.1), whose color ranges from yellow-orange to a pink-purple, was discovered around 1772 in Saramenha Hill, a district of Ouro Preto. Its existence was noticed by some of the first naturalists who visited

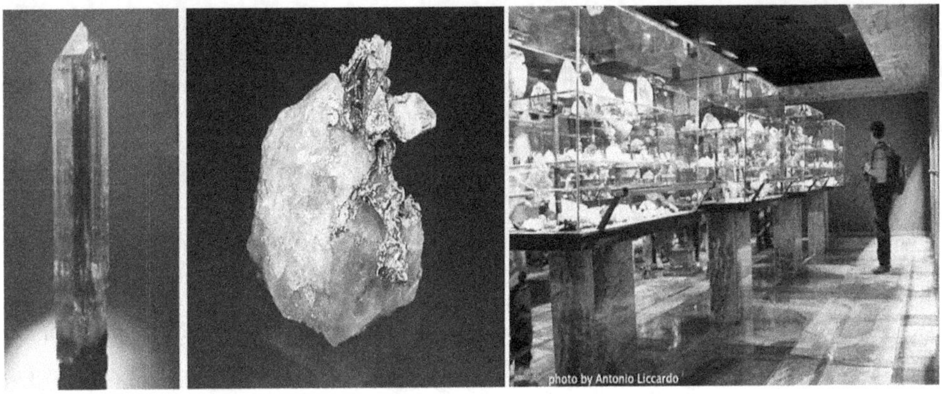

Figure 16.1 Images of the Museum of Science and Technology, including the largest imperial topaz crystal (22 cm) and an aggregate of gold and quartz from the region. Photos by Antonio Liccardo.

Brazil. John Mawe mentioned the mineral in his journey diary while visiting Minas Gerais in 1812. Although very limited, the production of imperial topazes drives an important jewellery trade in Ouro Preto, with more than 20 shops that sell crafted jewellery whose unique design reflects the local identity (AJOP, 2009). Topaz gemstones with good gemological properties can reach prices of up to US$1,500.00 per carat (US$ 7,500 per gram).

Soapstone is another mineral resource of Ouro Preto with touristic appeal. Soapstone, also known as steatite, is a metamorphic rock that has talc as one of its constituents. The presence of talc renders soapstone a soft rock with which sculptures of great beauty can be made. The works of Aleijadinho, a famous Brazilian artisan of the eighteenth century, are made of soapstone. Nowadays, soapstone lends itself to a stone masonry that is unique in Brazil (Figure 16.2). It has also played an important role in architectural composition, often in association with quartzite or schist (Pereira *et al.*, 2007).

The stone masonry brought by the Portuguese to Ouro Preto flourished during the period of thriving economy. The variety of rocks types available allowed their use for all kinds of

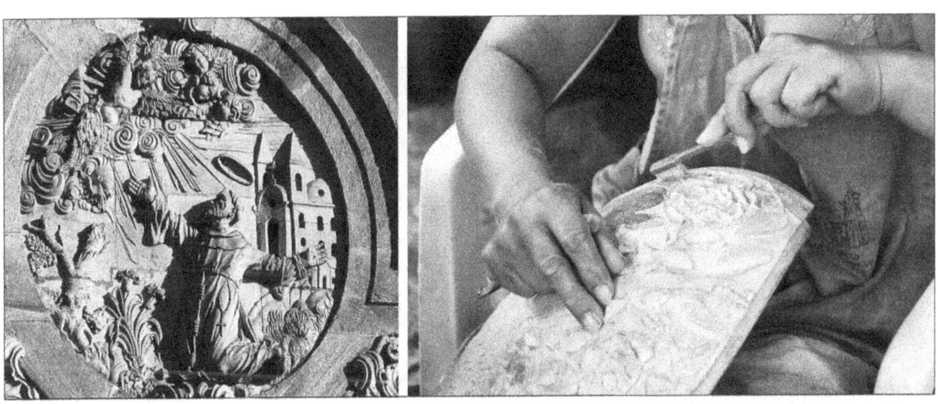

Figure 16.2 Soapstone work: medallion of the Saint Francis Church, carved in the 18th century by Aleijadinho, on the left, and the current work of a local artisan. Photos by Antonio Liccardo.

purposes: from the structural support of buildings to the decoration of church entrances, to religious sculptures, and also household items like pots and pans. This traditional craft was almost extinct, but some projects are now under way to recover, inventory, preserve and transmit it as an intangible cultural heritage (Pereira *et al.*, 2007).

Mining tourism in Ouro Preto

Owing to its historical, social and cultural importance, as well as its natural and architectural beauty, Ouro Preto inevitably has tourism as an alternative for investment. Available data show that tourism represented 10.4 per cent of the local gross product and 21.8 per cent of the tax revenues in 2004. Approximately 11 per cent of the region's workforce was directly or indirectly related to tourism in 2002 (Flecha *et al.*, 2008). However, most of the local population does not perceive the benefits of tourism, in part due to the lingering belief that mining should be solely responsible for the local development. Some of the existing mining heritage attractions located in Ouro Preto are described in Table 16.1. The photos in Figure 16.3 illustrate the tourist experience in one of the mines in Ouro Preto (photo on the left), as well as the traditional 18th century mining activity which is still carried out in this destination (photo on the right).

Opportunities and challenges

Ouro Preto is surrounded by a rich and varied natural environment with waterfalls, hiking trails and native vegetation partially protected as state parks. Parts of these resources are used for tourism. Paradoxically, this ecosystem contrasts with the human occupation of the region that produced, after centuries, a rich history and a cultural connection to mining, its oldest economic activity which triggered occupation. The region has an unlimited potential for tourism, especially in specific segments such as mining heritage tourism, in association or not with the existing ecotourism market. In fact, in Ouro Preto, tourism, history, geology and mining are often hard to distinguish; such is the inter-relationship between these segments.

For centuries, a major problem of mining has been the reuse of the affected areas. Modern mining projects proposed solutions to this problem right from the initial stages of operation, which did not happen until recently. As a result, most quarries and other old mining areas

Figure 16.3 Tourists on the trolley that descends into the 'Passagem' Gold Mine and artisanal gold washing, as done in the 18th century. Photos by Antonio Liccardo.

Table 16.1 Main attractions related to mining heritage tourism in Ouro Preto

Attraction	*Description*
The Inconfidência Museum	Presents historical aspects of the Brazilian Gold Cycle, including 18th century mining equipment and quartzite stone masonry in architectural elements that reflect the splendor of that period.
Casa dos Contos *(House of Tales)*	Former gold foundry house where taxes due to the Portuguese Crown were paid. The building dates from the 18th century and has stairs and details in quartzite masonry, in addition to content related to the Gold Cycle. Today, it houses the Center for the Study of the Gold Cycle and the Monetary and Fiscal Museum.
Museum of Science and Technology	Conserves a mineralogical collection formed since the 19th century, the largest one in Latin America, with more than 23 million items from all over the world (Figure 16.1).
Gold Mines	Given the chaotic early days of mining, the historic center of Ouro Preto has many former gold mine galleries. Some of them are superb attractions. A highlight is the 'Passagem' Gold Mine, the largest gold mine open for visitors in the world. Visitors learn about the mine's history and can observe some mining equipment. They can even try panning for gold from a tank with water, gravel, sand and some gold powder. It is also possible to go down by cable car to visit large galleries and a beautiful underground lake. The site also offers the possibility of diving into the underground galleries and tunnels flooded as a result of being below the water level (Figure 16.3).
Imperial Topaz	Visiting the centenary topaz mining district of Antônio Pereira is very easy. Access is made by a paved highway that runs past the mining area. Miners receive tourists and sell small topaz samples. This is the only imperial topaz mine open for visitation in the world. In Cachoeira do Campo, 'Capão' Mine only receives visitors sporadically, mainly for scientific purposes.
Queimada Hill	Region of the mountainside where Ouro Preto started to grow, Queimada Hill presents archaeological remains of early mining works. Stage of the first revolt against the Portuguese control, the formerly called Morro do Ouro Podre (*Rotten Gold Hill*), was completely burned by order of the then governor. Ruins and other vestiges are being considered for the implementation of an archaeological study site possibly open for visit.
Churches	Churches and chapels in Ouro Preto and the surrounding region have ornaments pigmented with gold that were common when the metal was abundant. They also display the most important stone masonry works in quartzite and soapstone. Typical Baroque sculptures in soapstone are also present in the churches of the region.
Soapstone Fair	Handicrafts in soapstone typical of the region are sold in this public exhibition place by dozens of artisans. This important activity is a strong cultural expression of the locals.
Gem Shops	The historic center of Ouro Preto concentrates over 20 gem shops, dozens of cutters and goldsmiths and one of the most intense gem trades in the country, especially at Tiradentes Square and Direita Street.
Stone Masonry	Stone masonry is an important cultural aspect recovered by the Federal University of Ouro Preto. An ancient craft in the history of mankind, stone masonry is currently an endangered activity. For its recovery, a comprehensive survey was carried out and the technique is now taught to new generations of stonemasons. A roadmap to the main works, which includes fountains, bridges, church entrances and other architectural details in quartzite and soapstone, is under implementation.

that do not have an appropriate destination represent serious environmental problems. Mining tourism utilizing exhausted mines is a source of employment and income. Tourism activities may even contribute to the recovery of degraded areas in various ways, such as reforestation for leisure purposes, or their transformation into history museums where aspects of local mining are interpreted.

Minas Gerais, and particularly Ouro Preto, provides the strong and rich cultural and historical content needed for the transformation of mining remnants into attractive tourism products, especially when combined with the existing cultural tourism of the region. Although mining tourism is explored in various parts of the world in extremely different social, economic, cultural and natural contexts, in Brazil it is still not a strategy readily adopted as an alternative for areas affected by mining activities.

The proposed installation of the Quadrilátero Ferrífero Geopark has Ouro Preto as its main geotouristic pillar. In this respect, the city concentrates not only mining examples representative of the entire region, but also hosts a university that is traditionally focused on mineral exploration, providing important scientific and technical support for the development of geotourism. The Museum of Science and Technology of the School of Mines, with the largest mineral collection in Latin America, is a great attraction for a tourist whose profile includes some mining interests. The presence of scientist-tourists such as geologists, geographers, mineralogists, historians and others is constant in the city, which indicates a potential for targeted tourism. Recently, Stonebridge Imports, a Canadian company, has launched a package tour that specializes in the visitation of mines and gem shops. Mining tourism represents an improvement opportunity for local receptive tourism operators by adding extra value to existing packages. Technical field trips to the region, for example, are among the most sought after options during specialized events. Some opportunities and challenges related to mining tourism in Ouro Preto are presented in Table 16.2.

In Ouro Preto, active mines and decommissioned ones like the Morro da Queimada can be considered historic mining, geological and archaeological sites. They form a perfect setting for the development of tourism. However, specific policies aiming at the implementation of tourism projects in these areas have not yet been adopted. As a result, the mines remain ignored and are often subject to inappropriate occupation, despite how important they are in the local context.

There are great possibilities of creating specialized routes in Ouro Preto which would increase the information content of urban attractions. Such routes would promote the (re)discovery of mining by tourists by combining past and present activities in the urban core and its surroundings. The possibility of a more comprehensive tourism circuit should also be considered. Important cities in the Quadrilátero Ferrífero are surrounded by mining activity, virtually all related to the historical context of the Gold Cycle and that can be considered geosites.

Another possibility of preserving natural and human-built heritage can be an ecomuseum. The city of Ouro Preto has an area of 1,200 km^2 that includes protected and inventoried natural and human-built heritages. Unlike numerous mining communities in Wales, where derelict mines have been transformed into tourism attractions (Edwards and Llurdés i Coit, 1996), Ouro Preto needs to promote its natural, human-built and visible heritage in order to become an open-air museum. A timely discussion is to what extent occupied environments are cultural heritage, such as the buildings of Portuguese heritage in Ouro Preto.

In the case of Ouro Preto, there is the opportunity of experiencing the unique scenario, used in all its fullness by the residents and that can be considered both an eco-museum and an open-air museum. Tourists and inhabitants mingle in the local rhythm. In fact, besides the

Table 16.2 Opportunities and challenges for mining tourism activities in Ouro Preto

Opportunities	*Challenges*
Geotourism	Geotourism in Brazil strides. Its implementation in Ouro Preto needs to be better promoted as a tourism product. The geological setting of Ouro Preto can be fascinating if properly presented as a tourism attraction.
Gemological Tourism	The production of gem minerals is mainly targeted at the European, Japanese, Chinese and American markets. Foreign tourists visiting Brazil are potentially interested in mining tourism that involves not only buying gems, as it already happens, but also other aggregated services, such as those proposed in this section.
Queimada Hill	Although Queimada Hill is already an inventoried heritage, no environmental protection actions have been taken so far. Some mines remain open but without any regular inspection. The hill is not considered by the municipal administration an attraction and, therefore, no orientation and protection is available to tourists.
The Quadrilátero Ferrífero Geopark	Creation of a geopark operated by a consortium supported by different institutions.
Iron and Aluminium Mines	Several iron and aluminium mines are worked in the surroundings of Ouro Preto which might be considered for tourism when decommissioned. Vale, the largest iron ore mining company in Brazil and one of the largest in the world has developed visitation programs in areas where it operates. This is done as a social contribution to the communities impacted by mining activities.
Placer Mining	The most rudimentary form of gold mining is still practiced along watercourses of the region. Some washers still survive from this activity, using the very same techniques of centuries ago. Mercury is not used, hence the activity being considered a form of artisanship (Figure 16.3). Gold washing is not explored as a tourism product in the area yet, but visitation would contribute to its preservation as a cultural heritage.
Communication	One of the greatest challenges of science is to shorten the distance between scientific content and common understanding. Geology and other mining-related scientists should use a less hermetic speech when communicating with the general public, so that the beauty of science can be more easily understood.

concepts discussed above, Ouro Preto also shows a mining heritage. Numerous sites open for visitation like the 'Passagem' mine, their equipment being exhibited at the Museu do Ofício (in English, *Craft Museum*). Today, the city also hosts a strong mining industry, with various other mineral resources being exploited such as aluminum and iron that can promote the development of a new type of tourism for these areas, such as industrial tourism.

The government's role in this process must be considered, as together with the local community, public policies define regulations, objectives and development strategies. The local population and private sector benefit from such actions, but the government is responsible for coordinating the process of tourism development.

Conclusion

Properly explored in a sustainable fashion, mining tourism thus brings several benefits to the community, including economic and cultural impacts. It can bring out distinctive tourism elements, adding geotourism to the historical-cultural tourism already sought by tourists. It can also ensure the maintenance of the city's traditional mining culture. Mining tourism provides, therefore, a fresher view of historical heritage and the ever changing socio-cultural dynamics.

Finally, a very important aspect that should be considered in further studies is that mining industries in Brazil are obliged to pay royalties to the municipalities where they operate. The use of those royalties in the development of activities that promote culture and heritage preservation would most definitively benefit mining tourism.

References

AJOP (2009) Roteiro Jóias e Gemas de Ouro Preto. Folheto de divulgação.

Barbosa, G.V. and Rodrigues, D.M.S. (1967) *Quadrilátero Ferrífero*. Belo Horizonte: Universidade Federal de Minas Gerais.

Eder, W. and Patzak M. (2004) 'Geoparks – Geological Attractions: A Tool for Public Education, Recreation and Sustainable Economic Development', *Episodes*, 27(3): 162–64.

Edwards, J.A. and Llurdés i Coit, J.C. (1996) 'Mines and Quarries, Industrial Heritage Tourism', *Annals of Tourism Research*, 23(2): 341–63.

Flecha, A.C., Amaral, H.F., Iquiapaza, R.A., Paula, S.F. and Ramon, D. (2008) 'Economic Impacts of Tourism in Ouro Preto, MG, Brazil', in First International Meeting in Tourism Management, Ouro Preto, *Annals of EIGTUR2008*. Ouro Preto: NUPETUR.

Hewison, R. (1987) *The Heritage Industry: Britain in a Climate of Decline*. London: Methuen.

Mantesso-Neto, V., Ruchkys, U.A., Nascimento, M.A.L., Liccardo, A. and Guimarães R.L. (2009) 'Quadrilátero Ferrífero, Brazil: General Information about a Potential Geopark', in *New Challenges with Geotourism – Proceedings of the VIII European Geoparks Conference*. Idanha-a-Nova, Portugal.

Pereira C.A., Liccardo, A. and Silva, F.G. (2007) *A Arte da Cantaria*. Belo Horizonte: ComArte.

Prentice, R.C., Witt, S.F.and Hamer, C. (1998) 'Tourism as Experience: The Case of Heritage Parks', *Annals of Tourism Research*, 25(1): 1–24.

Pretes, M (2002) 'Touring Mines and Mining Tourists', *Annals of Tourism Research*, 29(2): 439–56.

Ruchkys, U.A. de (2007) '*Patrimônio Geológico e Geoconservação do Quadrilátero Ferrífero, Minas Gerais: Potencial para Criação de um Geoparque da UNESCO*', Unpublished PhD dissertation, Federal University of Minas Gerais.

UNESCO (2008) *Guidelines and Criteria for National Geoparks seeking UNESCO's assistance to join the Global Geoparks Network*. Available online at: http://www.unesco.org/science/earth/doc/geopark/2008guidelinesJuneendorsed.pdf (accessed 10 October 2009).

17 Mining heritage, local development and territory identity

The case of Sardinia

Carlo Perelli, Paola Pinna and Giovanni Sistu

Introduction

This chapter will analyze the evolution of mining heritage management and promotion in Sardinia. The core theme will be the analysis of the changing perception of mines as an economic and development opportunity occurring in the shift from the productive phase to the transformation of mines into tourism attractions. In this analysis we will focus on the Geopark creation process and the inclusion in the related UNESCO list. The analysis will also be based on the interviews held with local decision makers, social and economic actors. Particular attention will be paid to identity issues in the background of the local development process.

Tourism products based on mining and geological attractions are nowadays commonly marketed. What is relatively new is the attempt to provide international coordination to the national initiatives. UNESCO took the lead in this process in June 2001, when the Executive Board mandated the development of a network aiming to support territories and tourism destinations interested in integrating Earth heritage into traditional tourism supply. The theoretical and operative framework adopted by UNESCO is the idea of the Geopark (Dowling and Newsome, 2006). It's by means of the Global Network of National Geoparks that UNESCO tries to promote and coordinate geological heritage preservation (UNESCO, 2008). The network strategies are integrated with other UNESCO centres of activities like the World Heritage Centre, the Man and the Biosphere Programme (MAB) and the World Network of Biosphere Reserves.

The basic assumption of the Geopark model is the need to integrate geological, historical and cultural heritage with environmental resources in a development process based on local communities and their economic activities (UNESCO, 2008). According to this approach geological heritage valorization finds its sense in relation to territories as the context of material, social and symbolic practices. Economic development strategies related to geological resources valorization are linked to an extensive notion of territorial culture and heritage. Interesting experiences of innovation in design, production and commercialization of traditional products emerge in Geoparks best practices as a result of innovating manufacturing and food sectors (Ruiz-Ballesteros and Hernández-Ramírez, 2007; see *www.europeangeoparks.org* for an overview of European best practices). Local communities are involved in an evolutionary process of transformation and adaptation of values, knowledges and practices. On the one hand this process recognizes economic significance of artefacts or traditional practices in crisis or perceived as marginal with respect to mass consumption markets. On the other, local knowledge can find new value in its originating *milieu*. Communities can preserve their particular heritage knowledge by managing the innovation processes and directly exploiting the benefits of such processes. As a consequence innovation processes can contribute to avoiding the risk of local knowledge trivialization by means of a continuous adaptation mechanism activated by local communities. Territories in the Geopark vision are the support for local based development process, shaped around community priorities and embodied in the lives of local people.

Reflections on the relationships between communities, territories and the tourism cross development debate have particularly involved geographers in the last decades (Soja, 1971; Raffenstin, 1981; Turco, 1988; Hall, 1994; Hall and Richards, 2000; Ruiz-Ballesteros and Hernández-Ramírez, 2007). The way the *milieu* (the simple space where communities imagine, project and perform their ideas on a piece of land) is related to *territories* (the result of communities' spatially determined vision and projects) is clearly developed in Raffenstin (1981). Space is conceived as a precondition of territories that are *produced* by the actors and, consequently, territories are by definition contested spaces, it being nearly impossible for all the actors to share the same imagination, vision and projects for that space. The concept of territorialization expresses the process of territories production. Tourism development projects can be interpreted in this way as territorialization elements, for the fact that they involve an idea, a vision of the projected space and of the actor's role in such a development project. The expected result is a process of definition of a space as a tourism territory.

Methodological notes

The chapter describes the results of the research activities carried on during the last two years by the authors on the Geopark experience in Sardinia. Particular attention is devoted to investigating the evolutionary process of the Geopark that in organized forms is now more than ten years old. To have a picture of the state of the art of the Geopark an analysis of the social, economic and governance patterns has been carried out with the regional and local actors directly or indirectly involved in the development process of the Geopark in Sardinia.

At the end of the 1990s extensive research focusing on 113 direct interviews with miners showed their perception of the transition phase (Boggio *et al.*, 2003). During the last two years a direct survey of 25 relevant actors (institutional representatives, enterprise managers, social and cultural associations) has been carried on during the Socio Economic Analysis of the Geopark (CRENoS, 2008). Semi-structured interviews have been realized with the aim of evaluating both the perception of the actors directly involved in the development processes but also to investigate how the Geopark communities are aware of the valorization processes in action.

At the same time desk analysis of the socio-economic data has been supporting the research design and analysis (CRENoS, 2008). The result is a view focusing mainly on territories, their actors and the dynamics related to mining activities heritage, viewed under the lens of tourism development analysis.

Mining territories in Sardinia

Historically, mining activities have been present in Sardinian territories. Traces of mineral extraction date back to the origins of the Nuragic civilization around 2000 years before Christ. Large scale territorial specialization in mining activities is not observable until the emergence of the modern age of the mine industry under the Savoy Kingdom during the eighteenth century. From the second half of the nineteen century the mining activities in Sardinia attracted the interest of North European capitalism. This was probably the beginning of the formation of a large scale mining industry and of a consistent specialized working class. From the 42,000 tons of minerals produced per year in the decade 1860–70 Sardinian production grew to 164,000 tons per year between 1890 and 1899. At the end of the nineteenth century around 11,200 people were employed in the mines and at least one-tenth of Sardinia population depended on the industry. Mining integrated farming activity revenues defined a new productive specialization but also marked the emergence of mining territories as they appear today.

Analysts agree on the strong colonial character of the Sardinian mining industry at this stage (Ortu, 1998). On the one hand the internal market was not able to sustain the birth of a modern extractive industry given the low population density and the subsistence character of the rural economy. On the other hand foreign capital controls nearly all of the mining sector and the raw minerals are destined for the European markets. Both the work organization model and the broad paternalism of the mines entrepreneurship produced enclavic economies with reduced multiplier effects on the regional scale. But during its development history, mining has represented at the same time an impressive modernization opportunity for Sardinia, for instance supporting workforce mobility flows at the transnational scale. In 1862, one-third of the Sardinian workforce in the mining sector (around 3,100 workers) was composed of people from abroad. At the end of the same decade three quarters of the 9000 people workforce of the regional mining sector was from Sardinia. In a few years, mining contributed to knowledge exchange, innovation in lifestyle and in the self representation of rural Sardinian people (Ruju, 2008; Salvadori, 1990).

Notwithstanding working under inexcusable conditions and marked inequalities in the local distribution of wealth, miners perceived their work as an improvement in life conditions. On the local scale mining has resulted in the consolidation of a specialized expertise and a revolutionary shift from farmhand working conditions to the mines salary. Territories started to experience internal mobility in search of better working opportunities and the highest level of salary. Moreover, for Sardinian workers the transition phase from the traditional rural economy to the mining sector appears more radically oriented than in other mining territories. Mining and farming multi-activities that are commonly present, for instance, in Spanish and Belgian history are uncommon in Sardinia. Return to the land is usually related to lack of employment during the recurrent falls in mineral price levels on the international markets. Miners act like a specialized work force with high levels of membership of mine workers' unions especially in the coal productions areas. Social protest even under the most anti-democratic political phases and under the Fascist Regime (1922–1943) has become part of the cultural background of miners and their families, and also of the territories involved. The identity of territories and communities is strongly dependent on the solidarity networks developed in the mining environment.

Mining industry priorities determined on the one hand a slow adaptation of territories to the reorganization of space and of the communities around the new production system. However, in some cases they produced impressively drastic territorialization and reterritorialization processes. This was the case in Sulcis and the coal district of Carbonia, an urban settlement that starting from the name (in English something like *coaltown*) embodied a new territorial identity based on coal production. In 1935, the Fascist Regime created a government controlled authority to manage coal production at the national scale and focused on Sulcis as one of the most relevant production areas in Italy. Industrial extraction on a large scale started in 1937 and the regime planned a new urban settlement to house the workers. In a territory characterized by dispersed settlement supporting the traditional farming economy a new 12,000 inhabitant city was built in less than 300 days. Carbonia was planned as a dispersed settlement in effect interfering with any form of social integration and spontaneous aggregation between the workers. More than 90 per cent of the residents were male. Spatial organization reflected the image and the sense the Fascist Regime envisioned for the newly born coal mining territory. A few years after the end of the Regime, Carbonia reinvented its identity, becoming the core area of miners' fights and workers' social claims, led by the Communist Party.

Mining activities marked in different ways territories with metallurgical mining industry specialization and coal mining areas. The traditional metallurgic areas were mainly characterized by private entrepreneurs, supporting paternalistic social policies and the professionalization of a specialized working class. At the end of the productive phase municipalities in some way tried to assist private entrepreneurs to provide alternative development options and to offer services formerly provided by mining companies. On the other hand, the coal industry was born as a public project around a factory city owned by the public company managing industrial sites. The state has been the miner's traditional counterpart both in its Fascist Regime form and, as a democracy, after the Second World War. The proletarization of a huge workforce with reduced technical skills characterized coal mines and created a strong solidarity network related to unions and the Communist Party, the base of successful social claims against the state.

During the transition from the productive phase to the contemporary attempts of heritage valorization mining territories have been experiencing a loss of identity. During the 1960s petrochemical and metal industry development, far from involving local communities in the elaboration of a shared development project, only partially absorbed the mining workforce. In 1957, the mining sector recruited the last miners and during the 1960s private entrepreneurs rapidly abandoned the mining sector that survived only through public intervention. For a decade, coal mines were supported by a public company aiming to produce energy from coal but in 1974 the majority of these mines closed. Mining territories had finished their productive phase and needed to reinvent themselves.

Previous research described the miners' perception of the transition phase between the last productive phase and the introduction of the Geopark idea (Boggio *et al.*, 2003). In this research, 90 interviews were carried out with miners receiving vocational retraining for environmental recovery projects. They clearly perceived the difficulties of a valorization project founded in territories marked by low entrepreneurship and a tradition of external (private or public) funding of economic activities. The feeling that the mining social tissue had nearly disappeared emerged clearly from the interviews together with a certain lack of confidence in the territories' capacity for self organization. It was like people still living in former mining territories were there only because "... they have not be able to escape before the fall down!", as one miner said during the interviews. On the other hand possibilities offered by the changed development paradigms emerged. What was useless for the extractive industry and was left untouched, now could be the core asset of a tourism development looking for a high level of nature, silence and unique landscapes. Together with the awareness of the exceptional territorial resources emerged the issue of a territory without entrepreneurial capability and private financial resources to support a local tourism industry.

The Geopark: imagined tourism territories

In October 1997, the Geomining Historic and Environmental Park of Sardinia as declared by UNESCO as the first park of what would become the World Geopark Network. The Italian Ministry of the Environment formally constituted the park in 2001 and it is now organized as a consortium of public authorities at the national, regional and local levels. The Geopark project at the moment involves 8 areas and 87 municipalities of the 377 in Sardinia. The population involved represents around 35 per cent of the total inhabitants of the region. It is one of the biggest and most differentiated parks in Italy, extending for around 3,800 km^2: approximately 15 per cent of the territory of Sardinia.

The territories involved show a strong element of heterogeneity linked by the common mining heritage and the need to define new development strategies. In the definition of mining heritage, territorialization processes identify their own limits and borders more in terms of the congruity of a space image and perception rather than on other material elements related to historical dynamics. Municipalities aiming to be involved in the Geopark project in some cases can claim very poor surviving elements of mining heritage but are strongly attracted by the growth opportunities related to Geoparks. Mining heritage is thus chosen and exploited as an economic factor, supporting a new development phase.

At the same time nearly all the traditional mining territories in Sardinia share some critical issues related to their socio-economic structure. Population growth rates are nearly stagnant or negative in the majority of Geopark areas outside of urban settlements. Residents are ageing and the percentage of active people is consequently reduced. Tourism supply is concentrated on the coastline with a strong presence of holiday houses and a high seasonality. The integration process between the traditional summer tourism supply and the inland area is reduced, and at most of the sites, being a distance of more than 40 km from the coastline makes it very difficult today to find a hotel to accommodate a bus tour with 50 people. In Geopark territories, the main issue is not to differentiate the tourism supply but to create or consolidate a tourism market.

The analysis of the Geopark potential tourism supply reveals the existence of an enormous number of sites (383) perceived as relevant local heritage resources both by local communities and decision makers. The majority of the attractions surveyed by the study can be defined as mining heritage sites.

Interestingly, expert perceptions of what can be attractive for a tourist largely differ from what communities perceive as representative of their heritage. Only 20 per cent of the sites selected by local communities present significant elements of interest to a potential tourism market in terms of accessibility, state of conservation and historic value. The selection process will represent a major issue for Geopark managers in the resources allocation processes.

Table 17.1 Conservation conditions of the main heritage sites of the Geopark (2008)

Area	*Abandoned*	*Open to public*	*Partially restored*	*Completely restored*	*Industrial activities*
Argentiera-Gallura	4	0	1	0	0
Funtana Raminosa	1	1	0	0	0
Orani-Guzzurra-Sos Enattos	4	1	0	0	1
Sarrabus-Gerrei	2	2	3	0	1
Monte Arci	0	4	0	1	0
Iglesiente	20	8	3	1	4
Sulcis	2	1	2	5	0
Guspinese-Arburese	6	4	0	0	1
Total	39	21	9	7	7

Source: Direct investigation (CRENoS, 2008).

The territorialization process addressed by planning for heritage tourism inevitably produces a hierarchy between competing resources, sites and territories.

At the moment only 21 sites are open to the public in the Geopark territory. The analysis of sites clearly shows the difficulties in rationalizing tourism site management and promotion activities. The research firstly evaluated the level of efficiency in the management of each site. A relevant element here is the management model. The sites are administered in some cases by the public company IGEA, in some others by associations or privates companies and municipalities. Only one site is directly managed by the Geopark. The variety of solutions adopted represent in some ways a limitation on the coordination attempts of the Geopark.

Visitors are relatively few. The total number of visitors for all the sites combined can be estimated at around 60,000 people per year. The most successful site, Porto Flavia, hosts around 17,000 visitors each year. Especially in the less relevant sites almost no marketing activities are supported due to budget limitations. Geopark site managers have only a vague perception of visitor characteristics. Little information is collected about tourist motivations and customer satisfaction levels. Some interesting elements resulted from research adopting Contingent Valuation Analysis techniques to evaluate the characteristics of possible mining heritage tourists and their visit experience (Balia *et al.*, 2003). The sample was composed of more than 400 interviews with *cultural tourists* in South Sardinia. Some questions tried to verify tourists' willingness to visit Porto Flavia, the most popular site of the Geopark. With respect to visitors' motivations, mining history and, more generally, mining environment experience were indicated as the most relevant. A quarter of the sample declared that participating in mining tourism would be an opportunity to escape the traditional coastal holiday in Sardinia. On the other hand tourists not interested in mine visits acknowledged their indifference as being mainly due to the difficult access to sites, the distance from the seaside and other motivations related to the organization of a visit. In the attempt to investigate the willingness to pay for a visit the research showed that cultural tourists interested in mining heritage declared themselves to be ready to pay from 50 per cent to 100 per cent more than the real price (€ 8) to visit Porto Flavia. Tourists clearly prioritized also a hierarchy in the services they expected to find during a visit experience. More than the half expected to find a guided tour of the galleries and, respectively, the opportunity to go trekking and the presence of minerals expositions at the sites.

Together with documenting the number of visitors, an analysis of revenues and costs has been carried on in the Geopark sites (CRENoS, 2008).

At the moment profits from attractions like mining gallery visits or museum entrance fees are the main source of income while other possible revenues from bookshops, rental for events, conferences, incentives and also catering services are largely underdeveloped. The total number of ticket sales in the Geopark sites is around ten times higher than the revenues from other activities. These data clearly show that in terms of tourism product creation and marketing, the Geoparks sites have a lot to learn from related European experiences in Spain, Belgium or the UK (Hose, 2007; Cole, 2004). The economic analysis of the sites shows that costs for raw materials and equipment are the highest expenditures together with personnel costs.

Table 17.2 Costs and revenues of the Geopark sites open to public (2008)

Costs			*Revenues*			
Employees	Other costs	Total	Tickets	Other revenues	Total	Total
€ 459.148	€ 669.764	€ 1.128.884	€ 305.362	€ 35.300	€ 340.500	–€ 788.384

Source: Direct investigation (CRENoS, 2008).

Territories development and the environmental challenges

Mining territories in Sardinia show some relevant environmental issues correlated to their industrial past that strongly limit tourism development projects. Industrial pollution and wastes are concentrated around the main extractive sites but also contaminate territories on a larger scale. In 2003, the regional authority evaluated 18.5 km^2 of lands occupied by different sorts of industrial wastes and dumps related to mining activities (Autonomous Region of Sardinia, 2003). All those areas are located in the Geopark territory. Sulcis, Iglesiente and Guspinese territories, where mining activities have been highly developed in the last two centuries, suffer from different forms of pollution. Aerial diffusion of minerals like lead, zinc, cadmium, and arsenic has been confirmed by the regional authority investigations. In the same areas groundwater and wells show high concentrations of lead and zinc even some kilometers distant from the pollution sources and acid mine drainage water due to sulphides is also present (Cidu *et al.*, 2007; Istituto Superiore di Sanità, 2004). During the last 10 years IGEA has been developing monitoring activities to define the level of pollution originated by mining activities on around 80 sites in the Geopark area. At present in only a few cases have recovery operations started. During the last 10 years resources have been concentrated in the Sulcis, Iglesiente and Guspinese territories that received, between 1996 and 2003, nearly all of the 47 million Euros available for environmental recovery of mines. In 1990, the Italian Government classified the Sulcis area as a territory with a high risk of environmental crisis and developed a Recovery Plan (Autonomous Region of Sardinia, 1994). Sulcis is also affected by the pollutants from industrial area of Portovesme: gas emissions, aerial powders, waste waters and industrial dumps. From 1995 to 2005 around 70 million Euros have been devoted to 60 depollution projects, more than half related to the industrial area activities. In 2005, the Sardinia Health Department published research also involving the Sulcis, Iglesiente and Guspinese territories, investigating the health conditions of residents in areas with a relevant presence of industrial, military or mining activities. The research considered epidemiological data between 1981 and 2001 and tried to delineate possible scenarios based on the data analysis. Concerning the Geopark area Sulcis municipalities showed illness trends higher than regional averages in relation to the proximity with Portovesme industrial sites. Respiratory diseases generally had a rate between 30–65 per cent higher for men and between 18–23 per cent higher for women than the regional average. Lung cancers had a rate between 26–62 per cent higher for men and between 16–54 per cent higher for women than the regional average, depending on municipalities. The other traditional mining areas of Iglesiente and Guspinese showed similar trends even though they were not affected by the same industrial development as Sulcis. Lung diseases were 149 per cent higher for men and lung cancer incidence was 28 per cent higher than the rest of Sardinia in Guspinese territories. Abnormal trends for both sexes were found for urinary and respiratory diseases, cancers and particularly lung cancers in Iglesiente.

Taking Geopark seriously as a strategic factor

Evidence from health investigations clearly shows that the environmental recovery of Geopark territories should not be postponed. Two main elements of complexity can be seen. On the one hand there is the confrontation between the different decision making levels over management of the big *green business* of environmental recovery. In 2008, the Italian National Government entrusted the President of the Sardinia Region with the management of the environmental recovery process in 34 municipalities of the Sulcis, Iglesiente and Guspinese territories. The regional government published in March 2008, a plan forming 6 macro areas of intervention with a budget cautiously forecast at around 310 million Euros (Autonomous Region of Sardinia, 2008). We can easily forecast that it will require more than 500 million Euros for the recovery of the whole Geopark area.

The second aspect is related to the level of investment required to support Geopark development. But this involves also the strategic vision of the mining territories. Regional economic and political elites show a contradictory attitude toward the Geopark. Even if the *mantra* of tourism diversification is a common argument in Sardinian debate, there's a clear diffidence towards the Geopark as a local development actor. This is only partially justified by the influence of coastal tourism lobbies (especially tourism developers and building sectors operators) on regional development strategies. Again, even if the *development for all* discourse constantly emerges from political debate, an analysis of decision makers' practices underlines the persistency of a different discourse. Considering Sardinia as a simple leisure space encourages the reproduction of a development model inevitably concentrating resources on the coastline, more and more represented as a place to consume. Furthermore, it is generally recognized that mining heritage tourism has low employment impact on the local and regional scale in the short term (Cole, 2004). Political discourses on employment and unemployment are a common rhetorical playground for Sardinian representatives of any political ilk. Geopark development is only marginally attractive in this sense. Difficult transport connections, poor tourism infrastructures and services, the open question of environmental recovery are among the conditions that limit tourism development and employment rate growth in the short term. In the background can be seen a gap between a theoretic (and sometimes rhetorical) adherence to Geopark development project concepts and the practice of local development planning that refuses to delegate to the Geopark enough power for it to become an effective development driver. And being that municipalities and their territories are the core actors of the Geopark development process a gap between local communities and regional development strategies clearly emerges.

The Sardinian provisions of regional law indicate municipalities as the final beneficiaries of the mining sites formerly owned by the regional government. Furthermore, the Geopark

Table 17.3 Forecast budget for environmental recovery of Sulcis, Iglesiente and Guspinese mining areas (2008)

Intervention area	*Analysis*	*Planning*	*Execution*	*Total costs*
Macro area Montevecchio Levante	1.500.000	2.100.000	59.000.000	62.600.000
Macro area Montevecchio Ponente	1.000.000	450.000	52.500.000	53.950.000
Macro area Masua	1.300.000	700.000	30.000.000	32.000.000
Macro area Malfidano	400.000	400.000	22.000.000	22.800.000
Macro area Barraxiutta	700.000	300.000	11.000.000	12.000.000
Macro area Valle del Rio San Giorgio -Iglesias	3.700.000	2.300.000	117.000.000	123.000.000
Orbai Mine Area	200.000	100.000	1.500.000	1.800.000
Su Zurfuru Mine Area	200.000	150.000	2.000.000	2.350.000
Ports	840.222	0	0	840.222
Total	9.840.222	6.500.000	295.000.000	311.340.222

Source: Autonomous Region of Sardinia, 2008.

is not included in the regional law between the possible owners of mining sites and municipalities are therefore facing enormous economic constraints in assuming control of the sites and justification of the maintenance costs in comparison with direct and indirect economic benefits. At the moment the transfer of ownership is only partially realized. IGEA Corporation has been created by the Autonomous Region of Sardinia (that is the only partner of the company) with the aim of coordinating the transmission of the mining sites property to local municipalities. IGEA is still in charge of the environmental recovery process, and of the implementation and maintenance of security standards in the sites but the company also manages the most relevant sites from a tourism perspective. Both economic reasons and the need to assure effective security standards, especially in mining galleries, are today the main limitations on including other actors apart from IGEA in the management of the sites. The mining technical knowledge is today in the hands of IGEA as a result of employing former mine workers with relevant skills.

Even imagining a separate management for galleries and the rest of the site attractions, it will be really difficult to integrate the different management strategies of a public body like IGEA and private actors involved in running mining sites. Overlapping responsibilities and a difficult role assignment process are the key questions to solve if regional government will aim to support this kind of public-private partnership. At the moment IGEA is granting the access to galleries and more generally to sites by the authority of the institutional mission the regional law defined for the company. Tourists seem to like the *authentic* style of the visit experience guided by formers miners who are still in charge of site maintenance, even if they have no specific tourism training. Nevertheless some critical issues in providing tourism services emerge mainly related to the public nature of IGEA. Sunday closure of the sites, no coordination with tourism private operators, reduced adaptability of working hours to visitors' needs and rigidity in the management of the ticket selling process are commonly reported as limiting tourism development in Geopark sites. Last year for instance IGEA introduced new rules for ticket sales and tourism operators had to reserve at least ten days before visits and pay in advance for tickets.

Another organizational model is nowadays strongly supported by Geopark management. Some consortium agreements between the Geopark and municipalities succeeding IGEA in the ownership of mining sites have been signed in the last ten years. Several key elements are shared by the different consortiums developed in former years. A board of directors is normally created under municipal control. At the same time Geopark cofinances at least 50 per cent of the costs of the consortium and private entrepreneurs start to be directly involved in consortium agreements. Furthermore, the Geopark is the very first case of a park accepted and supported by large majorities of local communities in Sardinia. Different from other park planning experiences in Sardinia, influenced in some ways by a *politics of backwardness* (Heatherington, 2001), the Geopark project has been obtaining the consensus of local communities. This process seems to imitate best practices observable in Europe and to potentially open up a strong involvement of local decision makers and entrepreneurs (Hose, 2007; Cole, 2004).

Together with the environmental recovery issues that transverse any development scenario in former mining territories, the main obstacle for an advance of the Sardinia Geopark is the uncertainty of the site transfer process from IGEA to municipalities. On one hand, the public company is the only Sardinian actor able to guarantee the technical support needed to maintain accessible mining sites and to preserve the mining knowledge heritage held by the few miners who are not yet retired. On the other the development of competitive tourism products needs the introduction of management mechanisms and expertise different from those of a

Public Body like IGEA. Regional decision makers are called to answer to territories' efforts in redefining development trajectories and to support what is probably the first attempt to empower local communities in Sardinia by means of the Geopark tool. In conclusion, a shallow and only formal adherence to the UNESCO Geopark network would result in another missed opportunity for Sardinia.

References

Atzeni, P. (1988) *Tra il dire e il fare. Cultura materiale della gente di miniera in Sardegna*, Cagliari, CUEC Editrice.

Autonomous Region of Sardinia (1994) *Piano di disinquinamento per il risanamento dell'area ad alto rischio di crisi ambientale del territorio del Sulcis*, Assessorato della Difesa dell'Ambiente.

Autonomous Region of Sardinia (2003) *Piano regionale di Gestione dei Rifiuti, Piano di Bonifica Siti Inquinati della Regione Sardegna*, Assessorato della Difesa dell'Ambiente.

Autonomous Region of Sardinia (2008) *Piano di Bonifica delle Aree Minerarie Dismesse del Sulcis – Iglesiente – Guspinese*, Assessorato della Difesa dell'Ambiente, March 2008.

Balia, S., Brau, R. and Strazzera, S. (2003) *La valorizzazione turistica dei beni di archeologia industriale: il caso del sito di Porto Flavia in Sardegna*, in Economia Pubblica, 33(6): 97–115.

Boggio, F., Pracchi, R., Asole, A. (eds) (1988) *Atlante economico della Sardegna*, Milano, Jaca Book.

Boggio, F., Sistu, G. and Stanzione, L. (2003) *Beni culturali e aree minerarie dismesse: il caso del Sulcis-Iglesiente (Sardegna sud-occidentale)*, in E. Dansero, C. Emanuel and F. Governa (eds) *I patrimoni industriali*, Milano, FrancoAngeli, pp. 79–98.

Cidu, R., Di Palma, M. and Medas, D. (2007) *The Fluminese Mining District (Sw Sardinia, Italy): Impact Of The Past Lead-Zinc Exploitation On Aquatic Environment*, in R. Cidu and F. Frau (eds), *IMWA Symposium 2007: Water in Mining Environments*, 27–31st May 2007, Cagliari, Italy.

Cole, D. (2004) 'Exploring the Sustainability of Mining Heritage Tourism', *Journal of Sustainable Tourism*, 12(6): 480–94.

Conesaa, H. M., Schulin, R. and Nowack, B. (2008) 'Mining Landscape: A Cultural Tourist Opportunity or an Environmental Problem? The Study Case of the Cartagena–La Unión Mining District (SE Spain)', *Ecological Economics*, 64: 690–700.

CRENoS (2008) *Piano economico-sociale del Parco Geominerario Storico e Ambientale della Sardegna*, Cagliari, September 2008, pp. 367.

Dowling, R. and Newsome, D. (eds) (2006) *Geotourism: Sustainability, Impacts and Management.* Oxford: Butterworth Heinnemann.

Edwards, J. A., and J. C. Llurdés i Coit (1996) 'Mines and Quarries: Industrial Heritage Tourism', *Annals of Tourism Research*, 23(2): 341–63.

Fernández, G. and Guzmán-Ramos, A. (2004) 'El patrimonio industrial-minero como recurso turístico cultural: el caso de un pueblo fábrica en Argentina, Pasos', *Revista de Turismo y Patrimonio Cultural*, 2(1): 101–09.

Hall, D. and Richards, G. (2000) *Tourism and Sustainable Community Development*, London: Routledge.

Hall, M. (1994) *Tourism and Politics: Policy, Power and Place*, Chichester: Wiley.

Heatherington, T. (2001) 'Ecology, alterity and resistance in Sardinia', *Social Anthropology*, 9(3): 289–306.

Hose, T.A. (2007) 'Geotourism in Almeria Province, Southeast Spain', in *Tourism: An International Interdisciplinary Journal*, 55(3/2): 259–76.

Hose, T.A. (2008) 'Towards a history of geotourism: definitions, antecedents and the future', in C.V. Burek and C.D. Prosser (eds), *The History of Geoconservation*, Geological Society, London, pp. 37–60.

Istituto Superiore di Sanità (2004) 'Heavy Metals Contamination Related to Disused Mining in Sardinia. Preliminary Study', *A cura di Sergio Costantini, Luciano Bodano, Rosa Giordano e Sonia D'Ilio*, iv: 72 pp. Rapporti ISTISAN 04/28.

Lee, J., Arnason, A., Nightingale, A., Shucksmith, M. (2005) 'Networking Social Capital and Identities in European Rural Development', *Sociologia Ruralis*, 45(4): 269–83.

Ortu G.G.(1998) 'Tra Piemonte e Italia. La Sardegna in età liberale (1848–96)', in L.Berlinguer and A. Mattone (eds), *Storia d'Italia. Le Regioni. La Sardegna*, Torino, Einaudi.

Raffestin, C. (1981) *Pour une géographie du pouvoir*, Les Libraires Technique, Paris.

Ruiz-Ballesteros, E. and Hernández-Ramírez, M. (2007) 'Indentity and Community – Reflections on the Development of Mining Heritage Tourism in Southern Spain', *Tourism Management*, 28: 677–87.

Ruju, S. (2008) *I Mondi minerari della Sardegna*, CUEC, Cagliari.

Sack, R. D. (1986) 'Human Territoriality: Its Theory and History', Cambridge Studies in *Historical Geography*, Cambridge University Press.

Salvadori, I. (1990) 'L'industria mineraria', in F. Boggio (ed) *Atlante economico della Sardegna*, Industria, Milano, Edizioni Universitarie Jaca, 2: pp. 117–34.

Soja, E.W. (1971) 'The Political Organization of Space', *Commission on College Geography, Resource Paper 8*. Washington, D.C.: Association of American Geographers.

Turco, A. (1988) Verso una teoria geografica della complessità, Unicopli, Milano.

UNESCO (2008) *Guidelines and Criteria for National Geoparks seeking UNESCO's assistance to join the Global Geoparks Network*, Global Geoparks Network, June 2008.

18 Geotourism, mining and tourism development in the Bay of Fundy Canada

Keith Dewar and Randall F. Miller

Introduction

People have always been fascinated by rocks, minerals and fossils. Specimens can be found in the homes of ancient peoples from China to Rome. Rocks and minerals as souvenirs have always been common and provide conversation pieces. Mining and mineral exploration open the earth to visitors attracting the serious collector and the curious. Industries that build up around mines and communities that grow and fade with the fate of the resource are important additional elements to the sites' appeal and an important part of a region's heritage. Mineral exploration, even if no mine is opened, leaves maps, specimens, descriptions, and stories that can interest heritage tourists and collectors.

This chapter sets the context and defines how incremental development is leading to a greater appreciation and use of mining and geological sites by the tourism industry in the Bay of Fundy Region of Canada. Further, it discusses the current state of geotourism and mining related tourism in the region and the part it is playing in the rejuvenation of the tourism industry. It is not a 'success story' nor a 'best practices' example but a work in progress that provides insight into the issues and challenges faced by governments, organizations and individuals attempting to expand and coordinate existing geological based attractions and possible strategies for a more stable and sustainable future for the tourism industry.

Bay of Fundy geology

The Bay of Fundy Region includes two of Canada's Maritime Provinces, New Brunswick and Nova Scotia, and a small part of the State of Maine in the USA. The Canadian provinces share 750 kilometres of Fundy coastline. The complex plate tectonics history of the region created the Bay of Fundy and combined with recent glaciations and the world's highest tides have led to the development of a highly eroded and spectacular coastline. Such tourist attractions as Cape d' Or, Hopewell Rocks, and Saint John's Reversing Falls are examples of major attractions that have a geological base.

Rich pockets of mining and mineral heritage could be used to enhance the tourism product and add heritage educational opportunities for visitors and local people alike. Future development and redevelopment of these and other related geologically significant places could give a much-needed boost to the variety of destinations and an understanding of earth history. The geology of the Bay of Fundy region includes a relatively complete and diverse geological record from the Late Precambrian to the Neogene, excepting only the Paleogene (Fensome, 2001). This is clearly recognized in the Fundy Tourism Partnership's literature (Figure 18.1) where one of the three tourism zones is labelled as 'Sea Cliffs and Fossils'.

Highlights in the regional geological record include:

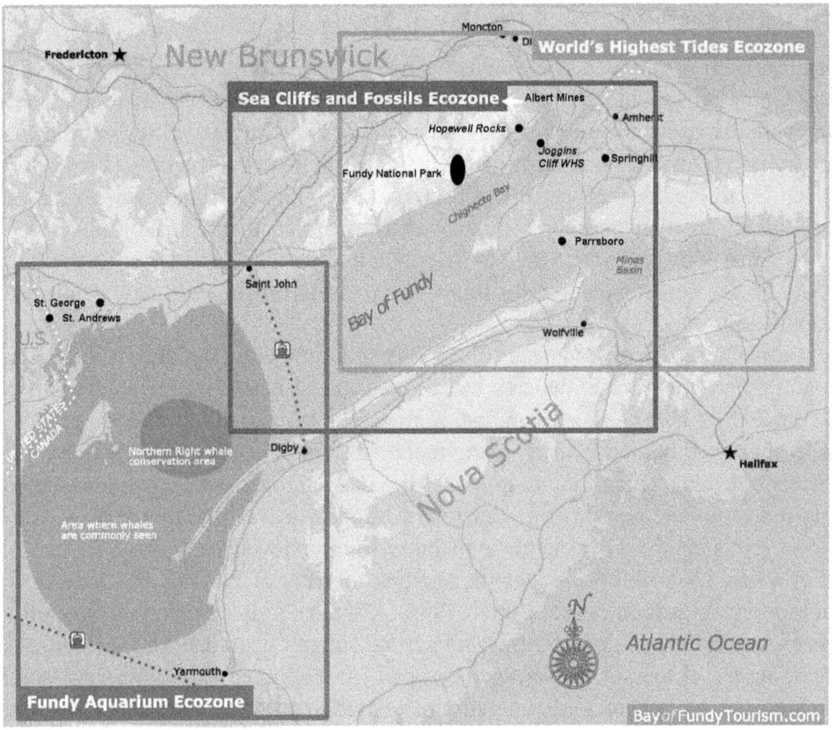

Figure 18.1 The Bay of Fundy Tourism Partnership provides a website and overview map which includes a Sea Cliffs and Fossil Ecozone outlining the geological importance of the region. (Reprinted with permission Bay of Fundy Tourism, 2009).

- the first described Precambrian stromatolites
- a classic Cambrian record
- Ordovician to Devonian record of the closing of the Iapetus Ocean and building of the Appalachian Mountains
- Mississippian deposits rich with terrestrial plant and vertebrate fossils
- a marine record with deposits of potash, salt, gypsum and limestone
- a Pennsylvanian record including the Joggins Fossil Cliffs UNESCO World Heritage Site and coal mining areas
- a Permian-Triassic-Jurassic-Cretaceous record of the opening of the Atlantic Ocean that includes Canada's oldest dinosaurs
- classic mineral collecting localities in Mesozoic volcanics and
- a Neogene interglacial-glacial record including rapid climate events such as the Younger Dryas cold interval.

Mining history dates back thousands of years with more recent activities including quarrying for lime production, graphite and metallic mineral mining, potash, salt and gypsum mining, quarrying of building stone, bitumen and coal extraction, oil and gas production and extraction of surficial sand and gravel.

Tourism and geotourism and the Bay of Fundy

Tourism has been a major part of the industrial mix of the region since the 1860s. By the 1880s the burgeoning railway industry gave mass access and the cool ocean environment attracted wealthy summer residents escaping the heat of major New England cities and the Montreal area of Canada. Famous summer residents included the Roosevelt family (Franklin Delano Roosevelt, 32nd President of the United States) on Campobello Island, New Brunswick and Hiram Walker (Distiller) and Sir William Van Horne (Canadian Pacific Railways) settling in nearby St. Andrews-by-the-Sea. The Algonquin Hotel and resort opened there in 1889 providing luxury accommodation for those who were not inclined to buy a summer home (Fairmont Hotels, ND).

The tourism industry on both sides of the Bay of Fundy grew rapidly after WWII as cars and road improvement made the area more accessible. Visitors reached a peak in the late 1990s. Since then there has been a gradual decline first as a result of the events of '9/11', followed by increased border security, rising fuel prices, and the general stagnation of the area as visitors took advantage of increasing alternate destinations. With exceptions, such as the thriving industrial and port city of Saint John with its cruise ship terminal, New Brunswick and Nova Scotia continue to struggle to maintain tourism levels.

Butler's Life Cycle Model would suggest that the Bay of Fundy area could be defined as being in the stagnation phase (Butler, 1980). Changes to the tourism destination system have been unhurried and governments and private entrepreneurs have been slow to react to the changing travel market. Slowness to respond can be linked to a number of factors that affect rural tourism destinations generally including; small 'c' conservatism, outmigration, and a small GDP resulting in fewer funds spent on tourism as aging urban infrastructure and healthcare absorb provincial budgets.

Other challenges such as the small size and lack of multiple activities at destinations reduce their effectiveness as attractions. This is compounded by a combination of the undeveloped nature of many destinations and the distance between them. The secondary road system is ageing and often in poor condition discouraging travellers from exploring many smaller 'out of the way' attractions. There is also a lack of information available at many of the sites. This is changing and a number of new initiatives are underway including several with a strong geotourism component. While geotourism (Dowling and Newsome, 2006) and more specifically the mining tourism concepts are not new, the development of the European Geoparks Network and the UNESCO Global Geopark Network are providing models for a more focused engagement of the public in the appreciation of mining tourism and earth science. The UNESCO Geopark concept is designed to be:

> . . . a platform of active cooperation between experts and practitioners in geological heritage. Under the umbrella of UNESCO, and through exchange between the global network partners, important national geological sites gain worldwide recognition and profit through the exchange of knowledge, expertise, experience and staff with other Geoparks.
> (UNESCO, 2009)

Since the beginning of this initiative in 2004, 58 UNESCO Geoparks have been inscribed. Of the 58 Geoparks, 15 see mining heritage as an element, and seven have mining as a major focus. The more familiar UNESCO World Heritage Site Program has similar goals and seven of these sites include mining as the major element in their inscription documents.

Figure 18.2 Joggins World Heritage site, a 14 kilometer stretch of coastal cliffs that represents the world's most complete coal age fossil record. (Photo: K. Dewar 2009).

In the Bay of Fundy region, there are several long-standing geotourism attractions and a number of recent developments that take advantage of the growing interest in geotourism. One of the newest and most important is Joggins Fossil Cliffs World Heritage Site (Figure 18.2). While fossils are the major focus of the site, mining was an early and important element (Falcon-Lang, 2008). The remnants of the wharves, mine props, as well as a small miner's display in the visitor centre, attest to the importance of mining.

In the upper Bay of Fundy a UNESCO Biosphere Reserve (inscribed June 2008, and managed by an NGO) has been created, in part due to the area's geology. The reserve is in the early stages of development. In addition to the reserve and the Joggins Fossil Cliffs World Heritage Site, there are several other new or developing destinations with geological components that can act as focal points including:

- The Fundy Trail Parkway, NB, opened in 1998 completion expected by 2011, managed by an NGO with funding from the Federal and Provincial governments (Fundy Trail Parkway, 2009).
- Harbour Passage in Saint John, NB includes geological interpretation (begun in 2006 and ongoing, local government, business and NGO partnership) (Saint John Waterfront Development, 2009).
- A potential UNESCO Geopark centered on the Saint John, NB region, is under consideration and could be the first Geopark in North America (government and NGO partnership).

In Nova Scotia, several excellent geology based tourist maps and routes are available to the public while New Brunswick has an outdated geological highway map. The provinces have formed a Bay of Fundy Tourism Partnership. This joint effort highlights many of the major geological features but at present no mining sites are highlighted in the organizations' marketing literature (Bay of Fundy Tourism Partnership, 2009). The potential for development and linking of these major points of interest with the smaller less visible sites is considerable, but there is a lack of coordination and cooperation. Much of the work to date to research mining and mining heritage has been spearheaded by provincial museums and geological survey departments and the Atlantic Geoscience Society.

Mining tourism potential in the Bay of Fundy region of New Brunswick

The history of mining in New Brunswick extends back thousands of years with people of the First Nations exploiting small deposits of stone and possibly copper and tin (Martin, 2003). Europeans noted copper occurrences near Beaver Harbour in 1604 and the first coal was mined from the Grand Lake area in 1639 (Martin, 2003; Falcon-Lang, 2008). Beginning in the early 1800s Abraham Gesner (1839), James Robb (1850), Loring Bailey (1864), Henry Hind (1865), and others produced detailed reports for the government outlining the potential for mineral and rock exploitation. On a brief visit to Saint John in 1852, Sir Charles Lyell commented on the nature of graphite mines at the Reversing Falls. Local newspapers followed Lyell's comments closely and eagerly wrote of great things to come for New Brunswick's mining industry (Miller and Buhay, 2007). In the late 1800s and early 1900s bitumen and oil were extracted in the upper Bay of Fundy region along with gypsum. During the same period Precambrian marble, Devonian granite and Pennsylvanian sandstone were exploited as building stone for the developing cities of New Brunswick and Nova Scotia and to supply New England. Potash was discovered in the Sussex area in the 1970s and has generated considerable mining investment. Recent discoveries of oil and gas have renewed interest in historic oil fields near Weldon and the related geological basin. Uranium exploration has also shifted the focus of the mining industry south toward the Bay of Fundy region.

Many of the mining operations surrounding the Bay of Fundy have long since closed as coal, building stone, and small industrial mineral and base metal mines have been replaced by petroleum, manufactured building materials, synthetic products and large scale mines. Although abandoned mines still dot the landscape (Ruitenberg, 1969), they were often small, and are a hidden part of the region's heritage. Overgrown stone quarries are difficult to distinguish, small graphite mines have disappeared into urban development, and entire mining communities have almost vanished. The Department of Natural Resources reports that there are more than 375 mine openings in New Brunswick at 63 separate mines sites, most on private land (New Brunswick Department of Natural Resources, 2004).

Regardless of the rich mining history, New Brunswick's list of more than 1,100 Heritage Places only includes two sites recognized for historic mining activities (Miller, in press); one at Albert Mines where a few remaining foundations of houses and the head frame exist, and one in St. George ('Granite Town') at the former site of the Milne-Coutts & Co. Granite Shed.

With one exception, museums and interpretations centres do not focus on mining history. On-site interpretation of mining activities does not exist, and few rock and mineral specimens or manufactured products are available for sale to the public. Although numerous modern and old mines and mining communities with great potential for tourism exist, their development has not been considered by either government or private parties. However, new geotourism sites are creating an awareness of the importance of mineral and fossil wealth and possible potential for future development.

Historic mining activities do receive some remote interpretation. The Albert County Museum and the Hillsborough and Salem Railroad Museum both note local albertite and gypsum mining as part of their larger focus. Local tour operators lead trips to gypsum caves in the area and the geological formations at Hopewell Rocks, an iconic destination, attracts thousands of visitors each year, but with a focus on tides. Nearby Weldon recognizes its oil history with a small pumpjack installed at a roadside picnic stop (Figure 18.3), where one of North America's first oil wells was drilled at Stoney Creek in 1859. Unfortunately, without interpretation, visitors are likely left wondering why it is there.

Figure 18.3 Pumpjack at Weldon, New Brunswick (Photo: R. Miller 2007).

Some *ad hoc* on-site interpretation occurs at the Albert Mines (Figure 18.4) where foundations for the head frame and community buildings still exist. This is an example of a site that could be a key 'point of interest' tourism attraction to increase the visitors' knowledge of the early days of the global petroleum industry (Brice, 2002). The bitumen 'Albertite' mined here was used by Abraham Gesner to distil kerosene. There is ample documentation to accompany the albertite story and the characters involved comprise a Who's Who of early geology including Sir Charles Lyell, Sir William Dawson, Hugh Miller, Charles Jackson, Joseph Leidy and others. There are great human-interest stories of fortunes made and lost, shots fired, and a landmark court case that helped found the province's legislation concerning mineral rights.

Westmorland and Albert County also produced much of the building stone used locally and as far away as New England (Figure 18.5). Many quarries and building foundations from the communities can still be seen. Good documentation, accompanying stories, and many existing buildings can be traced back to specific quarries (Martin, 1990). The reconstruction of Saint John after the Great Fire of 1877 provides a connection with the quarrying operations, as expressed in an interpretive book by the New Brunswick Museum (Miller and Hughes, 2009), an example of geology and mining incorporated into the larger tourism product.

Other sites that could provide anchors for regional networks include the proposed UNESCO Geopark near Saint John. The park would encompass one billion years of earth history including stories of plate tectonics, fossils, history of scientific investigation, and include mining and related activities as part of geological interpretation. Saint John is the province's energy hub with Canada's largest refinery, a newly commissioned LNG facility and a nearby nuclear power generating station. The proposed Norton Fossil Forest Interpretation Centre, part of

Figure 18.4 Top: Albertite mining site, Albert Mines, New Brunswick, 1860. Courtesy of Provincial Archives of New Brunswick, PANB Miscellaneous Collection - P1–21. Bottom: Geologists' Association, London at Albert Mines (Photo: R. Miller 2007).

Figure 18.5 Left: Old Post Office, 115 Prince William Street, Saint John, built 1877–1880. Right: Sackville Freestone Quarry, 1900–1910. New Brunswick Museum Collection (X11479).

the Geopark planning (MacKenzie, 2008) would focus on fossils, but also interpret regional mining including potash at one of the largest mines in Eastern North America, and recently discovered local gas reserves.

The Fundy Biosphere Reserve has a long history of mining. The four kilometres long Coppermine Trail in Fundy National Park tells the story of the Vernon copper mines. As well, both the Acadian and British settlers cut high quality grindstones and building stone from the Pennsylvanian sandstone beds in the reserve. There was a brisk trade in grindstones and gypsum with the Boston area all through the eighteenth and nineteenth centuries. Numerous quarries were dug and building stone shipped to Nova Scotia, Quebec, Ontario and New England.

Impediments to developing mining tourism include a lack of documentation of mining history and associated stories, safety, and trained interpreters. There are also issues surrounding the collection of rock, mineral, and fossil specimens that need to be addressed. Although there is a wealth of scientific literature, maps and geological reports, conducting historic research about mining and geology is time consuming and is hampered by competition for limited research dollars. Electronic compilations of heritage information and museum catalogues are making information more accessible. The New Brunswick Department of Natural Resources has developed an online resource that includes a mineral history database derived largely from transcribed newspaper records going back to the late 1700s. For anyone interested in the history of New Brunswick geology this is an invaluable resource for developing tourism products.

Visitor safety is another concern in developing mining tourism. Modern mines have necessary restrictions similar to any industrial site and many of the old mines that make interesting geotourism sites are remote and not suitable to receive visitors. The province has an Abandoned Mines Sites Policy (New Brunswick Department of Natural Resources, 2004) that addresses safety concerns associated with mine sites that was never intended to address the need of geotourism.

Mining tourism potential in the Bay of Fundy region of Nova Scotia

Nova Scotia's mining history on the Bay of Fundy also goes back thousands of years with the first European exploitation dating to the mid 1600s (Falcon-Lang, 2008). Abraham Gesner (1797–1864), one of Canada's most important geologists and inventor of the process for distilling oil from bitumen (kerosene), was born and lived in Parrsboro. He eventually became the first Provincial Geologist in neighbouring New Brunswick and opened one of the first public museums in Canada, the foundation for the New Brunswick Museum (Miller and Buhay, 2007). 'Gesner's Museum' included amethysts, chabazite and stilbite specimens found near his Nova Scotia home.

This mineral rich region has led to it being the centre for the collector and visitor interested in rocks, minerals, and fossils. The importance of rock and mineral collecting resulted in the Parrsboro Rock and Mineral Show, the only event of its kind in Eastern Canada. The two-day event draws approximately 3,500 visitors of which 80 per cent are tourists. It is estimated that each adult tourist spends an average of $50 per visit. This infusion of cash is small by some standards but an important addition to the economy of this rural community (Corbett, 2008). The event, along with two commercial 'rock and mineral shops', are the base of a geotourism destination. The shop run by Mr Eldon George is the oldest 'Rock Shop' (Figure 18.6) registered in Canada as a retail business; it was opened in September of 1948 by Mr George when he was eleven years old.

Figure 18.6 Eldon George's Rock and Mineral Shop, this is reputed to be the oldest rock and mineral retail business in Canada (founded 1948) and still in the hands of the original owner. Much of the local material sold here is found at old quarry sites in the region (Photo Li 2009).

The extractive industry in the Parrsboro area, the history of collecting, and the nearby discovery of dinosaur fossils led to the opening of the Fundy Geological Museum in 1993 greatly enhancing the Parrsboro destination. With an average visitation of 20,000 per year the museum is an important element in the tourism business and economy of the town. Tourists who visit the Fundy Geological Museum and/or show are also likely to visit the Joggins Fossil Cliffs World Heritage Site less than an hour east of Parrsboro.

As well as the exceptional fossil record the Joggins area coal mining history dates back to approximately 1710 and the area was worked until relatively recently. Remnant pit props and wharf supports can still be seen near the main entry point to the beach. The area was also important as a provider of grindstones. Their production began early perhaps as early as 1658 but was certainly in full swing by the mid 1700s (Falcon-Lang, 2008). Abandoned and broken stones can still be seen at Lower Cove some two kilometres from the Joggins coal measures.

Nearby, the town of Springhill is an important part of the mining heritage of Canada. Three major disasters in the Cumberland Mine make it one of the most tragic sites in Canadian mining. The first was the 1891 coal dust explosion that killed 125 men. A second coal dust explosion in 1956 killed 39 miners. Then on 23 October 1958 Canada's most infamous mine disaster occurred when a 'bump' rocked Collier No. 4. A hundred men were trapped and rescued but 74 died, and many still lie buried beneath what is now an industrial park. The Springhill 'bump' has obtained a special place in Canadian history not only for the tragedy and the bravery of the miners but also as the dawning of a new media era. It was the Canadian Broadcasting Corporation's (CBC) first live news broadcast from a remote site. The BBC with the aid of the British Air force who flew film footage from the site of the disaster to London by Vulcan bomber was also a first. The onsite visit by the Duke of Edinburgh and provincial premier Robert Stanfield undoubtedly gave the disaster special media attention. The final triumph was the rescue of twelve miners, trapped over three kilometres down (3,200 foot level), after six days with no food or water and, even more surprising, seven more were discovered alive after eight days. The disaster of 1958 was the first blow in the gradual closing of underground coal mining in Canada (Nova Scotia Archives and Records Management, 2006; Canadian Broadcasting Corporation, ND).

There is little evidence of colliers 2 and 4 today. Five ageing buildings are still scattered amongst the new small businesses of the industrial park. The site is transected by the Miners Memorial Drive. There are two National Historic Site Plaques erected in 2008 near the pithead and a small memorial park at the entrance to the Memorial Drive. Presently, other interpretation in Springhill is at the Syndicate Mine Site/Miners Museum in another part of the town. The now closed Syndicate Mine is the major tourism attraction related to mining in the Fundy Region. Open seasonally, visitors can descend about 100 metres into the mine with an interpreter who explains the workings. The shaft has been enlarged and the roof 'enclosed' to prevent 'coal falling on the visitors . . .'. All the mines are now flooded and serve as a geothermal resource heating many businesses and homes. There is considerable potential to develop this area as a tourist attraction related to mining including the restoration of some of the remaining Cumberland Mine buildings.

Discussion

The Bay of Fundy region has an exceptional geological heritage. This fact has long been understood but there has been a lack of coordination and integration into the general marketing of the Fundy Region by government, business and NGO's. Each tourism jurisdiction and organization has specific mandates that hamper opportunities to present the geology and its associated mining heritage as a whole. Although 'centre of interest' geological sites such as Hopewell Rocks, Reversing Falls and Joggins have long been exploited, no overreaching strategy or integrative approach has been developed for the entire region. New developments such as Joggins World Heritage Site and the Fundy Biosphere Reserve have helped give the geology of the area much greater exposure. The possible introduction of a UNESCO Geopark is also having a positive effect.

Most mine sites in the Fundy Region are small, scattered and not easily accessible. They are not yet interpreted directly on site or even in museum exhibits and visitor centres. Sites that do exist depend on the local or regional tourism organizations for marketing. Most lack permanent staff and there is no one organization responsible for the telling of the geological story of the Bay of Fundy. The Bay of Fundy Tourism Partnership is an attempt to bring the fragmented elements of the tourism industry together into a marketable tourist region centred on the bay. At present their regional boundaries largely follow coastal provincial highways with some exceptions (McCulloch personal communication, 1 July 2009). This regionalization leaves out such sites as Albert Mines, Weldon, and Springhill, some of the more important mining sites within the region. However this organization provides a communication and marketing fabric and has made a good start in bring disparate groups and attractions together in the immediate coastal areas. Although there is little enthusiasm amongst suppliers, regional government tourism organizations or NGO's to establish a new organization or network specifically for geotourism, continued progress towards the concepts integration into existing systems can form an important group of destinations that will attract and hold visitors and offer increasing educational and recreational opportunities for local communities. The network will depend on the larger point of interest attractions as anchors and markers. Although duration of visits to many sites may remain short cumulatively, if properly developed they will help retain visitors and provide recreation and educational opportunities for both the local communities and the visitors. There are other issues as well; most small mining sites are on private property which even if the owner was willing to allow access does not address the difficult issue of liability insurance. Nor does it address who and how interpretation will be managed. Some solutions might include purchase of the land of interest and the establishment of park areas by regional or provincial government organizations, conservation agreements, or encouraging local museums/visitor centres and entrepreneurs to establish small displays and sales points of geologically related material near or at some of the more visible and accessible sites. Accessibility can be addressed by the growing use of GPS points of interest systems and downloadable themed tourist maps from various websites, either free or for a small fee. Some local road signage would be beneficial.

Another major issue revolves around collecting. New Brunswick is developing policy and regulations that will control collecting and protect sites of significance. In Nova Scotia, legislation already exists but is outdated and various sections of the legislation conflict with one another. Management at Joggins World Heritage Site and others are presently lobbying for major changes to allow some controlled collecting. Care must be taken to protect valuable resources such as important fossil sites but to not be so restrictive that the amateur collector cannot take part in their recreational activity. Further, commercial collecting for resale of local fossils and minerals is a legitimate business that allows the small business person to maintain a profitable business or allow tourism businesses to sell specimens as part of their product line; however this must be balanced against the heritage value and Crown ownership of specimens. Public institutions should not have to compete with commercial interests in order to preserve heritage objects in the public domain. While museums and NGOs can use profits from the sale of surplus and scientifically redundant material to the public to help with the continuing issue of lack of funds, revenues from other tourism activities could be used to support geotourism infrastructure. The current guidelines for the UNESCO Global Geoparks Network states 'In order to be seen to be impartial in its management of the geological heritage, its managing body must not participate directly in the sale of geological objects within the Geopark (no matter from where they are) and should actively discourage unsustainable trade in geological materials as a whole, including short-sighted selling of Earth heritage, minerals and fossils.' Regulation and control remain the central issue. Designation of valuable areas, controlled collection, clear regulation and proper education of visitors can work together to avoid losing those who would enjoy owning a piece of the earth's rich history.

Future important work will involve joint activities such as regional geotourism packages; many of these can be GPS based and available through regional tourism websites and internet sources such as Google Earth. Educational and industrial tourism can also play an important role by providing additional tours and learning packages as well as information related to safety and legal collection of material. One example of this process could be provided by larger museums through outreach programs, small displays and temporary interpretive exhibits at smaller museums, tourism destinations and private attractions.

To ensure visitor destination quality and a credible educational component an increase in effort to research the history of the region's geological and mining history is required. Further tour guide, tourism information staff and interpreter training is essential. This can be accomplished by working with regional tour operators, Fundy National Park, the two provincial museums (Nova Scotia and New Brunswick) and the several universities and community colleges in the region. Destination Management Organizations also need to coordinate their training and activities and lobby the Departments of Tourism at the provincial level to increase marketing material related to geology and mining. It is these provincial level organizations that control the advertising dollars and their clear tourism focus make them essential players in the geotourism development of the region. The present stagnation of the tourism industry in most of maritime Canada is a problem that has several possible outcomes. As discussed above, if mining tourism and the more general umbrella of geotourism along with other initiatives are properly developed they will lead perhaps to a slower but more sustainable level of tourism growth.

Acknowledgements

Thanks to the New Brunswick Environmental Trust Fund and the Social Sciences and Humanities Research Council (CURA) for their support.

References

Bay of Fundy Tourism Partnership (2009) Official Site of the Bay of Fundy. Available online at: http://www.bayoffundytourism.com/ (accessed 27 June 2009).

Brice, W.R. (2002) 'Abraham Gesner (1797–1864) A Petroleum Pioneer', *Oil Industry History* 3: 72–80.

Butler, R.W. (1980) 'The Concept of a Tourist Area Cycle of Evolution: Implications for the Management of Resources', *Canadian Geographer*, 24: 5–12.

Canadian Broadcasting Corporation (ND) *Springhill Mining Disasters*. Available online at: http://archives.cbc.ca/economy_business/natural_resources/topics/111/ (accessed 27 June 2009).

Corbett, C. (2008 October) [Letter to Dewar, K.], Parrsboro, Nova Scotia.

Dowling, R. and Newsome, D. (eds) (2006) *Geotourism*, Heinemann, Oxford: Elsevier, Butterworth.

Fairmont Hotels (ND) *Fairmont Algonquin: Hotel History*. Available online at: http://www.fairmont.com/EN_FA/Property/ALG/AboutUs/HotelHistory.htm (accessed 30 May 2009).

Falcon-Lang, H.J. (2008) 'Earliest History of Coal Mining and Grindstone Quarrying at Joggins, Nova Scotia, and its Implication for the Meaning of the Place Name "Joggins"', *Atlantic Geology*, 45: 1–20.

Fensome, R.A.W.G. (2001) '*The Last Billion Years. A Geological History of the Maritime Provinces of Canada*'. Atlantic, Halifax: Nimbus Publishing.

Fundy Trail Parkway (2009) *Fundy Trail*. Available online at: http://www.fundytrailparkway.com/en/ (accessed 1 July 2009).

MacKenzie, C. (2008) *Fossil Forest Could Acquire Status*. Available online at: http://www.globalgeopark.org/publish/portal1/tab80/info2938.htm (accessed 4 July 2009).

Martin, G.L. (1990) *For the Love of Stone: Volume I: The Story of New Brunswick's Building and Stone Industry (Miscellaneous Report 8)*. Fredericton, New Brunswick: Department of Natural Resources and Energy Mineral Resources.

Martin, G.L. (2003) '*Gesner's Dream. The Trials and Triumphs of Early Mining in New Brunswick*'. Fredericton: Canadian Institute of Mining, Metallurgy and Petroleum New Brunswick Branch.

Miller, R.F. and Buhay, D.N. (2007) 'Gesner's Museum of Natural History: And Early Canadian Geological Collection', *Geoscience Canada*, 34(1): 37–48.

Miller, R.F. (in press) *Geoheritage 5: Geoscience Heritage in New Brunswick*. Ottawa: Geoscience Canada.

New Brunswick Department of Natural Resources (2004) *Abandoned Mine Sites Policy*, MRE 006.

Nova Scotia Archives and Records Management (2006) *Men in the Mines*. Available online at: https://www.gov.ns.ca/nsarm/virtual/meninmines/default.asp?Language=English (accessed 27 June 2009).

Ruitenberg, A.A. (1969) *The Old Mines of Mascarene*. New Brunswick Department of Natural Resources, Fredericton: Mineral Resources Branch.

Saint John Waterfront Development (2009) *Walk, Run, Skate, Stroll, Enjoy*. Available online at: http://www.sjwaterfront.com/ (accessed 1 July 2009).

UNESCO (2009) *Global Network of National Geoparks*. Available online at: http://portal.unesco.org/science/en/ev.php-URL_ID=6400&URL_DO=DO_TOPIC&URL_SECTION=201.html (accessed 23 May 2009).

19 From mining boom towns to tourist haunts

The ghost town life cycle

Bruce Prideaux and Dallen J. Timothy

Introduction

The term 'ghost town' conveys a sense of abandonment, desertion, dereliction, isolation, decay and even the supernatural. The popular image of a ghost town is one where the former inhabitants have moved away leaving only abandoned buildings as a testament to the community that once stood there. However, this is only one type of ghost town. Factors leading to abandonment include exhaustion of natural resources on which the settlement formerly depended, bypassing by railways or new highways (as in the case of several small towns along Historic Route 66 in the United States), sudden events or disasters, or the shifting of human economic activities to other areas. The most famous, or perhaps infamous, ghost town of the recent past is Chernobyl, Ukraine. A 1986 accident in the Chernobyl nuclear power plant forced the sudden evacuation of thousands of residents. Today, Chernobyl remains a city without residents, a ghost town where entry is tightly controlled because of high radiation levels and other safety hazards.

There are many ghost towns throughout the world that arose when resource extraction or manufacturing ceased. Some of these have become salient tourist attractions for a number of reasons. For example, ghost towns whose settlements were based on timber or mining in mountain regions are often located in areas of natural beauty. There are many such examples in the north-western United States and in the Australian outback and Tasmania. In other cases, they are well-known and visited because of their role as a venue for some famous event or period of history, or their connection to a legendary character (e.g. Tombstone, Arizona). Additionally, others are preserved for their architectural, aesthetic and artistic value, or because they remain relatively intact in their original form (e.g. Bodie, California).

In some cases, tourism has been able to resurrect ghost towns or bring dwindling resource communities from the brink of becoming complete ghost towns (Johansen, 1998). Several examples exist where towns on the verge of abandonment have been reinvented for tourists as ethnic villages (Frenkel *et al.*, 2000) or what Getz (1993) called 'tourist shopping villages' (Frost, 2006; Timothy, 2005), which design themselves as tourist and day-tripper destinations for shopping opportunities combined with heritage sites and resources.

In this chapter a mining ghost town is described as a mining settlement that was abandoned or nearly abandoned after resource extraction ceased. We examine the processes that lead to a settlement being transformed into a ghost town or reinvented into something else and consider the role that tourism may play in revitalizing the economies of resource-depleted communities.

A role for tourism

The mining landscape is dotted with abandoned towns, left by most or all of their inhabitants when mining resources were depleted. From an historical perspective, many mining towns were destinations in their own right as people travelled to prospect for gold, silver and other precious metals, and to conduct business or visit relatives. Even the most isolated mining towns catered to the needs of outsiders, offering accommodations, saloons, transportation, eating establishments, and attractions of vice, such as prostitution and gambling. When mining ceased, many mining towns of the American and Canadian west, as well as the Australian outback, underwent rapid economic decline, most notably in the late 1800s and early to mid 1900s. Some of these declining communities redefined themselves and reoriented their extractive economies into service economies and capitalized on their mining past as a tourism resource (Neil and Tykkyläinen, 1998). Other mining communities, unable to find an alternative economic base failed, resulting in permanent out-migration and eventual abandonment and dereliction. The remnants of these communities have popularly become known as ghost towns, with some having become salient tourist attractions and inscribed on various heritage protection lists. Others have been used as sets by the motion picture industry, often shoring up their historical importance in the eyes of policy makers, local residents, and tourists. Much of the 1969 hit movie, *Butch Cassidy and the Sundance Kid*, starring Robert Redford and Paul Newman, for example, was filmed in Grafton, Utah, a small ghost town, which has now been listed as an important heritage location.

Once a settlement's economic base has been lost, decline is irrevocable unless a substitute economy can be developed. As noted above, tourism has resurrected some ghost towns and provided this substitute economic base with the town's mining heritage becoming the focus of visitor attention. Tourism has also brought many virtual ghost towns back from the brink of complete demise. To understand how the process of decline occurs and to illustrate the role that tourism can play in providing a substitute economic sector the Ghost Town Life Cycle model illustrated in Figure 19.2 was developed. While this chapter is concerned with mining towns, and their potential as a tourism resource, the term 'ghost town' has a wider use and is often used to describe a process of urban decline that may include abandonment of specific urban localities in a larger urban area that is not itself in decline or the situation where visitor numbers to a specific site have declined to the extent that the site is empty or appears to be unused.

Surprisingly, the tourism literature has ignored ghost towns as an area of research despite their appeal as a novel experience, their heritage value or even their appeal as a site of dark tourism. Academic interest in the phenomenon has emerged in other disciplines including anthropology and geography. Thus, in the broader social sciences, ghost towns are often seen as examples of the rise and decline of places, and of the post-modern transformation from primary and secondary economies to tertiary (service) economies. In the non-academic literature there is a growing number of publications about ghost towns (Varney, 1994; Winn, 1985) and a growing number of websites that focus on ghost towns from a heritage perspective, offering advice to potential visitors (www.ghosttowns.homestead.com; http://www.legendsofamerica.com/LA-Ghosttowns.html; www.drivewa.com/poi/1287/ghost-towns.html).

Understanding the economic drivers that create, unravel and rejuvenate mining communities is a key requisite to understanding the function that the tourism industry may have in providing a substitute economic sector. The economies of many remote communities are built around the extraction of natural resources that are exported, often with little or no onsite processing. These natural resources, described as 'staples' are susceptible to boom and bust cycles linked to external market forces (Nelson and McKinnon, 2004; Schmallegger submitted). Economic and political control usually resides in distant metropolitan centres giving staples-dependent communities little ability to exercise control over their own destiny. Most mining communities exhibit an over-dependence on raw staples, and for this reason are susceptible to being trapped in their development paths by failure to diversify beyond their initial resource-driven export base. While tourism may appear to offer an escape path from mineral dependency there is potential for tourism to become the new staples trap if economic and political control is simply transferred from one group of metropolitan institutions to another with no local input. A number of authors (Schmallegger submitted; Müller and Jansson, 2007) have observed the tendency to view tourism as a relatively easy path to rescue regional communities undergoing decline. However in the absence of an ability to exercise control over local tourism development, communities remain staple-dependent and subject to the same forces of supply, demand and control experienced by the natural resource that tourism replaced.

In recent decades tourism has been identified as one of the few economic alternatives to mining in resource extractive communities (Edwards and Llurdés i Coit, 1996; Jensen and Blevins, 1992, 1995). Two key requirements that must be satisfied before a mining settlement can tap into the tourism market are accessibility and a resource base that will attract the interest of tourists. Access should not be confused with distance, for as McKercher (2006) observed in the case of Tap Mun Island located in Hong Kong's New Territories, a site may be a relatively short distance from a generating region but difficult to reach because of poor transport networks or difficult topography. Access is also a function of the time taken to reach the site as well as the actual cost of travelling from a generating region to an attraction. Access is a major problem for mining settlements located in remote and difficult to reach areas and is compounded by a rapidly deteriorating transport infrastructure once mining has ceased. The quality of tourist resources is also an important factor and includes accommodation, buildings, landscapes, streetscapes, museums or other site-specific activities. In many cases the uniqueness of specific ghost town resources including the landscape in which it is located and access to generating markets will determine if it has potential to be transformed into a viable tourism resource.

Difficulties posed by remoteness and access, usually defined as problems of peripherality, have attracted considerable interest by tourism scholars. Botterill *et al.* (2002) described peripheral areas as being socially, physically, economically, psychologically, politically and developmentally isolated from, and marginalized by, the core. Access rather than physical distance is often a major inhibiting factor when assessing the potential for a ghost town to be transformed into a tourist attraction.

Geographers have long had an interest in the relationships between social and economic cores and their peripheries. The Core Periphery Model introduced by Prideaux (2009) is an extension of that concept and highlights the difficulties of access and resource attractiveness that remote mining settlements face when attempting to attract tourists. Building on earlier research (Weaver, 1998; Oppermann, 1993) and the concept of distance-decay (Duval, 2007; McKercher *et al.*, 2009) the model highlights the strong relationship that exists between the level of tourist resource attractiveness and distance from an origin market. In this example a mining settlement located at point B has to compete with a more conveniently located attraction at A. In most cases B will need to offer resources that have sufficient additional value as an attraction shown in Figure 19.1 as the right hand vertical axis to entice visitors to justify the extra travel costs of money and time required to bypass A and travel to B. Prideaux (2009: 20) postulated that as the level of remoteness and/or peripherality increases, 'the scale of uniqueness' or 'differentness' of the attractions offered by destinations must increase in proportion to the increased distance travelled. Otherwise visitors will travel to destinations closer to home, a concept known as intervening opportunity. In the language of the push-pull model (Uysal and Jurowski, 1994) a mining settlement located on the periphery must develop a level of resource attractiveness that exerts sufficient pull attractiveness to enable it to match the push aspirations of target visitor groups, successfully compete with other appealing destinations and conquer the problem of access.

The attractiveness of a ghost town as a tourism resource is a function of the remaining built environment, how the site is presented and the form of interpretation that occurs. DeLyser

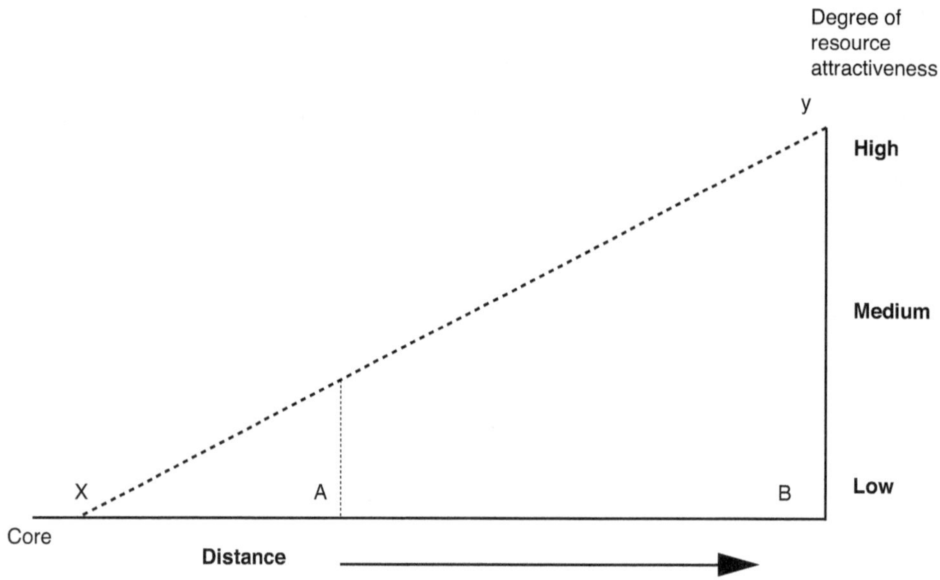

Figure 19.1 The core periphery model.

Source: Prideaux (2009).

Figure 19.2 The Imperial Hotel Ravenswood.

Photo courtesy of Bruce Prideaux.

(1995; 1999) in a study of authenticity in ghost towns raised a number of issues about how ghost towns are presented to tourists. While the structure of ghost towns can be expected to be authentic, the presentation of the past may be inauthentic and a construct that leaves impressions that are not an authentic representation of the reality of the community that lived in the settlement when it functioned as a working mining community. Commenting on the master plan for the Bodie State Historic Park, DeLyser (1999: 603) made the following observation: 'Thus Division personnel believed that by presenting Bodie as an "authentic ghost town," devoid of civilization's "oppressive elements" visitors to the park would be able to see, to feel, and therefore to understand the lives of our "courageous" and "resourceful" pioneering American forebears. In other words, to Division personnel, the "authentic ghost town" of Bodie represented mythic notions about the American West that could be experienced through the landscape'. In a commentary on nineteenth century Arizona lawman Wyatt Earp, Barra (1998) made similar comments about how the legendised life of Wyatt Earp has become part of the myth of the American West and central to the image of the Old West offered for tourist consumption in towns such as Tombstone. In some circumstances the inauthentic presentation of the authentic may actually add to the attractiveness of the resource (Martens and Timothy, 2006).

As DeLyser (1999: 606) noted, ghost towns such as Bodie have become living museums that do not convey the past per se, but rather show 'how people in the present think about the past', which is the underlying ethos of heritage conservation and heritage tourism today (Graham *et al.*, 2000; Timothy and Boyd, 2003). By this mechanism the power of an external visual landscape can be 'linked to an internal experience or cultural belief (in this case the mythic West and contemporary American values)' (DeLyser, 1999: 608). In reality it may be argued that authenticity is a construct with different meanings, in different places to different people and even the same people but at different times (Till, 1999; Timothy and Boyd, 2003; 2006).

Creating a ghost town experience is therefore a function of suitable access, appealing physical resources based on streetscapes, surrounding landscapes, quality of infrastructure and interpretation. Facilitating the successful transformation of the local economy from mining dependence to tourism dependence and in the process avoiding a new staples trap is a further factor. In the case of Bodie, the state park system has taken control of the town, disempowered the small number of remaining residents and created a new staples trap – tourism.

Methods

The study reported in this chapter has two objectives: to identify the process of decline that occurs after the cessation of mining and to examine the role that tourism may have in arresting decline by creating demand for heritage experiences. The principle research method employed in this investigation was historical research using a range of archival and printed histories of 20 Australian mining ghost towns. Similarities and differences in growth and decline patterns were recorded and incorporated into the model. In the following discussion the process of decline that commences with abandonment is referred to as *ghostification*.

The general pattern of growth and decline over all the settlements was very similar, except for the time scales involved. When plotted in graph form the pattern of growth and decline suggested that a life cycle process could be observed. In the past the life cycle theory has been used to explain the phenomena of growth, decline and even rejuvenation with the best known application in the tourism literature being Butler's (1980) Tourism Area Life Cycle used to explain the forces that create, sustain and cause decline in tourism resorts. Applying the same concept to ghost towns, a specific Ghost Town Life Cycle could be identified although the term 'life' hardly seems appropriate when the discussion focuses on settlements that have effectively died.

When applied to ghost towns the life cycle concept demonstrates a capacity to explain why growth occurs, why decline occurs and the process that follows if the economic base of the settlement is not rejuvenated by industries such as tourism. It was also apparent that the process of physical decline, described as ghostification, occurs relatively quickly if alternative economic staples are not available. To demonstrate the capacity of the Ghost Town Life Cycle model (Figure 19.3) as a method for explaining the economic processes at work in mining communities, two former mining towns that were never fully abandoned are used as case studies.

Case studies

To identify the function of tourism in arresting the process of ghostification and provide a declining mining settlement with an alternative economic base two mining towns were examined as case studies. Neither town can be classed as a ghost town in the sense that they are completely devoid of population or were completely abandoned in the past. In both cases the process of decay after cessation of mining was arrested and both towns were revitalized as tourist centres. The first case is the former gold mining settlement of Ravenswood located in North Queensland, Australia. The second case is Tombstone located in Arizona, USA.

Ravenswood, Queensland

The area around Ravenswood was first settled by grazers in 1861 and alluvial gold was discovered in 1869. The initial alluvial gold phase passed quickly and a number of publicly listed mining companies commenced underground mining. By 1871 two stores, numerous grog shops and a court house were operational (Hooper, 2002). With futher discoveries of silver and gold the town took shape adding a state school in 1873, the School of Arts in 1876, Post Office in 1878 and Catholic Convent in 1885, and a rail link was opened to the coast in 1884. Further gold discoveries supported the opening of new underground mines, and in the first

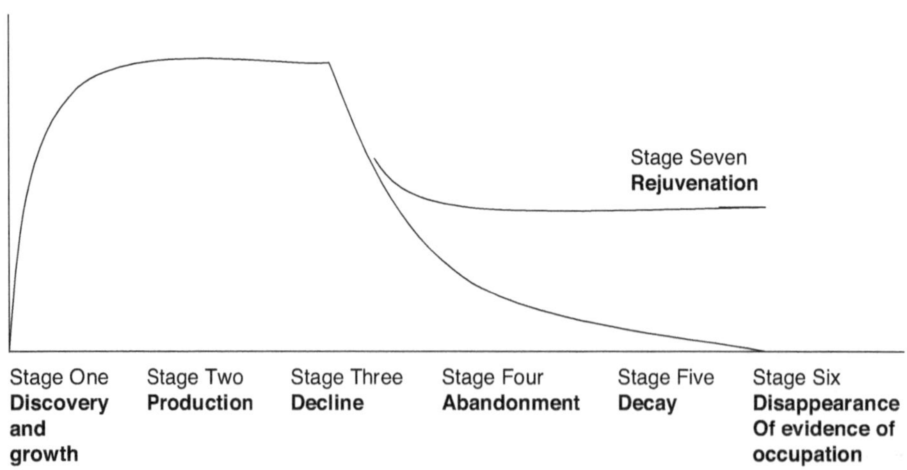

Figure 19.3 The ghost town life cycle.

decade of the twentieth century the town's central business district was rebuilt (Winn, 1985) with the Ambulance Headquarters and the Railway and Imperial Hotels (Figure 19.2) being typical examples. These buildings were constructed of brick and remain standing in 2009. At the peak of its production the area supported a local newspaper and 42 licensed hotels. The closure of the Sunset-Grant mine in 1909 was the first sign that the gold field was reaching the end of its economic life. Industrial unrest in 1912, the volunteering of many miners for service with the Australian Defense Forces during the First World War (1914–1918), and the closure of more mines saw further declines until most mining had ceased by 1917 (Mather and Cox, nd). Many buildings were demolished or dismantled and moved to other settlements in the region. The Catholic convent, for example, was built of timber. It was dismantled and removed to Giru on the coast. In 1930 the rail link closed. Not all businesses left because the now much scaled-down community continued to function as a service centre for the surrounding pastoral industry. In the 1960s interest in the town saw efforts made to preserve the remnants of its past as the basis for a tourism sector. Today the town, which currently has a population below 100, is a minor tourism attraction unable to command more than passing interest from tourists because much of its heritage streetscape has been lost, and the community is fairly isolated.

Tombstone, Arizona

Best known as the site of the famous gunfight at the OK Corral, Tombstone, Arizona, was established in 1881 as a base for silver mining. The town grew quickly, attracting miners, businesses and entrepreneurs who supported the mining industry and its workers including hotels, casinos and brothels. When the town's ore was depleted, many residents moved away but the fortuitous location as the seat of Cochise County prevented the town from being fully abandoned. However, the relocation of the county seat to nearby Bisbee in 1929 resulted in further decline of the town. Its cowboy heritage, however, previously celebrated in various Hollywood western movies, prevented its complete decline. Following the move of the county seat, the old Cochise Council Courthouse and the adjacent gallows yard was saved from dereliction by being turned into a museum. The construction of state highways, the post-war growth in car ownership, increased interest in heritage, and its celebratory status as the site of Wyatt Earp's famous 1881 gunfight at the OK Corral saved the town from following the path of other nearby silver towns towards complete decay. Increasing interest in its heritage provided a viable basis for a growing and, some would say, successful tourism industry. Today, the town has some 1500 residents who are mainly occupied servicing approximately 450,000 day trippers who visit the town each year.

The ghost town life cycle

The life cycle of this type of newly abandoned ghost towns exhibits six specific stages commencing with initial development and concluding with the settlement's absorption into the landscape after a period of decay. An alternative to stages five and six is also shown and is based on rejuvenation after an alternative economic base is developed. The life cycle is illustrated in Figure 19.3. The initial stage (Discovery and Growth) describes the period of time when the resource is first developed. Growth is often rapid but will be capped by the size of the workforce required to exploit the resource, the size and type of mineral reserve, its location relative to other settlements, access to transport networks, the company structure of the mining sector and any other economic functions that develop, such as providing service centre functions for nearby rural communities. The size of the workforce required to work the ore body and any associated smelting, and the projected life span of the mine will be the key factors that determine the size of the town that is established and the type of buildings that are erected.

Stage Two (Production) describes the period during which the mining sector is at its highest level of activity. If the ore reserve is expected to support mining operations over a long time period, substantial buildings are usually constructed. The building activity that occurs in this stage will define the future potential value of the settlement as a heritage tourism destination. If the buildings that comprise the streetscape were built for long-term use there is high degree of probability that they will remain intact for some time even in an abandoned state and will eventually constitute the core of the settlement's future ghost town attractiveness as a tourism resource. The quality of the buildings and the streetscape they form will ultimately determine the heritage value of the location after mining ceases. One factor that may influence the type of streetscape that is built is the corporate structure of the mining sector. Where two or more companies are operating, responsibility, building infrastructure is more likely to be controlled by a local municipal authority and result in a streetscape that will differ from that which might be built by a company dominated town.

If the ore body has a short life span, as was the case with many of the gold rush settlements in Australia, New Zealand, and the USA (including Alaska), timber was the preferred building material and if not demolished or relocated will quickly decay after abandonment. In North Queensland, dozens of short-lived mining settlements were established in the nineteenth century and now can only be located as a map reference. For example, Hightville was established to support the Wee McGregor Copper mine (Hooper, 2002). The first buildings were constructed some years before Hightville was officially surveyed in 1913. The McGregor Hotel was known to be operating from 1909. Aside from the hotel, a gathering place of miners found in almost all mining towns, the settlement had a post-office that operated between 1909 and 1914 and a public school. The cemetery has six graves with the first and only marked grave dated 1912 and the last containing Mr Tom Price who was buried in 1918. Hightville was abandoned about 1918 and the only evidence that a settlement existed on the site is rubble from a decayed miner's hut and a few concrete foundations. Its timber buildings quickly succumbed to termites and bush fires. In a more recent example of this process, the town of Wittenoon, Western Australia, was established in 1937 to support asbestos mining and was closed in 1996 because of health concerns about asbestos. Most residents moved away and unoccupied buildings rapidly deteriorated. In 2006, the town was degazetted and its name removed from official signs and maps. The few residents who continue to live in the town have found it increasingly difficult to remain as shire roads are closed and public services withdrawn. Devoid of even a map reference, a fate shared by most of the ghost towns that were surveyed for this research, the town will linger on until the last resident moves away after which the process of decay will then take over eventually removing almost all trace of the town from the landscape.

While evidence of Hightville was quickly erased from the landscape, the more substantial gold ore body at Ravenswood was able to sustain a substantial mining operation for nearly four decades. During this time substantial buildings were constructed, and today the remaining structures constitute the core of the settlement's tourism appeal.

Stage Three (Decline) describes the period when mineral production begins to decline. This may result from exhaustion of the ore body, a decline in demand for the mineral or cost escalation that makes further mining unprofitable. Decline may be rapid if the ore body is controlled by one company or take longer if there are a number of companies operating, as was the case with Ravenswood. During this stage the settlement's inhabitants face a dilemma: do they stay or move. As demand for labour falls there is a parallel fall in demand for housing causing house prices to decline rapidly. Those who move may have some opportunity to recover some of the investment made in their house while those who stay risk losing a large proportion of their investment as the downward spiral of real-estate values gathers pace. In a similar manner, the demand for goods and services will fall forcing shops and other businesses to close further adding to the downward spiral of the local economy. Just as the attainment of a threshold will made it profitable for a business to be established and new services provided by the public sector, breaching of thresholds will lead to disestablishment and disinvestment. As thresholds are breached businesses close and the population declines until the process becomes almost irreversible. If the settlement acquires a secondary function such as a service centre for the surrounding regional economy the downward economic spiral may be arrested but not without significant population loss. Both Ravenswood and Tombstone had secondary functions that prevented their complete decline.

Once mining activity ceases there will be a rapid out migration of the remaining residents and without new business opportunities emerging, the settlement is likely to be totally abandoned (Stage 4, Abandonment). At this point it is likely that a few residents, often retired or unable to find work elsewhere, will remain behind living in the shell of what was a once thriving settlement. At this point, if not earlier, any buildings capable of being transported elsewhere will be removed, others will be demolished for scrap, and the remaining infrastructure abandoned unless alternative uses such as heritage tourism are identified. At this stage the process of ghostification is very apparent.

Following abandonment, the process of decay (Stage 5) commences and reflects the cessation of maintenance on buildings and infrastructure. After abandonment, the remaining buildings and the construction materials and techniques used, will decay at a rate determined by environmental processes. Desert settlements and settlements where stone and concrete were the main building materials used will last longer than settlements located in wet climates or where timber was the major building material. Ultimately, buildings will collapse and be incorporated into the landscape as rubble and almost all traces of the settlement except, perhaps, for earthworks and concrete foundations will disappear. This process is illustrated by Figures 19.3 and 19.4.

Abandoned buildings will undergo physical decay that without intervention will lead to eventual disappearance from the visual landscape (Stage 6). Figures 19.4 and 19.5 illustrate the process of decay that typically occurs to abandoned buildings. In Figure 19.4, a hotel in an abandoned town in the North Flinders Ranges of South Australia exhibits the first major visual evidence of decay that will speed up once the roof collapses. Once exposed to the elements, timber structures in the building's interior rot or are attacked by termites. In the next stage of building decay the interior walls collapse as illustrated in Figure 19.5 eventually leaving only rubble and foundations. In the final two stages of the life cycle, ghost towns and their remnants are not resurrected as living, vibrant communities as those that are re-invented for their shopping or cultural value are. Their ruins, though, may function as significant tourist attractions – heritage sites that are preserved (or not) by local or state conservation entities and visited by heritage enthusiasts (DeLyser, 1999).

The potential value of the settlement as a tourist attraction will be determined by actions taken late in Stage 3, or during Stage 4 and the beginning of Stage 5. If the built environment is remarkable enough to generate tourist interest, it's intact preservation is essential. Where elements of the streetscape are lost, the level of tourist interest can be expected to decline. For example, Ravenswood's peripheral location and substantial loss of built streetscape has

Figure 19.4 A former hotel in the early stage of decay following abandonment.
(Photo courtesy of Bruce Prideaux).

Figure 19.5 A former residence in an advance stage of decay after abandonment.
(Photo courtesy of Bruce Prideaux).

seriously eroded its potential as a ghost town that is able to hold visitors' attention. In comparison, Tombstone's demise was arrested early on, albeit at the cost of some of its authenticity (Martens and Timothy, 2006), and now thrives as one of Arizona's top tourist destinations.

Conclusion

Ghost towns are intriguing heritage places both for what they are today and what they once were. It is unlikely that the fascination that ghost towns exert will wane as new generation of tourists are attracted to them and as new ghost towns are created. Commercializing a ghost town is not, however, an easy task and will largely depend on the state of decay it has reached when tourism is first introduced, the distance the town is from other attractions and access points (e.g. highways), the degree to which the ghost town experience can be commodified to generate revenue for investors, and the regulations imposed by various public sector bodies that will have some interest in the town.

This chapter has introduced the concept of ghostification and the associated ghost town life cycle as concepts that can be used to identify the opportunities for developing declining towns to take an alternative path from decay. Tourism in this case is able to replace mining or other forms of economic activity as the new staple, but in the process faces the danger that the town will enter a new staples trap.

Today some abandoned towns have undergone the process of rejuvenation outlined in this chapter and have successfully tapped into the tourism market. Others, in more peripheral areas or suffering from advanced states of decay have remained abandoned and declined to the extent that only foundations and earth works remain as a visible testament to the previous history of their settlement.

References

Barra, A. (1998) *Inventing Wyatt Earp: His Life and Many Legends*. New York: Carroll and Graf.

Botterill, D., Owen, R., Emmanuel, L., Foster, N., Gale, T., Nelson, C. and Selby, M. (2002) 'Perceptions from the Periphery: The Experience in Wales', in F. Brown and D. Hall (eds) *Tourism in Peripheral Areas*, pp. 7–38, Clevedon: Channel View.

Butler, R.W. (1980) 'The Concept of a Tourist Area Cycle of Evolution and Implications for Management of Resources', *Canadian Geographer*, 24(1): 5–12.

DeLyser, D. (1995) 'Preservation with a Grain of Salt: The Fantasy Heritage of California's Ghost Town', *Small Town*, 25(5): 4–13.

DeLyser, D. (1999) 'Authenticity on the Ground: Engaging the Past in a California Ghost Town', *Annals of the Association of American Geographers*, 89(4): 602–32.

Duval, T. (2007) *Tourism and Transport: Modes, Networks and Flows*. Clevedon: Channel View.

Edwards, J.A. and Llurdés i Coit, J.C. (1996) 'Mines and Quarries: Industrial Heritage Tourism', *Annals of Tourism Research*, 23: 341–63.

Frenkel, S., Walton, J. and Anderson, D. (2000) 'Bavarian Leavenworth and the Symbolic Economy of the Theme Town', *Geographical Review*, 90(4): 559–85.

Frost, W. (2006) 'From Diggers to Baristas: Tourist Shopping Villages in the Victorian Goldfields', *Journal of Hospitality and Tourism Management*, 13(2): 136–43.

Getz, D. (1993) 'Tourist Shopping Villages: Development and Planning Strategies', *Tourism Management*, 14(1): 15–26.

Graham, B., Ashworth, G.J. and Tunbridge, J.E. (2000) *A Geography of Heritage: Power, Culture and Economy*. London: Arnold.

Hooper, C. (2002) *Angor to Zillmanton Stories of North Queensland's Deserted Towns*, 4th (edn), Townsville: Colin Hooper.

Jensen, K. and Blevins, A. (1992) 'Lead, South Dakota: The Remaking of a Company Mining Town', *Small Town*, 22(6): 4–11.

Jensen, K. and Blevins, A. (1995) 'Gambling on the Lure of Historic Preservation: Community Transformation in Rocky Mountain Mining Towns', *Journal of the Community Development Society*, 26(1): 71–92.

Johansen, H. (1998) 'Mining to Tourism: Economic Restructuring in Kellogg, Idaho, USA', in C. Neil and M. Tykkyläinen (eds) *Local Economic Development: A Geographical Comparison of Rural Community Restructuring*, pp. 251–68. New York: United Nations University Press.

Martens, T. and Timothy, D.J. (2006) 'Burying Tombstone: Fading Authenticity in a Wild West Tourist Destination', *Paper Presented at the Association of American Geographers Annual Conference*, Chicago, Illinois, March 7–10.

Mather, J. and Cox, J. (nd) *Ravenswood*. Charters Towers: Dalrymple Shire Council.

McKercher, B. (2006) 'Living on the Edge', *Annals of Tourism Research*, 33: 508–24.

McKercher, B., Chan, A and Lam, C. (2008) 'The Impact of Distance on International Tourist Movements', *Journal of Travel Research* 47: 208–24.

Müller, D.K. and Jansson, B. (2007) 'The Difficult Business of Making Pleasure Peripheries Prosperous: Perspectives on Space, Place and Environment', in D.K. Müller and B. Jansson (eds) *Tourism in Peripheries: Perspectives from the Far North and South*, pp. 3–18. Wallingford: CAB International.

Neil, C. and Tykkyläinen, M. (eds) (1998) *Local Economic Development: A Geographical Comparison of Rural Community Restructuring*. New York: United Nations University Press.

Nelson, R. and McKinnon, R. (2004) 'The Peripheries of British Columbia: Patterns of Migration and Economic Structure, 1976–2002', *Canadian Journal of Regional Science*, 27(3): 353–86.

Oppermann, M. (1993) 'Tourism Space in Developing Countries', *Annals of Tourism Research*, 20: 535–56.

Prideaux, B. (2009) *Resort Destinations: Evolution, Management and Development*. Oxford: Butterworth Heinemann.

Schmallegger, D. and Carson, D. (2010) 'Is Tourism Just Another Staple? A New Perspective on Tourism in Remote Destinations', *Current Issues in Tourism*, 13(3): 201–21.

Till, K. (1999) 'Authenticity', in L. McDowell and J. Sharp (eds) *A Feminist Glossary of Human Geography*, pp. 8–9. London: Arnold.

Timothy, D.J. (2005) *Shopping Tourism, Retailing and Leisure*. Clevedon, UK: Channel View.

Timothy, D.J. and Boyd, S.W. (2003) *Heritage Tourism*. Harlow: Prentice Hall.

Timothy, D.J. and Boyd, S.W. (2006) 'Heritage Tourism in the 21st Century: Valued Traditions and New Perspectives', *Journal of Heritage Tourism*, 1(1): 1–17.

Uysal, M. and Jurowski, C. (1994) 'Test the Push and Pull Factors', *Annals of Tourism Research*, 21: 844–46.

Varney, P. (1994) *Southern California's Best Ghost Towns: A Practical Guide*, Norman: University of Oklahoma Press.

Weaver D. (1998) 'Peripheries of the Periphery: Tourism in Tobago and Barbuda', *Annals of Tourism Research*, 25: 292–13.

Winn, J. (1985) *Ravenswood into a Golden Light*. Brisbane: Boolarong Publications.

PART VI

Lessons learned

20 Lessons in transforming mines into tourism attractions

Lee Jolliffe and Michael Conlin

Introduction

The previous nineteen chapters have presented what is likely the most comprehensive review to date of mining heritage tourism and related attraction development. Some 37 academics with expertise related to the subject area have discussed a wide range of mining site developments, both established and emerging, and have examined them in terms of an equally wide range of issues. For the most part, the organization of the book identifies the major categories of themes in the fields – preservation, interpretation, environmental impact, attraction development, sustainable management and future growth. Not surprisingly, several issues have come to the fore and are clearly pervasive throughout the field. These issues, again not unexpectedly, mirror the organization of the book as well spanning virtually all chapters – the fundamental conflicts between preservation and tourism attraction development strategies, the impact of governmental authorities on the planning and management of sites, the never ending challenge of funding acquisitions and more importantly the operations of sites, the need for and use of committed communities in terms of creation and sustained operations, and the challenge of raising the profile and attractiveness of mining heritage tourism for consumers.

The examples presented in the previous chapters offer some lessons for the sector and also present some models and tools to support the sustainable development of mining heritage tourism attraction development. What is clear from the narrative to date is how important the heritage encompassing the mining industry is for people and communities. In many areas of the world, mines and their resources have traditionally had a historically close relationship with their community. As discussed in Chapter 1, as society progresses and transitions from industrial to service economies these mines, both operational and defunct, with their rich industrial and social histories become valued as heritage assets. These generally large sites with industrial values and a strong cultural relationship to their communities are therefore highlighted, as regions begin to investigate these heritage assets and their potential use.

Heritage assets are recognized as being ideally situated to be developed as demand generators for tourism (McKercher and du Cros, 2002), and mines with their distinct site characteristics and rich histories are well situated to be considered for tourism development. Such developments, rather than being regarded as being a deterrent in the development of tourism (Edwards and Llurdés i Coit, 1996) can be repackaged for the purposes of tourism heritage (Graham *et al.*, 2000) appealing to tourists who appreciate industrial heritage. These developments reflect a trend in post industrial societies for regions to use industrial remains for heritage tourism (Timothy and Boyd, 2003).

According to Goodall and Beech (2006) industrial heritage comprises of three broad elements: the buildings used for the manufacture of products, the power sources used by the machinery in the factory and the means of transporting raw materials and finished products.

This means that mining related sites may include not only the buildings on site but also the actual mine (open pit or underground), the processing and power plants used to run the mine and the railways (tracks, cars, stations) that were used to transport the mine products. To some extent many former mining sites have tended to be abandoned once mining has ceased (Edwards and Llurdés i Coit, 1996) and in other cases when heritage values are recognized, these former mines can be transformed for the purposes of recreation, leisure and tourism, into parks, visitor attractions and other touristic assets. As demonstrated by a number of cases in this volume, mining sites available for tourism can contribute to the economic and social well-being of mining communities by generating new revenue streams. Mining heritage can also be reflected in local museums that keep alive and commemorate the mining heritage of communities.

Keeping all of these subjects of concern in mind this final chapter revisits the challenges for the communities that are in the various stages of transitioning from mining as production to mining as attraction. The structure of the chapter is to first review the various forms of visiting mines and experiencing mining heritage as an evolution of mining related tourism, to discuss typologies of mining attractions, then to briefly review the central themes of the book ultimately linking them together into a possible direction for future research.

Visiting operating mines

Visits to operational mines are part of industrial heritage tourism. Here the visitor can have the opportunity to see aspects of mining production. This can be an important public relations activity for the mining company, especially where there is criticism of the negative environmental impacts of mining operations. An example is found in the oil sands region of northern Canada at Fort McMurray, Alberta, an area where oil production is somewhat contested due to environmental impacts but where Fort McMurray Tourism promotes packaged bus tours of the Oil Sands. The package, promoted under the byline 'experience the energy' includes accommodation for two nights with breakfast, a bus tour of Suncor Energy – the world's first commercially successful oil sands plant, and visits to the Oil Sands Discovery Centre and Heritage Park (Fort McMurray Tourism). However, like mining tourism packages offered in other parts of the world, this is a very specialized and a very seasonal tourism product.

In other situations visits to operating mines can offer an opportunity for the narration of local indigenous mining history to visitors, as is the case with tours to the Posti silver mines led by former miners in Bolivia (Pretes, 2002). Buultjens *et al.* (2010) also identifies the relationship of the mining sector and indigenous tourism in Australia, but while noting the potential of mine pits and associated mining equipment as tourist attractions, limited indigenous involvement in mining heritage tourism is found here.

Developing defunct mines as attractions

When no longer required for industry or economically viable, Goodall and Beach (2006: 503) note that the fate of many mines is to be reopened as 'living museums'. As such their operation moves from profit generating for the mine owners to not for profit as agencies with social objectives guided by the principles of cultural heritage management (McKercher and du Cros, 2002) are established. This type of operation should be guided by a site-specific management plan, developed to take into consideration a number of factors, including the nature of the site, the mission of the operating organization, the carrying capacity for visitation, the type of interpretation to be provided, visitor services and marketing.

There are however recognized difficulties in developing former mining sites for tourism. The first is their size, which makes redevelopment costly. The second is the degradation of the surrounding natural environment that results from mining, making it somewhat difficult to attract any kind of new activity. A third issue is the often peripheral location of mines to normal tourist routes and circuits noted by Edwards and Llurdés i Coit (1996). However, this drawback could also be an advantage for heritage attraction development, as noted by Dyebdal (1998) who examined the development of flagship attractions in peripheral locations in Scandinavia. It is also possible that local authorities may be interested in investing in mining attractions to stimulate local tourism and the economy, as Swarbrooke (2002) notes has happened in Norway.

A further challenge is that visitor numbers to mining attractions (that may include mining museums, above ground and underground mining tours and even mining towns and mining regions) may be low, as Swarbrooke (2002) reports of Norway's mining attractions. Because of their outdoor nature, many of these mining attractions are also seasonal, some operating for a short summer season and others receiving only a few thousand visitors per annum. However in the island of Svalbard, Norway, Viken (2006) reports that mining heritage has been part of the transition of the main town Longyearbyen from mining company town to tourist resort, with remnants of mining being declared as heritage and one former mine being open as a museum. A benefit of mining attractions to communities and to visitors is the employment of former mine workers as interpreters, creating employment for the latter and contributing to an authentic visit for the former for example as in the case of the Cape Breton Miners' Museum in Canada (Brown and Geddes, 2007).

Local mining museums

Mining history is to a large extent community history and this is often reflected in local museums. Mining communities are those where the community is significantly affected by mining operations in the community or nearby. Planning for mining operation to sustainability needs to include both the life of the mine and the post-mining closure period (Veiga *et al.*, 2001). The authors (ibid) indicate that the legacy left to a community is emerging as an important consideration in mining operations. Sustainability involves improving environmental performance and mitigating environmental impacts as well as co-operative decision making with the community and diversification.

Tourism offers an opportunity for mining communities, both those with operating and defunct mines, to develop in a sustainable direction. It offers an important method of supplementing community incomes in the post-mining period. Government assistance is often available to allow communities to transition to other industries such as tourism by establishing mining related museums. Cole (2004) examines mining heritage tourism as a sustainable niche within industrial heritage tourism, in particular. Both tourism and mining depend on the natural environment as a key resource, in fact Murphy (1983: 3) referred to tourism as 'Canada's other resource industry'. In addition the resource base for tourism is cultural heritage, a heritage that in the case of mining communities is affected by the interaction between the mine and the community.

Towards a typology of mining attractions

One method of studying visitor attractions is to look at them by type. A number of authors have previously discussed typologies of attractions (Prentice, 1993; Williams, 1998; Swarbrooke, 2002; Timothy and Boyd, 2003). Williams (1998) noted a change in the scale, diversity and extent of heritage based attractions with heritage tourism being responsible for introducing tourism into locales previously unknown as tourist places. Heritage sites related to mining, for example in the form of what Prentice (1994) terms primary production attractions, are only one example of this phenomenon. Timothy and Boyd acknowledged (2002) that the foundation of heritage tourism is found in heritage attractions, as exemplified by those related to mining profiled in the chapters in this volume. Turning more specifically to heritage attractions based on industry, Edwards and Llurdés i Coit (1996) had proposed a typology of industrial heritage attractions, as productive, processing, transport and socio-cultural attractions.

Swarbrooke (2002) notes that while there is no clear definition of attractions, they can generally be split into four categories. The first category includes features within the natural environment, the second is human-made buildings, structures and sites designed for a purpose other than visitation, the third is human-made buildings, structures and sites that are designed for visitation and the fourth is special events. In terms of the mining related attractions surveyed in this volume, all of these types are present. Building on Swarbrooke's (2002) classifications a useful typology of mineral and mining related attractions can be proposed, incorporating concepts of theme and scale in the literature from the various typologies of visitor attractions reviewed (Table 20.1).

The value of this typology is that it can be used to classify mining related tourism assets in a region for the purpose of planning mining tourism, including the bundling of individual components into itineraries, thematic routes or touring products. However, given that many regions with mining heritage resources are located in rural and peripheral areas, this could be somewhat dependent on the necessary infrastructure for the planning, promotion and marketing of such tourism products. As Swarbrooke (2002: 8) indicated a typology is only a tool, as types may overlap, but it can be useful in 'gaining an understanding of this complex and potentially confusing subject' of visitor attractions. In the case of mining related visitor attractions, as documented by the previous chapters in this book, many mining attractions are based on heritage assets not originally designed for visitation and the issues of scale, access and safety mixed with mining identities and their communities likewise make the study of mining attractions a similarly confusing subject. However, it is hoped that the chapters in this book have shed light on the way forward by providing a global snapshot of the development of mining heritage attractions through their relevance to communities and visitors, regional partnerships and innovations in operations and development.

Table 20.1 The four categories of mineral and mining related attractions

Natural	*Human-made not originally designed primarily for visitation*	*Human-made and purpose built for visitation*	*Special events*
Caves Rock Faces Landforms	Mines Quarries Mining Railways Mining Communities Mining Ghost Towns	Mining Theme Parks Mining Museums Mining Open Air Museums Mining Community Museums Mining Interpretation Centres Mining Routes	Mining Festivals Mining Commemorations Mining Anniversaries

Source: After Swarbrooke, 2002.

Lessons for the future

As demonstrated by the contributions to this volume, the motives for transforming mines into attractions will differ according to local circumstances, needs and priorities. Past history and current circumstances will determine the possible conversion and interpretation of both existing and defunct mines and their related built heritage, landscape and settlements into attractions for tourists, as Cloke *et al.* (1996) indicate moving from 'wasteland to wonderland'. The process is not unique to the mining tourism product as Swarbrooke (2002) acknowledges that a number of factors will come together in the future to bring about changes in the attraction product, the market and the way that attractions are marketed.

It is however possible to identify a continuum for this process of mines moving from industrial sites to heritage attractions in terms of the role of mines, in both industry and the service based attractions industry. As discussed above, it is proposed that mine visitors and visitor attractions form a continuum from operating mines with visitation to defunct mines open to visitors, to local museums recreating and commemorations of mines and their history. As noted earlier, a number of complex relationships come into play here, for example the operating mine has a strong relationship to the community in terms of employment, economic impact and the relationship between the mining company and the community. All of these types of relationships are illustrated in this volume. The lessons on the current development of mining heritage for the purpose of tourism provide a basis for understanding the current state of mining attractions around the globe.

Conclusion

The global reach of the volume reinforces the fact that mining heritage is complex, and the potential for mining related tourism, including tourism related to geology, is enormous. The ways in which mining heritage is valued, preserved and then interpreted to the public are diverse. The value of this book in part is that it has provided some baseline information on the way in which mining heritages are being employed for tourism around the world. More research is needed in the key areas of interpreting mining heritage, transforming mines into heritage attractions and mining tourism destinations, as reflected by the organization of the book. Particular research priorities building on this work could include:

Interpreting mining heritage

- Investigation of how new media (web, Twitter, etc.) could be utilized to expand the audience base for mining attractions, in particular developing new audiences.

Transforming mines into heritage attractions

- Establish best practice guidelines for development and operation.
- Examine impacts of mining attractions on their communities, in terms of social and economic effects, as well as develop best practice ideas for increasing community involvement in the development and operation of mining attractions.

Mining tourism destinations

- Identify strategies for achieving a higher profile within national/regional tourism development and marketing agendas.
- Examine network development of mining sites (such as the European Geopark Network).
- Study the bundling of smaller sites into regional products, developing mining routes and tours.
- Identify mining attraction potential for potentially defunct mines, well before their closures, so that appropriate re-use options can be explored and adopted.

Mines and mining communities for the most part, at least the majority of those surveyed within this volume, are at an early stage in their development as attractions. Many are not yet part of established tourism circuits nor are they fully developed for destination tourism (Gunn, 1979). These research priorities will have, in varying degrees, relevance and value for any potential, emerging or established mining heritage visitor attraction.

Finally, visiting mining sites is to some extent unlike other forms of tourism in urban spaces which one author (Edensor, 2005: 95) describes as '. . . typically organized around the capture of familiar, famous sights, undisturbed and seamless forms of mobility, the dissemination of digestible, pre-packaged information and reified discourses about places and cultures, a series of "appropriate" activities centred upon consumption, and regularized itineraries which sew these sites together.' Visiting mining sites and their diminished or sometimes abandoned communities may be more like what Edensor describes as having the characteristics of visiting industrial ruins, namely that '. . . movement is rough, disrupted and potentially perilous, replete with sensations other than the distanced gaze; no one is there to enforce performative norms, and indeed, there are no preparations for entry into, and performance upon, such a stage'. Visiting these types of locales, which could well include abandoned mines, former mining villages and their landscapes, could be likened to Edensor's observation that this is a form of 'anti-tourism' lacking the commodification and the formality of contemporary tourism space.

Visiting mining heritage attractions, be they single sites, communities or regions may contain some of the elements of this kind of underdeveloped tourism. However, the examples cited in this volume do demonstrate that mining heritage attractions of various types are becoming commodified and packaged for tourism, moving into the mainstream of tourism activity. The challenge for these attractions, as has been discussed throughout this volume, is to balance the desire for historical context and relevancy along with the unstructured sense of adventure described by Edensor with the need to make the sites more attractive to larger numbers of visitors which, in turn, contributes to the sustainability of the sites.

In summary, the unique challenges in making mining heritage and mining communities accessible to tourists are related to the character, size and location of these heritage assets as well as to the perceptions of host communities and potential visitors. Several cases in this volume have highlighted the fact that forward planning and assessment of tourism potential should be part of any planned mine closure. It is clear that the areas of future research focus are many and there is in particular an opportunity for applied research that will assist communities around the globe in their journey to preserve and benefit from their mines once they have transitioned from industrial to heritage capital.

References

Brown, K.G. and Geddes, R. (2007) 'Resorts, Culture and Music: The Cape Breton Tourism Cluster', *Tourism Economics*, 13: 129–41.

Buultjens, J., Brereton, D., Memmott, P., Reser, J., Thomson, L., O'Rourke, T. (2010) 'The mining sector and indigenous tourism development in Weipa, Queensland', *Tourism Management*, 31(5): 597–606.

Cloke, P., Milbourne, P. and Thomas, C. (1996) 'From Wasteland to Wonderland: Opencast Mining, Regeneration and the English Natural Forest', *Geoform* 27(2): 159–74.

Cole, D. (2004) 'Exploring the Sustainability of Mining Heritage Tourism', *Journal of Sustainable Tourism*, 12(6): 480–94.

Dybedal, P. (1998) *Theme Parks as Flagship Attractions in Peripheral Areas*. Bornholms Forskingscenter: Bornholm.

Ednesor, T. (2005) *Industrial Ruins – Spaces, Aesthetics and Materiality*. New York: Berg.

Edwards, J. and Llurdés i Coit, J. (1996) 'Mines and Quarries: Industrial Heritage Tourism', *Annals of Tourism Research*, 23(2):341–63.

Fort McMurray Tourism. Available online at: http://www.fortmcmurraytourism.com/index.php?page_id=1023 (accessed October 24, 2009).

Goodall, G. and Beech, J. (2006) 'The Management of Heritage and Cultural Tourism', in J. Beech and S. Chadwick *The Business of Tourism Management*, pp. 485–505. Harlow, England: Prentice Hall.

Graham, B., Ashworth, G.J. and Tunbridge, J.E. (2000) *A Geography of Heritage, Power, Culture and Economy*. London: Oxford University Press.

Gunn, C. (1979) *Tourism Planning*. New York: Crane Russak.

McKercher, B. and du Cros, H. (2002) *Cultural Tourism: The Partnership between Tourism and Cultural Heritage Management*. New York: Haworth.

Murphy, P.E. (1983) 'Tourism: Canada's Other Resource', in P. E. Murphy *Tourism in Canada: Selected Issues and Options*, pp. 3–19. Victoria, British Columbia: University of Victoria Department of Geography.

Pretes, M. (2002) 'Touring Mines and Mining Tourists', *Annals of Tourism Research*, 29(2): 439–456.

Prentice, R.C. (1993) *Tourism and Heritage Attractions*. London: Routledge.

Swarbrooke, J. (2002) *The Development and Management of Visitor Attractions*. Oxford: Butterworth-Heinemann.

Timothy, D.J. and Boyd, S.W. (2003) *Heritage Tourism*. Harlow, Essex: Pearson Education Limited.

Veiga, M.M., Scoble, M. and McAllister, M.L. (2001) 'Mining with Communities', *Natural Resources Forum*, 25: 191–202.

Vicken, A. (2006) 'Svalbard, Norway', in: Baldacchino, G. (ed.) *Extreme Tourism: Lessons from the World's Cold Water Islands*, London: Elsevier, 129–44.

Williams, S. (1998) *Tourism Geography*. London: Routledge.

Index

Page numbers in **bold** refer to illustrations.

For Product Safety Concerns and Information please contact our EU
representative GPSR@taylorandfrancis.com
Taylor & Francis Verlag GmbH, Kaufingerstraße 24, 80331 München, Germany

www.ingramcontent.com/pod-product-compliance
Lightning Source LLC
LaVergne TN
LVHW080648011026
849160LV00014B/1361
* 9 7 8 1 1 3 8 8 8 0 6 9 6 *